Fundamental Behavior of Unsaturated Widely-Graded Soil

Foreword

It is a great pleasure to write this foreword for the book titled, *Fundamental Behavior of Unsaturated Widely-Graded Soils*" authored by Xu Li, Hongfen Zhao, and Limin Zhang. I have followed with interest the ongoing unsaturated research program at the Hong Kong University of Science and Technology under the direction of Prof. Limin Zhang. Xu Li and Hongfen Zhao were two Ph.D. students guided by Prof. Zhang. I have been fortunate to be able to follow their research studies on this unique class of unsaturated soils. Colluvial soils form a significant part of the natural terrain in many countries and have become the subject of rainfall-induced landslides. These soils pose a serious risk to human safety and economic development worldwide. The engineering profession has desired to better understand the risks faced by frequent, intensified rainstorms. Colluvial soils typically comprise a mixture of coarse and fine-grained particles, which are often widely-graded and exist under unsaturated conditions. The behavior of coarse widely-graded soils is variable and complex, resulting in difficulties in evaluating the stability of colluvial soil slopes. The seepage and strength characteristics of widely-graded soils are critical in the assessment of landslide hazard mitigation and engineering construction. The authors have conducted fundamental studies into these scientific problems for two decades and have become a well-qualified research team to synthesize the research material on this topic. It is an opportune time to write a book that presents the mechanisms of behavior and evaluates the methodologies that the scientific community needs for prudent applied engineering practices.

This book focuses on the seepage and strength characteristics of widely-graded soils. It presents the fundamental behavior of these soils covering multimodal features of grain-size distribution and pore structure. The book covers the evolution of soil fabric under drying–wetting cycles as well as the characteristics of seepage and shear strength. The mechanisms of rainfall-induced colluvial landslides are clearly presented. It also presents creative laboratory test techniques and numerical simulation methodologies for the study of seepage and stability related to unsaturated coarse, widely-graded soils.

I am pleased to witness the timely publication of this book. Students, engineers, and researchers who are concerned about the seepage behavior of unsaturated soils and the stability of unsaturated soil deposits can benefit from this book. I am confident that this book will serve as an important source of reference for future studies on this topic.

Dr. Delwyn G. Fredlund
Professor Emeritus
University of Saskatchewan
Saskatoon, Canada

Preface

Colluvial soils are widely distributed on natural terrains and often serve as fill materials for earth structures, including embankments, foundations, and slopes. Colluvial soils are characterized as coarse, widely-graded, and often under unsaturated conditions. Related geotechnical problems dealing with natural hazards and disasters such as dam failures, ground settlement, and landslides have demanded increased attention. Most current studies on unsaturated soils primarily focus on fine-grained soils. How water infiltrates into such coarse soils and how the shear strength of these soils decreases during water infiltration are not well understood. Also, it is not clear how coarse granular particles affect water infiltration and groundwater flows into the soil deposits. To answer these questions, several challenging scientific issues need to be addressed on the soil-water characteristics, hydraulic conductivity, and shear strength of these coarse soils.

The primary objectives of this book are to investigate the hydraulic and mechanical behavior of unsaturated widely-graded soils and the mechanisms of rain-induced failures of loose, coarse-grained soil deposits. These issues have been addressed through extensive laboratory tests, theoretical studies, numerical analyses, and case studies. This book advances our engineering understanding of the fundamental behavior of loose coarse soils. The results will facilitate landslide hazard mitigation and engineering constructions around the world.

Beijing, China — Xu Li
Guangzhou, China — Hongfen Zhao
Hong Kong, China — Limin Zhang

Acknowledgments

In the process of writing this book, the authors received enthusiastic help and invaluable assistance from many individuals. The authors would like to express their special thanks and respect to Prof. Delwyn G. Fredlund, who gives valuable guidance, constructive suggestions, and continued encouragement throughout the authors' research studies. The first two authors are grateful to their Ph.D. supervisor, Prof. Li-min Zhang for years of supervision and help. Prof. Zhang's wisdom, insight, and enthusiasm for the research always guide us to go further.

The authors are grateful for the editing help provided by Mrs. Fei Wang, Dr. Yong-kan Wu, Mr. Xiao-kang Li, Mr. Yu-xin Zhao, and Mr. Shuang-fei Zheng.

The support from the National Natural Science Foundation of China (No. 51109003; 51479001; 51979002; 51909287) and Guangdong Provincial Natural Science Foundation (No. 2020A1515010872) is gratefully acknowledged.

All authors contributed equally to this book.

Contents

Chapter 1
Introduction

1.1 Background

Widely-graded soil deposits are very common in nature, including colluvial soil landslides, residual soil slopes, the natural loose-packing dams of barrier lakes, glacial till deposits, etc. For example, quite a number of the natural slopes in the Three-Gorges Reservoir Zone (TGTZ), landslides in Hong Kong, and colluvial soil deposits formed after the Wenchuan earthquake are composed of colluvial soils. The grain-size analysis of colluvial soils reveals that such soils are often widely-graded with a small sand fraction and show a bimodal feature of grain-size distribution (Zhang and Chen 2005). Most of these widely-graded soil deposits are under an unsaturated state, and are very sensitive to the influence of wetting actions, such as rainfall, or reservoir water level fluctuation (Fig. 1.1).

Landslides in widely-graded soil deposits are the main threat in the Three-Gorges Reservoir Area of China. As many as 684 individual slopes larger than 100,000 m^3 and numerous smaller slopes have been identified in TGRZ in China (The Yangtze Water Conservancy Commission 1997). These slopes have been partially submerged by the reservoir water since 2003. Landslides were identified as being a major hazard of the Three-Gorges project. As shown in Fig. 1.1a, the Qianjinangping landslide occurred as the water level was being raised in July 2003, killing 24 people, destroying 346 houses and causing the loss of four factories (Yang and Wang 2009). In the yearly operation of the reservoir, the water level is lowered rapidly from 175 to 145 m for flood control during the wet season. The drawdown of the water level can impose a large hydrodynamic force on the slopes and reduce the stability of the slopes. In addition, rainstorms are also likely to occur during the reservoir level drawdown. The compounding effects of reservoir water infiltration, drawdown of reservoir level, and rainfall on the stability of the slopes could be extremely severe. To prevent the landslide disaster and its surge hazard, over 200 slopes in the TGRZ have been stabilized before the reservoir water level rose to 175 m.

Rainfall-induced landslide of widely-graded soil deposits is also a major concern in Hong Kong, China (GEO 2004; Emily et al. 2015). In 1972, a disastrous landslip

X. Li et al., *Fundamental Behavior of Unsaturated Widely-Graded Soil*,
https://doi.org/10.1007/978-981-19-3402-5_1

(a) Qianjinangping Landslide occurred in Three-Gorge Reservoir Zone

(b) Po Shan Road landslide of Hong Kong in 1972 (https://hkss.cedd.gov.hk/)

(c) loose landslide soil deposits along Province Road 303 after the Wenchuan earthquake

Fig. 1.1 Natural widely-graded soil deposits are sensitive to the influence of wetting actions such as rainfall, or reservoir water level fluctuation occurred on Po Shan Road in the middle of the mountain, which destroyed three buildings, including a 12 storey building, and killed 67 people, as shown in Fig. 1.1b. More than 1,000,000 visible natural terrain landslides have been identified in Hong Kong (Evans et al. 1999). Failures of fill and natural terrain slopes killed over 140 people in the year 1972 alone and still continue to pose a major hazard to human life and properties in Hong Kong. From 1984 to 2002, 513 major landslide incidents (volume equal or greater than 50 m^3) and 5340 minor landslide incidents (volume less than 50 m^3) occurred in Hong Kong. Only 4.5% of the major landslide incidents and 0.7% of the minor landslide incidents occurred during the slope construction (Sun and Tsui 2003). Most landslide incidents instead were triggered by rainfall. For example, more than 800 slope failures were triggered by a rainstorm in November 1993 on Lantau Island, Hong Kong (Dai and Lee 2002).

At the same time, rainfall-induced landslides of widely-graded soil deposits are a major secondary disaster after strong earthquakes. For example, the Ms 8.0 Wenchuan earthquake in Sichuan in 2008 induced approximately 50,000 landslides covering a wide area of 285 × 20 km (Zhang 2009; Dai et al. 2010). A large number of hillslopes are composed of colluvial soils. After the Wenchuan earthquake, loose colluvial soil deposits are densely distributed on steep terrains (Fig. 1.1c). These soil deposits are marginally stable under normal climate conditions. During rainfall infiltration, the negative pore-water pressure (suction) in an unsaturated soil slope decreases gradually. Consequently, the contributions of soil strength from the negative pore-water pressure decrease, which may lead to the failure of numerous deposits. Since these soil deposits pose hazards to human safety and economic development, any hazardous soil deposits must be identified and their associated risks minimized.

1.2 Fundamental Behavior for Assessing the Stability of Widely-Graded Soil Deposits

1.2.1 Mechanism of the Trigging of Widely-Graded Soil Landslides

Widely-graded soil landslides are often triggered by rainfall and pose significant risks in many parts of the world, especially in tropical mountainous areas. The mechanisms of rainfall-induced landslides have been extensively studied through field monitoring, numerical simulation, and laboratory investigations (e.g., Lumb 1975; Wu 2003; Chen et al. 2004). Two distinct failure mechanisms have been observed and analyzed for rainfall-induced landslides (Collins and Znidarcic 2004). Wang and Sassa (2001) observed that significant positive pore water pressures build up in a low area on a slope or along with the soil/bedrock interface. Movements along the sliding surface lead to liquefaction along this surface, resulting in rapid movements, long run-out distances and finally a complete liquefaction of the failed mass. It is established that the rise of pore pressure as a result of rainfall decreases the mean effective stress in the slope, which lowers the soil strength and eventually triggers landslides. Fourie (1996) and Fredlund and Rahardjo (1993) noted that the soil is initially in an unsaturated state. During a rainstorm, water infiltration into a soil slope can cause a dramatic increase in the pore-water pressure in the slope. Consequently, the contribution to soil strength from the negative pore-water pressures decreases and the sliding forces increase. This can finally lead to a loss of stability of the slope.

During the dry season, water evaporates from the soil surface and the soil keeps unsaturated until the next rainy season. The widely-graded soil tends to have higher shear strength in an unsaturated state during the dry season, but a lower value when fully saturated. When the rainwater begins to infiltrate the unsaturated soil, only a certain amount of water can enter the soil during a storm and only a limited zone of soil is wetted. Field measurements of soil suction show that the suctions at shallow depths (within the top 3 m) show a gradual reduction during the rainy season, but at 3–17 m depths, the soil suctions do not vary much throughout the year. This is the reason why shallow-seated failures are the main failure mode in colluvial soil deposits during rainstorms (Dai et al. 2003; Fannin et al. 2005; Cascini et al. 2010). Such failures generally have sliding depths varying between 0.5 and 2.5 m on the slope surface within the wetted depth (e.g., Au 1998). The corresponding confining stresses are in a low range of less than 25 kPa. Hence, this book focuses on the behavior of saturated and unsaturated colluvial soils under very low confining pressures to provide a better understanding of shallow-seated failures in colluvial soils.

1.2.2 Fundamental Behavior for Assessing the Infiltration Process of Widely-Graded Soil Deposits

Evolution in time of slope stability is controlled by the varying rates of infiltration which in turn depend on rainfall intensities, the slope geometry (arrangement of the different layers, the existence of open vertical cracks, and so on) and soil characteristics (Alonso et al. 1995). The soil properties that determine the evolution of the infiltration process are the soil hydraulic conductivity function (SHCF) and the soil water characteristic curve (SWCC). Therefore, for a given rainfall regime, slope geometry, and initial boundary condition, these hydraulic properties control the evolution of slope stability, which is usually evaluated as the safety factor.

The total energy at one point can be expressed as energy per unit weight, which is called a potential or a hydraulic head. This hydraulic head, h_w, (i.e., energy) will drive water to flow in the soil. h_w at one point has three components, the gravitational energy, the pressure energy, and the velocity energy:

$$h_w = y + \frac{u_w}{\rho_w g} + \frac{v_w^2}{2g} \tag{1.1}$$

where g is the gravitational acceleration, which is a constant and equal to 9.8 m/s^2; y is the elevation of that point above a referenced, certain elevation; u_w is the pore-water pressure at that point; ρ_w is the density of water, which is a constant and equal to 1 g/cm^3; v_w is the flow velocity of water at that point. In the unsaturated flow through soil, the velocity head $[v^2{}_w/(2g)]$ is negligible in comparison with the gravitational and pressure heads. So the hydraulic head in Eq. (1.1) can be simplified as,

$$h_w = y + \frac{u_w}{\rho_w g} \tag{1.2}$$

The seepage flows in saturated/unsaturated soils are commonly assumed to be governed by Darcy's law and the mass conservation. Considering an unsaturated soil element with two-dimensional transient water flow in the x-direction and y-direction, as shown in Fig. 1.2, the mass conservation gives,

$$\left(v_{wy} + \frac{\partial v_{wy}}{\partial y}\mathrm{d}y - v_{wy}\right)\mathrm{d}x\mathrm{d}z + \left(v_{wx} + \frac{\partial v_{wx}}{\partial x}\mathrm{d}x - v_{wx}\right)\mathrm{d}y\mathrm{d}z \\ + q(t)\mathrm{d}x\mathrm{d}y\mathrm{d}z + \frac{\partial \theta_w}{\partial t}\mathrm{d}x\mathrm{d}y\mathrm{d}z = 0 \tag{1.3}$$

where v_{wy} is the flow velocity of water in the y-direction; v_{wx} is the flow velocity of water in the x-direction; $q(t)$ is the applied boundary flux; θ_w is the volumetric water content; dx, dy, dz are the size of the unsaturated soil element in the x-direction, y-direction, and z-direction separately. Darcy's law gives,

$$\begin{aligned} v_{wy} &= -k_{wy}\frac{\partial h_{wy}}{\partial y} \\ v_{wx} &= -k_{wx}\frac{\partial h_{wx}}{\partial x} \end{aligned} \tag{1.4}$$

where k_{wy} is the water coefficient of hydraulic conductivity of the soil in the y-direction; k_{wx} is the water coefficient of hydraulic conductivity of the soil in the x-direction; $\partial h_{wy}/\partial y$ is the hydraulic gradient in the y-direction; $\partial h_{wx}/\partial x$ is the hydraulic gradient in the y-direction.

One of the major differences between water flows in saturated soils and flows in unsaturated soils is that the hydraulic conductivity is not a constant but a function of the degree of saturation (or soil suction) in an unsaturated soil. Another major difference between water flows in saturated soils and flows in unsaturated soils is that the water contents in the soil change with the variable soil suction in unsaturated soils. Considering the seepage problem in isotropic unsaturated soil, the soil hydraulic conductivity can be expressed as,

$$k_{wx} = k_{wy} = k(\psi) \tag{1.5}$$

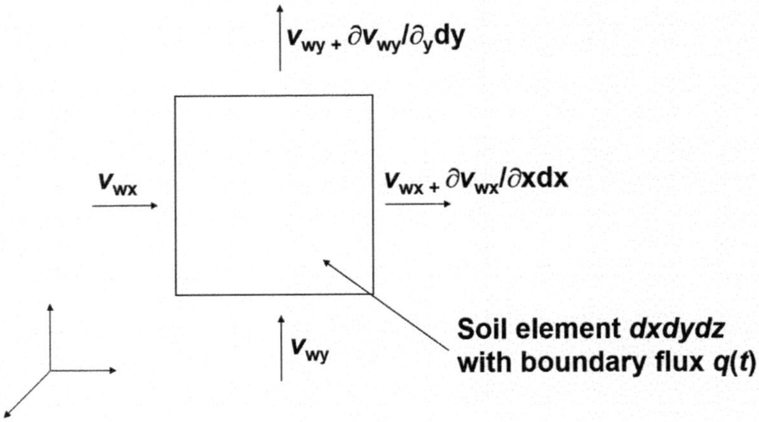

Fig. 1.2 Unsaturated soil element undergoing two-dimensional transient water flow

For a given soil water characteristic curve, the soil water content can be expressed as,

$$\theta_w = \theta(\psi) \tag{1.6}$$

where ψ is the matric suction, which is equal to $(u_a - u_w)$; u_a is the pore air pressure. Substituting Eqs. (1.2), (1.4) and (1.6) into Eq. (1.3),

$$k(\psi)\frac{\partial u_w}{\partial x} + k(\psi)\left(1 + \frac{\partial u_w}{\partial y}\right) + q(t) + \frac{\partial \theta(\psi)}{\partial t} = 0 \tag{1.7}$$

Equation (1.7) reveals that the soil hydraulic conductivity function $k(\psi)$ and the soil water characteristic curve, $\theta(\psi)$ are three dominative soil property functions in the infiltration process.

The SWCC establishes a relationship between water content or degree of saturation and suction of soil. The SWCC is not only a key hydraulic property function which determines the infiltration process in unsaturated soil, but also a key factor which controls the shear strength of the unsaturated soil. The SWCC has become basis for the evaluation of slope stability and is a key factor in the investigation of mechanisms of slope failure (Lumb 1962; Ng and Shi 1998; Collins and Znidarcic 2004). The colluvial soils are considered to have a significant change of water content at suctions smaller than 2 kPa. A new SWCC device needs to be developed for measuring the SWCC at extremely low suctions (smaller than 2 kPa).

The measurement of soil hydraulic conductivity function (SHCF) includes the measurement of saturated soil hydraulic conductivity and the measurement of unsaturated soil hydraulic conductivity. Existing unsaturated soil hydraulic conductivity function measurement methods are often time-consuming, limited in the available measurement ranges of hydraulic conductivity or dependent on selected soil hydraulic conductivity function models. It is desirable to develop a direct method for the measurement of the soil hydraulic conductivity function that is time-saving and applicable to a wide range of hydraulic conductivity and a wider range of soil suction.

1.2.3 Fundamental Behavior for Assessing the Instability Process of Widely-Graded Soil Deposits

Experimental studies on rainfall-induced landslides have been conducted by many researchers (Anderson and Sitar 1995; Lade 1992, 1993; Riemer 1992). When a slope becomes unstable, its movement starts to accelerate. Therefore, in the geotechnical context, the word instability should be defined as the behavior in which large plastic strains are generated rapidly (Chu et al. 2003).

In the laboratory, the instability behavior of sand has been investigated through undrained triaxial tests for loose granular soil (Yang 2002; Lade 1993). It has been consistently observed in undrained triaxial tests that, subjected to monotonic loading, very loose sand exhibits a peak strength at a small shear strain and then collapses rapidly to large strains and low strengths. The type of behavior described above is now recognized as flow liquefaction, which may produce the most devastating effects of all the liquefaction-related phenomena. Flow liquefaction failures are characterized by a sudden loss of strength and rapid development of large deformation. Based on their observations, instability occurring in loose sand under undrained conditions has been considered as the main factor causing the failure of loose granular soil slopes.

Chu et al. (2003) conducted constant shear drained tests (CSD) under constant shear stresses but decreasing mean effective stress to study the instability of sand. This stress path simulated the field stress path (Fig. 1.3). For loose sand, prefailure instability in the form of a rapid and sustained increase in the axial strain rate can occur under fully drained conditions when a specimen is sheared under a CSD path. The conditions for drained instability are the same as those for undrained conditions. The instability line defined by Lade and Pradel (1990) can be used to specify the conditions for drained instability as well. For dense sand, prefailure instability also occurs under drained conditions along a stress path of decreasing mean effective stress when the stress state is above the instability line.

However, many natural soil deposits encountered in engineering practice are often unsaturated. The stress paths derived from the conventional triaxial compression tests such as the consolidated drained tests, the consolidated undrained tests, and constant shear drained tests (Fig. 1.3) fail to replicate the field conditions. The conventional triaxial compression tests are conducted with the application of increasing axial load at constant displacement rates. As a result, these strain-controlled conventional tests are mainly used for the determination of the shear strength of slope materials.

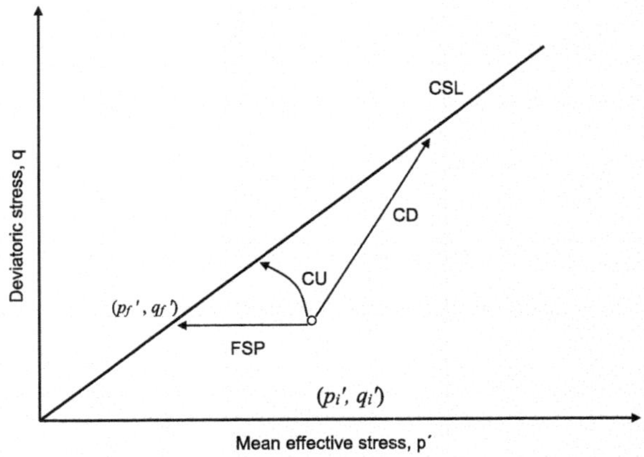

Fig. 1.3 Comparison between field stress path (FSP) and conventional consolidated undrained (CU) and drained (CD) stress paths (after Brand 1981)

As the second mechanism of rainfall-induced landslides shows, failure of unsaturated soils is often caused by a reduction in the soil suction. During rainfall infiltration, a reduction in suction causes a decrease in the shear strength, which leads to the development of volumetric and deviatoric strains and ultimately to the instability of the soil. This process can be idealized as a wetting path along which the shear stress and the net mean stress keep constant but the suction decreases over time. However, the behavior of soils along the wetting path has seldom been investigated.

In order to analyze the stability of colluvial soil slopes comprehensively, the shear strength and hydraulic properties and distinct deformation and slippage properties (Brand 1981; Mair 1993; He and Wang 2006) need to be determined. Among these properties, the hydraulic properties of colluvium soils are required in the seepage analysis, which is helpful for the investigation of the triggering mechanisms of failure for structures involving colluvium soils, such as colluvial slopes, and is the basis for evaluating the evolution in time of the stability of structures involving colluviums. The shear strength of colluviums is also required in the stability analysis of structures involving colluviums.

1.3 Outline of This Book

To prevent the failure of widely-graded soil deposits under extreme hydraulic conditions (such as reservoir water level fluctuations, long-period rainfall, or their combinations), a deep understanding of the fundamental behavior of widely-graded soil is desired. The knowledge of hydraulic property functions and unsaturated shear strength for such widely-graded soil deposits is required in seepage analysis and stability analysis, and is the foundation for evaluating the evolution in time of the stability of colluvial soil deposits. Further, this knowledge can be used to guide the stabilization of dangerous soil deposits, the maintenance of stabilized deposits, and the monitoring of potentially dangerous deposits.

This book focuses on the hydraulic-mechanical properties and the shear strength of widely-graded soils. This book will improve the understanding of the unsaturated flow characteristics and shearing behavior of widely-graded soils and help evaluate the geo-hazards caused by widely-graded soil landslides. This book can also improve design methods for the stabilization of natural terrain slopes and for landslides risk management.

The structure of this book is as follows:

(1) Chapter 2 summarizes the properties of natural colluvial soils. The basic properties of the colluvial soils obtained from laboratory experiments are presented.
(2) In Chap. 3, the microporosity structures of colluvial soils have been studied. The variation of pore size distribution induced by wetting/drying is investigated. Furthermore, a mathematical model was developed to describe the bimodal PSDs.
(3) In Chap. 4, the techniques for measuring the hydraulic properties of unsaturated coarse granular soils were introduced. A new device for measuring soil water characteristic curves at very low suctions is presented. A "wetting front advancing method" was developed to measure the unsaturated hydraulic conductivity function.
(4) In Chap. 5, the measured soil–water characteristic curves and soil hydraulic conductivity functions for different colluvial soils are presented and discussed. The new bimodal soil hydraulic conductivity function model proposed in Chap. 4 is used to fit experimental data.
(5) Chapter 6 presents theoretical relations among soil fabric, water content, void ratio and soil suction. Using these relations, the PSD is used to estimate the water characteristic curve and the soil shrinkage curve.
(6) In this Chap. 7, the effect of coarse content on the compressibility and shearing behavior of saturated and unsaturated coarse and widely-graded soils under very low confining pressures are investigated. The critical state characteristics of saturated and unsaturated granular soils are also investigated.
(7) In Chap. 8, the bimodal features for coarse widely-graded soil are summarized. A new bimodal SWCC model is proposed. Then, a bimodal soil hydraulic conductivity function model and bimodal shear strength model are proposed.
(8) Chapter 9 describes the seepage analysis technique for unsaturated soils. Through theoretical analysis and numerical tests on unsaturated seepage, selection criteria are proposed to choose the critical mesh size and critical time step that lead to a convergent solution in different nonlinearity degrees.
(9) In Chap. 10, the instability of unsaturated granular soils along the wetting path is investigated experimentally. A method for evaluating the instability of saturated and unsaturated granular soils is proposed. Then, the infiltration process and the stability of colluvial slopes are investigated by numerical simulation.

References

Alonso E, Gens A, Lloret A, Delahaye C (1995) Effect of rain infiltration on the stability of slopes. In: Proceedings of the 1st international conference on unsaturated soils. Paris, France, 6–8 Sept 1995, pp 241–249

Anderson SA, Sitar N (1995) Analysis of rainfall-induced debris flows. J Geotech Eng 121(7):544–552

Au SWC (1998) Rain-induced slope instability in Hong Kong. Eng Geol 51(1):1–36

Brand EW (1981) Some thoughts on rain-induced slope failure. In: Proceedings of the 10th international conference on soil mechanics and foundation engineering, vol 3, London, pp 374–376

Cascini L, Cuomo S, Pastor M, Sorbino G (2010) Modeling of rainfall-induced shallow landslides of the flow-type. J Geotech Geoenviron Eng 136(1):85–98

Chen H, Lee CF, Law KT (2004) Causative mechanisms of rainfall-induced fill slope failures. J Geotech Geoenviron Eng 130(6):593–602

Chu J, Leroueil S, Leong WK (2003) Unstable behavior of sand and its implication for slope stability. Can Geotech J 40(5):873–885

Collins BD, Znidarcic D (2004) Stability analyses of rainfall induced landslides. J Geotech Geoenviron Eng 130(4):362–372

Dai FC, Lee CF, Wang SJ (2003) Characterization of rainfall-induced landslides. Int J Remote Sens 24(23):4817–4834

Dai FC, Lee CF (2002) Landslide characteristics and slope instability modeling using GIS, Lantau Island, Hong Kong, Geomorphology 42(3-4):213-228

Dai FC, Xu C, Yao X, Xu L, Tu XB, Gong QM (2010) Spatial distribution of landslides triggered by the 2008 Ms 8.0 Wenchuan earthquake, China. J Asian Earth Sci 40(4):883–895

Emily YY, Yeung MP, Sharon TT (2015) Hong Kong's emergency and disaster response system

Evans NC, Huang SW, King J (1999) The natural terrain landslide study: phases I and II. Geo report no. 73. Geotechnical Engineering Office, Hong Kong

Fannin RJ, Eliadorani A, Wilkinson JMT (2005) Shear strength of cohesionless soils at low stress. Geotechnique 55(6):467–478

Fourie AB (1996) Predicting rainfall-induced slope instability. Proc Int Conf Inst Civ Eng Geotech Eng 119(4):211–218

Fredlund DG, Rahardjo H (1993) Soil mechanics for unsaturated soils. Wiley, New York

GEO (2004) Geotechnical manual for slopes, 2nd edn (1984)

He KQ, Wang SJ (2006) Double-parameter threshold and its formation mechanism of the colluvial landslide: Xintan landslide. China Environ Geol 49(5):696–707

Lade PV (1992) Static instability and liquefaction of lose fine sandy slopes. Geotech Geol Eng 118(1):51–72

Lade PV (1993) Initiation of static instability in the submarine Nerlerk berm. Can Geotech J 30:895–904

Lade PV, Pradel D (1990) Stability and flow of granular materials: experimental investigation. J Eng Mech 116(11):2532–2550

Lumb P (1962) Effect of rain storms on slope stability. In: Proceedings of the symposium on Hong Kong soils. Hong Kong, pp 73–87

Lumb P (1975) Slope Failures in Hong Kong. Quart J Eng Geol Geol Soc Lond 8:31–55

Mair RJ (1993) Developments in geotechnical engineering research. Proc Inst Civ Eng Lond 93:30–37

Ng CWW, Shi Q (1998) Numerical investigation of the stability of unsaturated soil slopes subjected to transient seepage. Comput Geotech 22(1):1–28

Riemer MF (1992) The effects of testing conditions on the constitutive behavior of loose, saturated sands under monotonic loading. Ph.D. thesis. University of California, Berkeley, CA

Sun HW, Tsui HM (2003) Review of notable landslide incidents during slope works. Geo report no.171. Geotechnical Engineering Office, Hong Kong

The Yangtze Water Conservancy Commission (1997) Engineering geology of the three gorge project. Hubei Science and Technology Press, Wuhan

Wang G, Sassa K (2001) Factors affecting rainfall-induced flowslides in laboratory flume tests. Geotechnique 57(7):587–599

Wu TH (2003) Assessment of landslide hazard under combined loading. Can Geotech J 40(4):821–829

Yang J (2002) Non-uniqueness of flow liquefaction line for loose sand. Geotechnique 52(10):757–760

Yang HP, Wang JS (2009) Geological features and cause analysis of Qianjiangping landslide of July 13, 2003 on three gorges reservior. J Eng Geol 17(2):233–239 (in Chinese)

Zhang LM (2009) Challenges in multi-hazard risk assessment and management: geohazard chain in Beichuan Town caused by Great Wenchuan earthquake. In: Geotechnical risk and safety. Taylor and Francis Group, London, pp 237–244

Zhang LM, Chen Q (2005) Predicting bimodal soil-water characteristic curves. J Geotech Geoenviron Eng 131(5):666–670

Chapter 2
Particle-Size Distribution and Density of Natural Widely-Graded Colluvial Soils

2.1 Introduction

Landslides are an important type of natural hazard that causes catastrophic consequences in terms of human causalities and economic losses. In Hong Kong, more than 100,000 visible natural terrain landslides have been identified (MFJV 2007). In the Three-Gorge Reservoir Zone in China, 684 large slopes with volumes larger than 100,000 m^3 and numerous smaller slopes have been identified (The Yangtze Water Conservancy Commission 1997). The Ms 8.0 Wenchuan earthquake in Sichuan in 2008 induced approximately 50,000 landslides covering a wide area of 285 $\times$ 20 km (Zhang 2009; Dai et al. 2010). A large number of the aforementioned natural slopes are composed of colluvial soils. The grain-size analysis of colluvial soils reveals that many colluvial soils are coarse and widely distributed ranging from clay contents to meter-scale large particles (Zhang and Chen 2005).

After the Wenchuan earthquake, colluvial soil deposits are densely distributed on steep terrains. These soil deposits are marginally stable under normal climate conditions. During rainfall infiltration, the negative pore-water pressure (suction) in an unsaturated soil slope decreases gradually. Consequently, the contributions of soil strength from the negative pore-water pressure decreases, which may lead to failures in numerous soil deposits. Since these soil deposits pose risks to the human safety and economic development, any hazardous soil deposits must be identified and their associated risks minimized. Hence, there is a need to clearly understand the behavior of coarse, widely-graded colluvial soils and to evaluate the instability of such soil deposits under rainfall infiltration.

In this chapter, the properties (i.e., particle-size distribution and density) of natural colluvial soils are summarized. To investigate the behavior of different colluvial soils, laboratory experiments were conducted on artificial soils, which were obtained from natural colluvial soils. For best presenting different natural colluvial soils, the GSDs and density of natural colluvial soils are characterized in this chapter. The representative PSDs of natural soils were used to prepare artificial soils. Some basic

X. Li et al., *Fundamental Behavior of Unsaturated Widely-Graded Soil*,
https://doi.org/10.1007/978-981-19-3402-5_2

soil properties, such as optimum water content and maximum dry density, were measured and are presented here.

2.2 Properties of Natural Colluvial Soils

2.2.1 *Segregation of Soil Particles During a Landslide*

Colluvial soils are widely distributed on natural terrains. Colluvial soils have been subject to transportation and particle segregation (Fig. 2.1). Generally, flow selectively sort large grains upwards and small grains downwards. In other words, coarse particles move to the front and the top surface while fine particles accumulate at the bottom and in the rear part of the following granular body (Iverson and Vallance 2001; Wang and Hutter 2001). As grain-size segregation produces concentrations of large grains at the debris surface, where the velocities are typically the fastest, large grains preferentially migrate to the surge fronts. In the case of a growing heap on an infinite planar surface or within a cylinder of finite radius, the large particles are concentrated onto the lower and outer parts of the heap whereas the small particles are concentrated onto the upper part of the heap and in its core. Particle segregation or sorting occurring in the particle flow process of a landslide induces heterogeneity in the fabric, structure, and properties of landslide deposits. Particle segregation leads to a tremendous number of uncertainties in geological features when managing the risks imposed by large landslides (Costa and Schuster 1988; Schuster et al. 1992; Lacasse et al. 2004; Tang et al. 2006). The mechanism of particle segregation is very complex due to many influencing factors involved, i.e., particle size, density, shape, resilience, etc.

Fig. 2.1 Coarse-grained soils in a landslide deposit containing different coarse contents

2.2.2 *Particle Flow and Segregation in a Landslide Triggered by Wenchuan Earthquake*

2.2.2.1 The Landslide

Zhang et al. (2011a, b) and Zhao et al. (2013) reported an extensive field investigation on a giant fresh deposit of the Donghekou landslide, which was triggered by the Wenchuan earthquake (Fig. 2.2). The Donghekou landslide (105° 6′ 31″ E, 32° 24′ 23″ N) is located at Donghekou Village, Qingchuan Country, Sichuan, which is approximately 3 km to the north of the Yinxiu-Beichuan-Qingchuan fault that triggered the Wenchuan earthquake. It is a middle and low mountainous area in geomorphology, consisting mainly of dolomitic limestone, dark gray and dark siliceous shale, and carbonaceous sericite quartz phyllite. The upper part comprising dolomitic limestone has a sharp slope with gradients of 70°–80° due to a gully cutting. In contrast, the middle and lower part comprising siliceous shale and carbonaceous phyllite has a relatively gentle slope with gradients of 20°–30°. The V-shaped valley at the slope toe provides a favorable topography for the occurrence of potential flow-type landslides, which are often associated with long run-out distances.

At Donghekou, the earthquake lasted approximately 80 s. The prolonged ground shaking caused a shattering-ejection type of landslide (Fig. 2.2) with a total volume of approximately 30 million m^3 (Huang 2008), which caused the loss of approximately 700 people (Xie et al. 2008). It is a typical rapid and long run-out landslide with a height between the scar and toe of approximately 700 m, a length of 2700 m, a width of 150–600 m, and a covering area of 1.09 km^2. The thickness of the landslide deposits varies greatly, from several meters to dozens of meters. According to the characteristics of movement and deposition, the Donghekou landslide can be divided into three main areas: sliding source, flying and traveling area, and deposition and run-out area (Figs. 2.2b and 2.3a). In general, the landslide has a carrot shape with a wide sliding source and a narrow but long deposition and run-out area due to the constraint of topography. The sliding source is located at elevations between 1080 and

(a) (b)

Fig. 2.2 **a** Full view; **b** remote sensing image of the Donghekou landslide

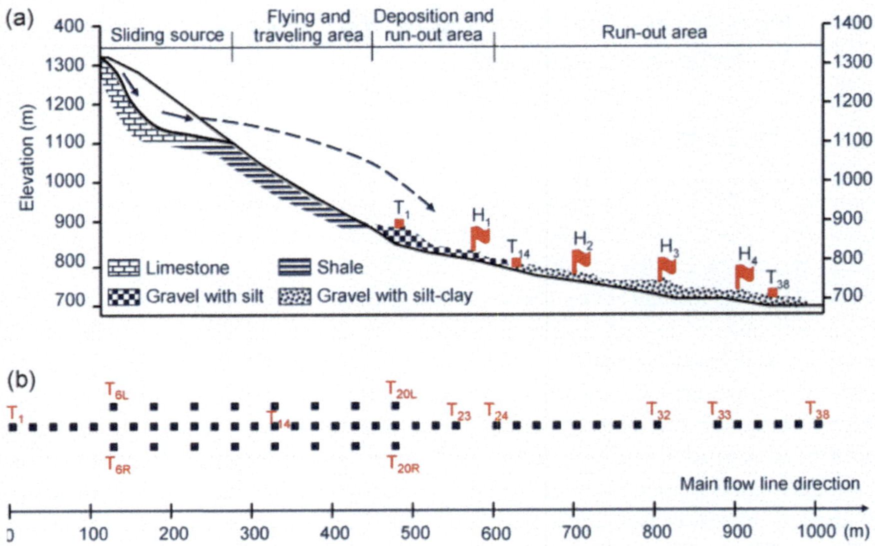

Fig. 2.3 **a** Schematic cross section of the Donghekou landslide (four local highlands of H1–H4 are labeled with flags); **b** layout of sampling locations along the main flow line

1330 m, consisting of highly weathered dolomitic limestone on a shale bedding that is a favorable contact interface for the slip zones. During the earthquake, approximately 6×10^6 m^3 of dolomitic limestone detached from what became the scar with a shape of convex in the middle but concave on the lateral sides (Fig. 2.2). The sliding mass was thrown under the vertical force of the earthquake and crushed in front of a platform formed at the top (Yin et al. 2009). Much of the rock mass was projected for a distance and crashed onto the valley floor in the initial deposition and run-out area at elevations of 800–900 m (Fig. 2.3a). Due to the scraping effect of the high-speed mass movement, the original loose deposits and vegetation on the valley floor were ripped off the slope as shown in Fig. 2.2 as the flying and traveling area. The debris then ran down the hill and deposited over a distance of approximately 1500 m.

The direction of the main mass movement changed twice along the travel path at two obstructive mountainsides (Figs. 2.2 and 2.4). The debris flow with high-speed hit and shaved the left side of the first obstructive mountain, then changed the route of movement (Fig. 2.4a). Unlike the scenario at the first obstructive mountain, part of the flowing debris stopped after hitting the right side of the second obstructive mountain and climbed up about 20 m (Fig. 2.4b). That might be caused by two factors: (1) the deposit traveled at a moderate speed not as high as that at the first obstructive mountain; and (2) the deposit comprised more broken soils and rocks than that at the first obstructive mountain. Then the mass flow changed the direction again and entered the Qingzhu River, resulting in a huge wave. A witness was dragged by the wave into the river but luckily rescued on the downstream. The witness heard a tremendous air explosion all of a sudden at the time of the failure. This illustrates

Fig. 2.4 **a** Hitting area 1; **b** hitting area 2 for the main mass movement along the travel path

that the landslide flow was extremely rapid, which has also been reported by Wang et al. (2009).

2.2.2.2 Test Plan and Procedures of Field Investigation

Shortly after the earthquake, we conducted field surveys at the Donghekou landslide deposit amidst numerous aftershocks. The surveys included identification of particle flow paths and measurements of the particle size, density, and fabric of the landslide debris along the flow paths. Among various factors that can produce particle segregation under certain circumstances, particle size is found to be the most important (Williams 1976). Therefore, great efforts were made to investigate the particle size of the landslide deposit in this study. However, the investigation of the particle-size distribution of landslide debris has never been standardized and is usually assessed by a partial estimation through the analysis of the finer fraction because landslide debris materials often range in size from blocks of tens of cubic meters to clay particles (Casagli et al. 2003). In our field surveys, three methods were adopted, including sieve and hydrometer analysis of fine particles in the laboratory, field sieve analysis of the fine-to-medium fraction (Fig. 2.5) and grid-by-number analysis of the very coarse fraction (Fig. 2.6). The samples were tested at a temporary test site on the landslide deposit. A small amount of soil from each block sample was taken for parallel testing in the laboratory. The field surveys were completed before the originally fresh soil deposit was disturbed by an intensive storm (180 mm of precipitation) on September 23–24, 2008.

In the field surveys, the grain-size distribution of the soil particles smaller than 100 mm was measured using the sieve analysis method. The grain-size distribution and dry unit weight of each of the samples were measured following the standard methods of ASTM (2000). Figure 2.5 shows the field testing procedure for sieve analysis. To obtain a relatively accurate result of soil sieving, double drying of sample soils, including sun-drying and gasoline-burning drying, were carried out to make sample soils as dry as possible. Special large-size sieves and gasoline-burning devices

Fig. 2.5 Sieve analysis for the fine-to-medium fraction of the deposit materials. **a** Digging of block samples; **b** sun-drying of sample soils; **c** re-drying sample soils by gasoline burning; **d** soil sieving

were prepared and used to assist in testing approximately six tones of soil samples on the site. Such a non-standard drying method was adopted to dry approximately 6500 kg of soil on site. The testing of such a large amount of soil would have been impossible if the soil had to be transported to a laboratory for oven drying.

The starting and ending sampling locations (T_1 and T_{38}) for field tests and surveys are shown in Figs. 2.2 and 2.3. Sampling location T_{14} is located at the boundary between the zone of gravel with silt and the zone of gravel with silt–clay (Figs. 2.3 and 2.8b). In total, 54 block samples, 400 × 400 × 400 mm in size, were taken manually at the surface along the main flow line and two side flow lines. Figure 2.3 shows the sampling layout along with the run-out distance of 1000 m. The distance between two neighboring sampling locations was 25 m, except the distance (50 m) between T_{23} and T_{24} due to the presence of a breach (Fig. 2.8d) and the distance (75 m) between T_{32} and T_{33} where a temporary emergency road was constructed after the earthquake. The particle-size distributions of these 54 samples were obtained using sieve analysis and the corresponding in situ bulk densities were measured.

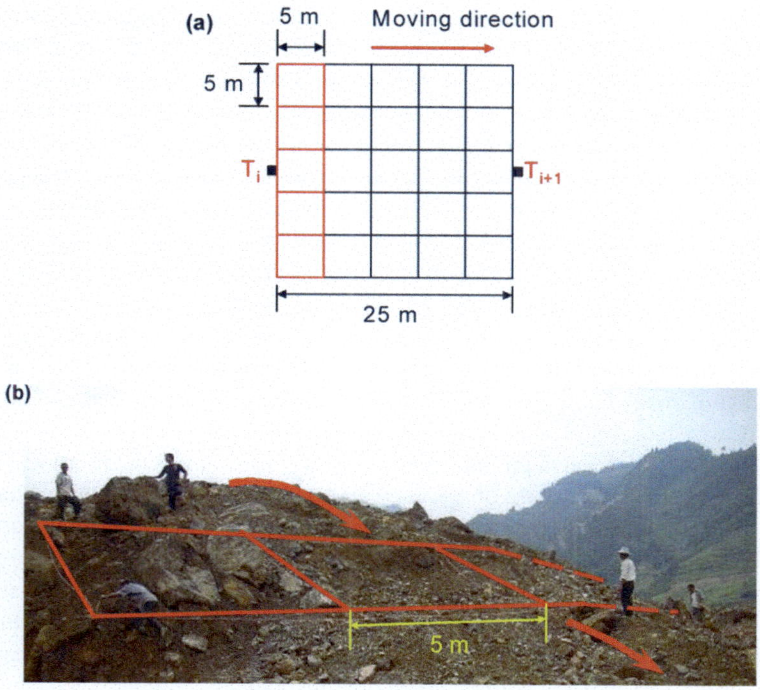

Fig. 2.6 Grid-by-number analysis for the coarser fraction of the deposit materials along a 1000 m long segment of the flow line. **a** Schematic grid for surveying large particles; **b** surveying large particle

Due to the sample size limitation, the grain size of particles larger than 100 mm was determined in situ through a grid-by-number analysis, with the smallest grid being 5 × 5 m in size. Rectangular girds of 25 m × 5 m, each containing five cells of 5 m × 5 m as shown in Fig. 2.6, were surveyed sequentially along a 1000 m long flow line. Within each cell, five groups of particles were identified according to their diameter ranges (11–30 cm, 31–50 cm, 51–100 cm, 101–300 cm, >300 cm) and the numbers of different-sized particles were counted. The total numbers of different-sized particles within a square grid of 25 m × 25 m are used for statistical analysis. The surveys were conducted sequentially over a 1000 m range along the main flow line from the starting point (T_1) to the ending point (T_{38}) as shown in Fig. 2.3 at an advancing interval of 5 m.

2.2.2.3 Segregation of Fine Particles

The grain-size distribution of a soil can be represented by several characteristic particle diameters of the soil, namely D_{60}, D_{50}, D_{30}, and D_{10} (particle diameters at 60, 50, 30 and 10% finer by weight). The values of these characteristic diameters at 38 sampling locations along a 1000 m long segment of the main flow line are presented in Table 2.1. The variations in the mean particle size, D_{50}, along the main flow line are shown in Fig. 2.7a and discussed to illustrate how these characteristic diameters vary along the flow line. The values of D_{50} are mostly in the range of 10–60 mm and exhibit a weakly increasing trend with increased run-out distance. The fluctuations of D_{50} become progressively larger with increased run-out distance. In an ideal particle flow without crushing, D_{50} would increase with increased run-out distance more markedly (Jullien and Meakin 1990; Iverson and Vallance 2001). The reason for the weaker trend in the present study is that the landslide debris underwent disintegration, particle crushing, and mixing with local landslide materials as it moved down the valley. This can be clearly seen by comparing the particles in the initial deposition area (Fig. 2.8a) and the frontal area (Fig. 2.8f). The deposit in Fig. 2.8a is mainly comprised of rocks and gravel and it is more uniform in particle size than that in Fig. 2.8f, which is comprised of silt with scattered rocks and gravel.

In Fig. 2.7a, four local maximum values of D_{50} correspond to the four local highlands, H_1–H_4, in Fig. 2.3a. This observation confirms that the segregation mechanism works by inverse grading, with large particles moving to the top and small particles accumulating at the bottom (Miyamoto et al. 2007). The percolation of fine particles and the rise of coarse particles took place (Iverson and Vallance 2001). For instance, Fig. 2.8e shows the third local highland, H_3, with the feature of inverse grading on a 20 m high exposure. The thin surface layer with a cluster of boulders lies on a layer of primarily sandy silt with scattered floating boulders. Similar inverse grading phenomenon was also observed in a large landslide that occurred in Siaolin Village in southern Taiwan on September 9, 2009 (Lee et al. 2009). In that catastrophic landslide event, the landslide volume was 23.9 million m^3 and the deposition and run-out distance was over 1600 m. Reversely graded bedding was found in the deposits in the traveling area. Zhou (2010) summarized that two stages might exist for particle segregation in a deposition process. In the first stage, most of the coarse particles flow upwards and concentrate in the top layers while fine particles are squeezed into the bottom layer, which is referred to as a squeeze expulsion mechanism. In the second stage, strong shearing occurs in the layers near the free surface and fine particles are observed penetrating into the granular body through the voids formed by coarse particles, which is referred to as a kinematic sieving mechanism. Therefore, both mechanical (squeeze expulsion) and geometrical (kinematic sieving) effects contribute to particle segregation in the flow deposition process.

To understand the particle sorting in the initial accumulation area, the variations of D_{50} along the left and right sides of the main flow line (Fig. 2.3b) are compared in Fig. 2.7a. Two features are observed: the value of D_{50} on the left is larger than that on the right at a small distance; after a critical distance of about 350 m (the transitional zone in Figs. 2.3 and 2.8b), the values of D_{50} on the left, in the middle, and on the right become less differentiable. This observation agrees well with the material composition of the landslide. The wide initial accumulation area can be divided into two parts involving two distinct types of deposits with an obvious borderline shown in Fig. 2.8c. The left part is comprised of gravel with a small fines content while the right part is comprised of soil with some gravel. As the deposit traveled down the

Table 2.1 Particle-size distributions of the landslide debris finer than 100 mm at the 38 sampling locations along the 1000 m flow line investigated in this study

Sampling point	Distance from T_1 (m)	Particle diameter (mm)			
		D_{60}	D_{50}	D_{30}	D_{10}
T_1	0	39.0	28.5	12.0	2.9
T_2	25	26.0	16.5	5.9	1.0
T_3	50	40.0	29.0	12.5	2.7
T_4	75	42.0	29.0	11.5	2.9
T_5	100	31.0	20.0	6.5	1.3
T_6	125	29.5	21.5	9.0	2.3
T_7	150	43.0	30.0	12.5	2.9
T_8	175	28.5	18.5	6.8	2.0
T_9	200	40.0	26.0	9.5	2.3
T_{10}	225	49.0	34.0	13.5	3.2
T_{11}	250	47.0	34.0	14.0	2.7
T_{12}	275	61.0	47.0	15.0	3.0
T_{13}	300	30.0	21.0	9.4	2.2
T_{14}	325	41.0	25.0	8.0	1.4
T_{15}	350	20.5	13.0	5.4	1.2
T_{16}	375	38.0	25.0	10.0	2.1
T_{17}	400	37.0	24.0	8.0	1.5
T_{18}	425	27.0	19.0	8.0	1.3
T_{19}	450	40.0	25.0	8.2	1.6
T_{20}	475	42.0	25.0	7.5	0.9
T_{21}	500	63.0	52.0	14.0	1.5
T_{22}	525	43.0	26.0	8.0	1.8
T_{23}	550	50.0	34.0	13.5	3.5
T_{24}	600	45.0	32.0	13.0	3.0
T_{25}	625	60.0	46.0	16.5	3.7
T_{26}	650	66.0	54.0	21.0	4.3
T_{27}	675	61.0	46.0	13.0	1.0
T_{28}	700	23.0	15.0	5.5	0.8
T_{29}	725	47.0	37.0	19.0	4.7
T_{30}	750	57.0	41.5	14.0	2.3
T_{31}	775	56.0	38.0	11.5	1.4
T_{32}	800	23.0	15.5	5.6	0.65
T_{33}	875	54.0	35.0	8.0	0.7
T_{34}	900	71.0	62.0	20.0	2.1

(continued)

Table 2.1 (continued)

Sampling point	Distance from T_1 (m)	Particle diameter (mm)			
		D_{60}	D_{50}	D_{30}	D_{10}
T_{35}	925	36.0	23.0	7.0	0.63
T_{36}	950	54.0	38.0	14.0	2.3
T_{37}	975	54.0	35.0	9.3	0.8
T_{38}	1000	22.0	13.0	4.3	0.62

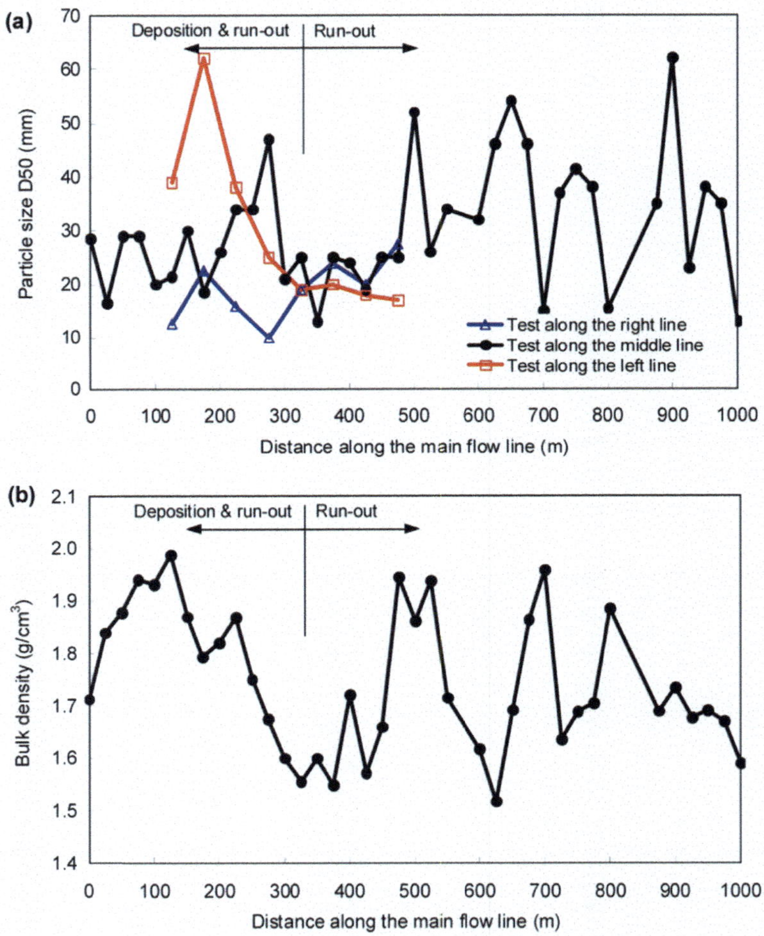

Fig. 2.7 **a** Variation of particle size D_{50} along three flow lines; **b** variation of bulk density along the main flow line

Fig. 2.8 Local geographic and soil conditions along the main flow line. **a** The initial zone with large quantities of rocks and gravel; **b** transitional zone; **c** the zone with the right part lying over the left part; **d** the breach associated with Quake lake 1 in Fig. 2.2; **e** the third local highland; and **f** the frontal zone with limestone from the landslide scar distributed along the flow line

slope, the right part tended to lay over and mix with the middle and left parts due to the topographical constraints (Fig. 2.8c), and hence, the material variability along the transverse line became less significant.

The variations in the soil bulk density with the run-out distance along the main flow line are shown in Fig. 2.7b. The bulk density ranges from 1.5 to 2.0 g cm^{-3}. Two locations of low bulk densities are notable in the transitional zone at a distance of 325–375 m (Fig. 2.8b) and in the breach zone at a distance of 625 m (Fig. 2.8d). In the transition zone, the materials were mainly from "free fall" and not subjected to much transportation and mixing. The breach zone, on the other hand, was a potential lowland filled with relatively fine particles. The bulk density values in Fig. 2.7b do not correspond well to the D_{50} values in Fig. 2.7a as D_{50} only represents a part of the landslide composition.

2.2.2.4 Segregation of Coarse Particles

The variations in the numbers of five groups of coarse particles within a 25 m × 25 m cell (diameter = 11–30 cm, 31–50 cm, 51–100 cm, 101–300 cm, >300 cm) with the run-out distance are shown in Fig. 2.9. The characteristics of the mass movement are different in the initial deposition and run-out area and the frontal run-out area (Figs. 2.2 and 2.3). In the initial deposition and run-out area, the rock mass projected by the action of the earthquake landed on the valley floor at high speed. The grids in which the maximum numbers of different-sized large particles occur (Fig. 2.9) are all in the initial deposition and run-out area. The locations of these grids for the

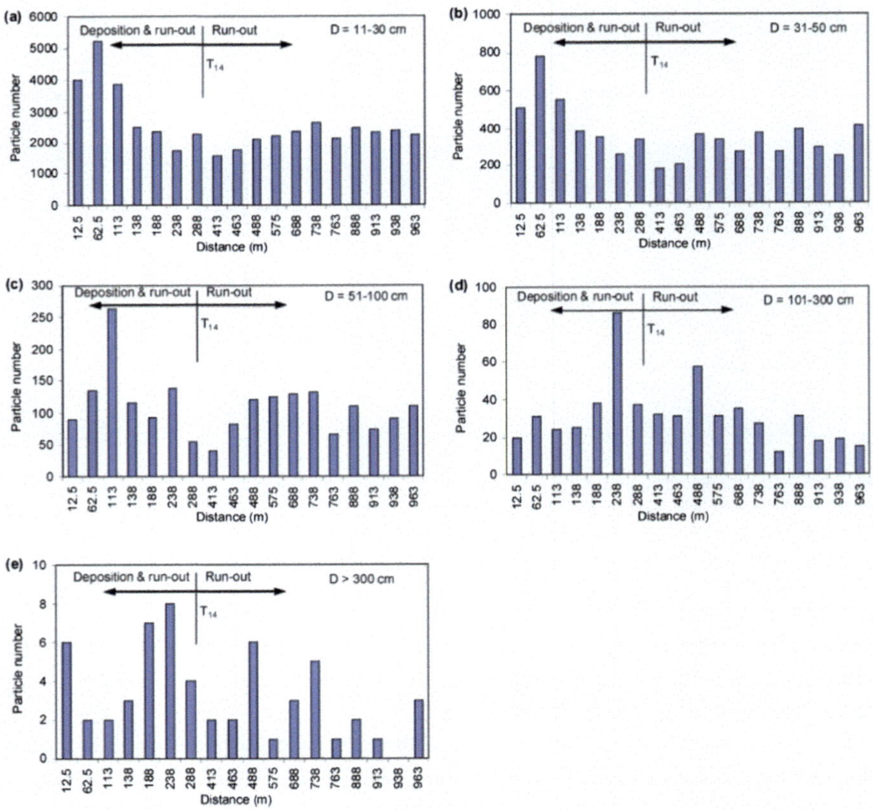

Fig. 2.9 Variations in the numbers of different-sized particles within each 25 m × 25 m cell with distance along the main flow line. **a** Particles with diameters of 11–30 cm; **b** particles with diameters of 31–50 cm; **c** particles with diameters of 51–100 cm; **d** particles with diameters of 101–300 cm; and **e** particles with diameters larger than 300 cm

five particle groups are at 62.5 m, 62.5 m, 113 m, 238 m, and 238 m, respectively, from the starting sampling point, T_1. The flying distance of particles in a particular particle-diameter range tends to increase with the mean particle diameter. This can be explained by the mechanism of trajectory segregation (Anderson and Bunas 1993), in which the distance that a particle travels after being projected, termed as the stopping distance, is proportional to the square of the particle diameter.

In the run-out zone starting from T_{14} (325 m from T_1), no projected debris was deposited and the materials simply flowed down the valley. Near the transition line at T_{14}, the large particles in all the five diameter groups appear to be minimal and the soil density is low. The numbers of large particles in the first two diameter groups (diameter = 11–30 cm and 31–50 cm) increase with the run-out distance along the main flow line and reach relatively constant values (Fig. 2.9a, b). The run-out distance from T_{14} at which the maximum particle number is found is defined as the characteristic flow distance. The characteristic flow distances for the above two particle groups are 338 m and 138 m, respectively. The trend in the run-out zone in Fig. 2.9a, b agrees with that observed in typical particle flows (Anderson and Bunas 1993): after a characteristic flow distance, the deposition of large particles tend to be steady. However, the observed characteristic flow distances do not increase with the particle diameter as expected in an idealized particle flow. The numbers of very large particles in the last three diameter groups (diameter = 51–100 cm, 101–300 cm, and >300 cm) do not increase with the run-out distance. Instead, they appear to decrease along the main flow line (Fig. 2.9c–e). There are two mechanisms at work in this landslide event. On the one hand, particle flow caused larger particles to travel further away from the scar. On the other hand, the released rock mass/aggregates disintegrated and large particles broke along the flow path due to internal interactions and impacts with the boundaries, leading to more small particles (Figs. 2.7a and 2.9a, b) and fewer large particles (Fig. 2.9c–e) along the flow path. The variations in the volume percentages of the five groups of particles with the run-out distance are shown in Fig. 2.10. A specific-sized particle is assumed as a sphere for volume calculation.

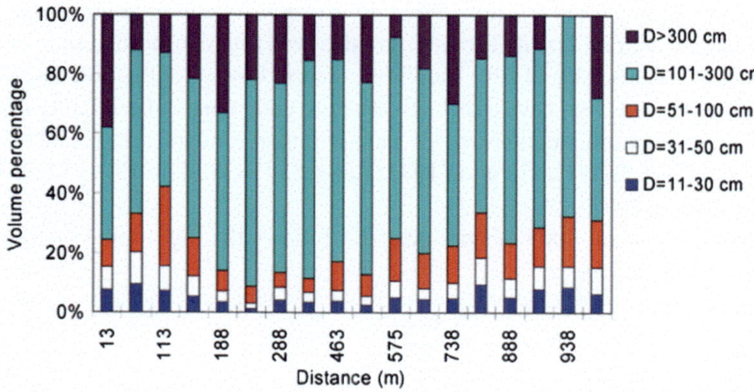

Fig. 2.10 Variations in volume percentages of the five groups of particles

The representative diameter values are assumed to be 20 cm, 40 cm, 75 cm, 2 m, and 3 m for the five groups of particles, respectively. As indicated in Fig. 2.10, the volume of particles with diameters of 1–3 m is the largest along the main flow line. Note that the volume percentages reported here are for the particles on the surface of the landslide deposit, not for those particles within a unit volume of material in the deposit.

Based on the above surveying results for both fine and coarse particles, it is found that the process of particle sorting and fabric changes during a large landslide differs from processes occurring in an ideal granular mass flow in a number of ways. First, the large landslide process involves the detachment of a vast amount of rock mass from the landslide scar. The rock mass disintegrates and large particles break gradually along the movement path. Second, the grain sizes of landslide materials are widely distributed, ranging from nanometers to a few meters. Third, the mass movement in an earthquake-induced landslide involves projection, long distance flying, sliding, rolling, trajectory segregation, inverse grading, etc. It is extremely challenging to simulate the entire process of particle sorting and fabric changes, which requires further large-scale field tests/investigations and robust theoretical studies.

2.2.3 *Grain-Size Distribution Analysis of Natural Colluvial Soil*

Figure 2.11 shows the cumulative grain-size distributions and grain-size density curves of 32 colluvial soils from a slope in the Three-Gorge reservoir zone, a natural slope in Hong Kong, and a recent earthquake-induced landslide in Sichuan, China. The colluvial soils are coarse and the size of colluvial soil grains is widely distributed, ranging from clay contents to meter-scale large particles. The grain-size density curves show the amounts of particles over intervals of grain sizes and reveal the grain-size distributions more clearly. These colluvial soils are widely and gap graded with a limited sand fraction; the grain-size density curves are bimodal. Such bimodal feature of grain-size distribution has also been found in the materials forming part of rockfill dams, glacial tills, mudflows, debris flows, solifluction sheets, and residual soil deposits that have a distinct structure and consist of a mixture of large (gravel or hard clay fragments) particles and a soft matrix of soil (clay, sand or sand-silt–clay mixture) (Deere and Patton 1971; Legget 1976; Vallejo 1979, 1989, 1997; Loganathan et al. 1992).

Assume the general GSD of unimodal soil follows the log-normal distribution, which can be written as:

$$P(d) = \frac{1}{\sigma\sqrt{2\pi}} e^{-0.5\times[(\log d - u)/\sigma]^2} \tag{2.1}$$

where d is the particle diameter, $P(d)$ is the differential weight proportion at particle diameter of d, μ is the arithmetic mean, and σ is the standard deviation.

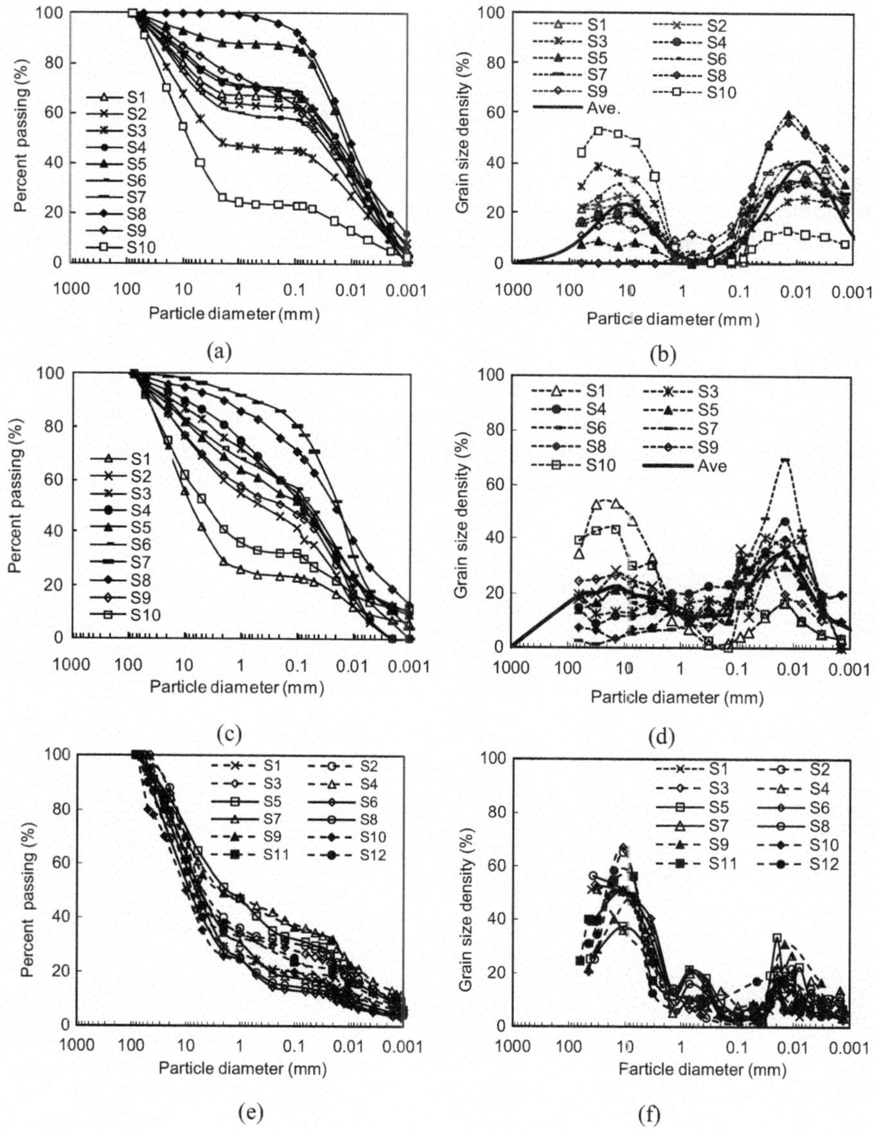

Fig. 2.11 Grain-size distributions and density curves of soils in **a**, **b** soils from a slope in the Three-Gorge reservoir zone; **c**, **d** soils from a slope in Hong Kong; **e**, **f** soils from a new landslide in Sichuan

The GSD of a bimodal or widely-graded soil can be viewed as a combination of two unimodal soils, which follows log-normal distributions.

$$P(d) = wP_1(d) + (1 - w)P_2(d) \tag{2.2}$$

where $P(d)$ is the differential weight at particle diameter of d within 1 g of solid, w is a weighting factor for the unimodal soil, which has a differential weight proportion $P_1(d)$. The weighting factor of the unimodal soil, which has a differential weight proportion $P_2(d)$, is $(1 - w)$. $P_1(d)$ and $P_2(d)$ follow Eq. (2.1).

Equations (2.1) and (2.2) were used to characterize the GSDs of soils from six natural deposits located in Hong Kong and one natural landslide deposit located in TGRZ. These landslide deposits are listed below:

(1) Landslide on the Shum Wan Road on August 13, 1995 (GEO 1996).
(2) Landslide at Kwun Lung Lau on July 23, 1994 (N.R. Morgenstern 2000).
(3) Landslide in the Three-Gorge area (The Yangtze Water Conservancy Commission 1997).
(4) Landslide above South Lantau Road on August 24, 1999 (GEO and Fugro Scott Wilson Joint Venture 2001).
(5) Landslide at the Leung King Estate on April 14, 2000 (GEO and Halcrow China Ltd. 2002).
(6) Landslide near 2B Bowen Road on August 24, 2000 (GEO and Halcrow China Ltd. 2002).
(7) Landslide at Queen's Hill on July 5, 1998 (Fugro Scott Wilson Joint Venture 2001).

The parameters describing the best-fit GSDCs for the soils from seven natural landslide deposits are shown in Table 2.2. The best-fit GSDCs and grain density function (GDF) curves of these natural soils are plotted in Figs. 2.3 and 2.4. Figure 2.12 shows bimodal grain-size density curves, which contain two dominant particle-size zones, whereas Fig. 2.13 shows unimodal grain-size density curves, in which only one dominant particle-size zone is found.

2.3 Soil Samples for Representing In-Situ Soil Behavior

2.3.1 Grain-Size Distribution Design of Artificial Soils for Laboratory Experiments

To simulate the behavior of natural soils, a soil with GSD similar to that of natural soil should be chosen for laboratory experiments. A coarse soil and a fine soil were separately used to simulate the coarse parts and the fine parts of four bimodal soils. Figure 2.14a shows the GSDs of the fine parts of four bimodal natural soils. The GSD of a design fine soil, which has a mean diameter, d_{50}, of 0.0185 mm, is also

Table 2.2 Characterization of grain-size distributions of soils form seven natural landslides using log-normal distributions

Deposits	Feature	Particle diameters (mm)	Parameters	Values (in log-scale)	Mean particle diameter (mm)
TGRZ	Bimodal	Coarse (0.5–80)	μ	1.069	11.73
			σ	0.5424	
		Fine (0.001–0.5)	μ	−1.954	0.0111
			σ	0.547	
Bowen Road	Bimodal	Coarse (0.5–80)	μ	0.33	6.76
			σ	0.352	
		Fine (0.001–0.5)	μ	−1.797	0.016
			σ	0.657	
South Lantau Road	Bimodal	Coarse (0.5–80)	μ	0.138	1.375
			σ	0.574	
		Fine (0.001–0.5)	μ	−1.545	0.0226
			σ	1.317	
Kwun Lung Lau	Unimodal	(0.001–80)	μ	−0.62	0.24
			σ	1.19	
Leung King Estate	Unimodal	(0.001–80)	μ	−0.712	0.194
			σ	1.06	
Shum Wan Road	Unimodal	(0.001–80)	μ	−1.77	0.017
			σ	0.73	
Queen's Hill	Bimodal	Coarse (0.5–80)	μ	0.753	5.66
			σ	0.382	
		Fine (0.001–0.5)	μ	−1.539	0.289
			σ	0.66	

μ is the arithmetic mean of the log-normal grain-size distribution function; σ is the standard deviation of the log-normal grain-size distribution function

shown in Fig. 2.14a and can fit the GSD of natural soils well. Figure 2.13b shows the GSDs of the coarse parts of several bimodal natural soils. The GSD of a design coarse soil, which has a mean diameter, d_{50}, of 6 mm, is also shown in Fig. 2.13b and seems to be similar to those of the natural soils.

To investigate the behavior of the colluvial soils, the fines and the coarse soil were further mixed according to five different weight proportions (as shown in Fig. 2.15a): 20% coarse soil +80% fines; 50% coarse soil +50% fines; 68% coarse soil +32% fines; 75% coarse soil +25% fines and 90% coarse soil +10% fines. The particle-size distributions of these test soils represent field widely and coarse-grained soils and are shown in Fig. 2.15b. The mean particle diameter, D_{50}, ranges between 0.015 and 3.5 mm. The grain-size density curves in Fig. 2.15b exhibit bimodal or trimodal features. The five design soils are classified as GW-GM with sand, SM with gravel,

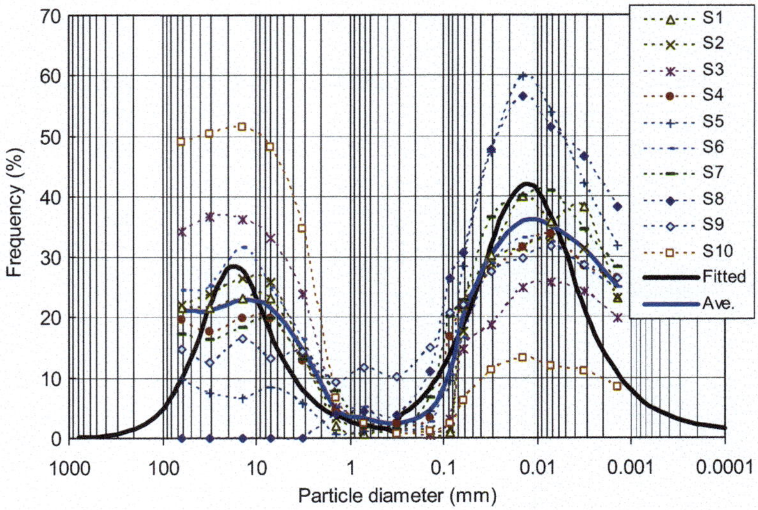

(a) soils in a landslide in the Three-Gorge Reservoir zone (The Yangtze Water Conservancy Commission, 1997)

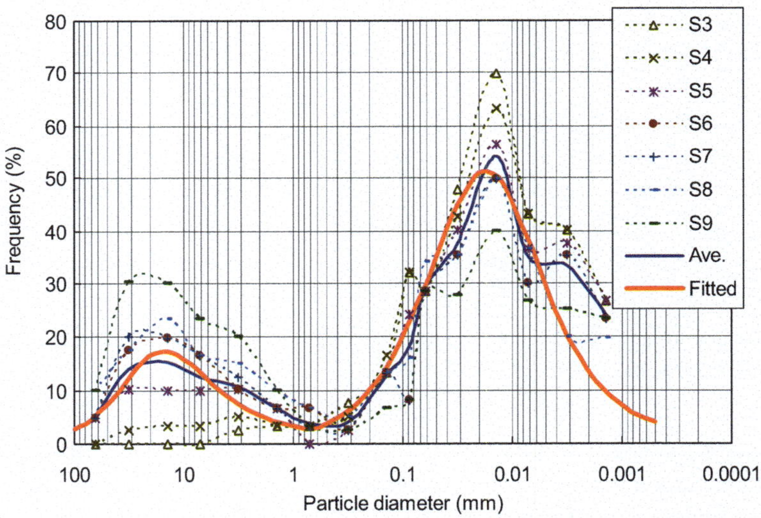

(b) soils in a landslide in Bowen Road on 24 August 2000 (GEO and Halcrow China Ltd., 2002)

Fig. 2.12 Bimodal GSDs of colluvial soils

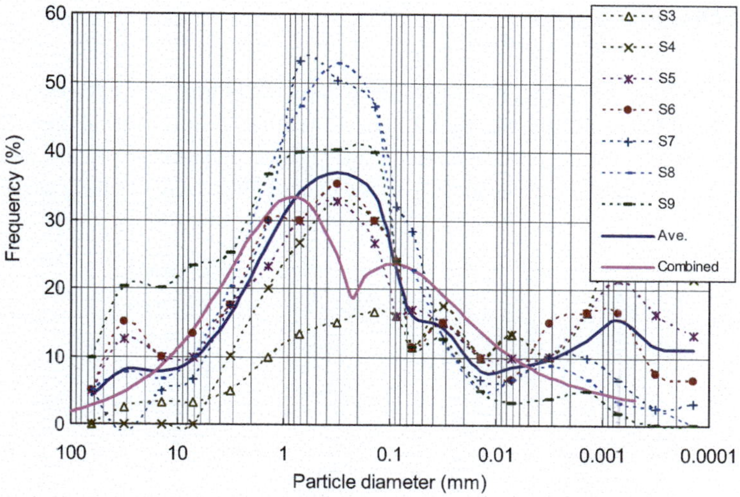

(c) soils form a landslide above South Lantou Road on 24 August 1999 (GEO and Fugro Scott Wilson Joint Venture , 2001)

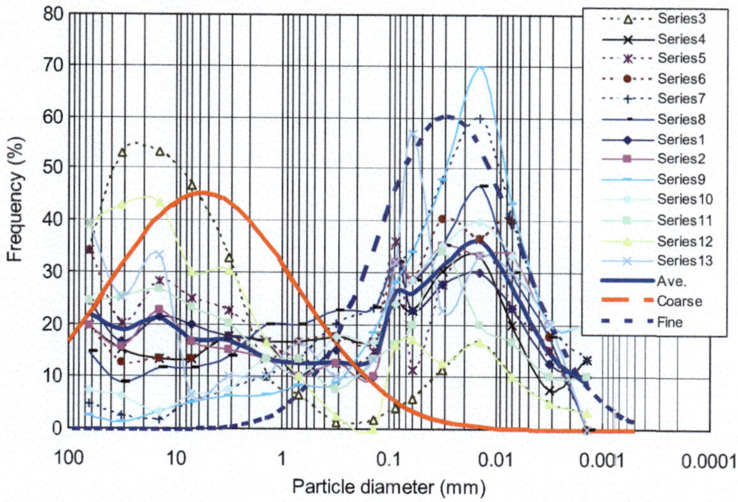

(d) a landslide at Queen's Hill on 5 July 1998 (Fugro Scott Wilson Joint Venture, 2001)

Fig. 2.12 (continued)

SC with gravel, sandy ML, and CL with sand (as listed in Table 2.3). The five designed soils have different gravel (particles with diameter in range of 4.75–75 mm according to ASTM D2487) proportions, i.e., 58%, 44%, 29%, 15%, and 0%, respectively.

2.3.2 *Soil Density Design*

Most fill slopes in Hong Kong, specifically those constructed before the establishment of the Geotechnical Controlling Office (GCO), used weathered soils as fill materials, for example completely decomposed granite (CDG). The typical range of void ratio of these fill soils is from 0.6 to 1.3 according to the data of 68 pre-GCO filling slopes

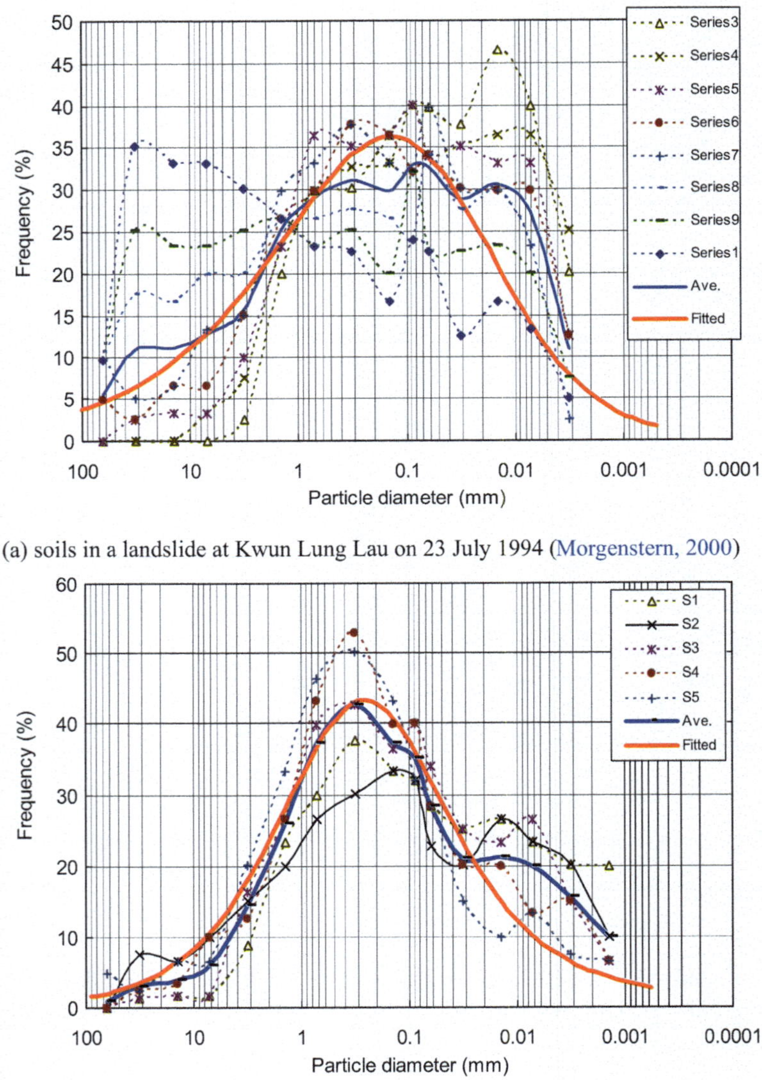

(b) a landslide at Leung King Estate on 14 April 2000 (GEO and Halcrow China Ltd. , 2002)

Fig. 2.13 Unimodal GSDs of colluvial soils

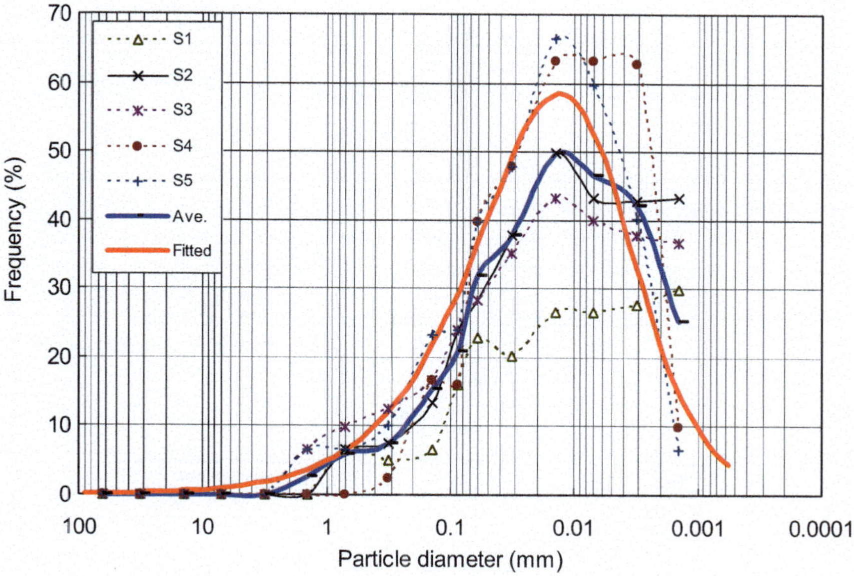

(c) a landslide on Shum Wan Road on 13 August 1995 (GEO, 1996)

Fig. 2.13 (continued)

(Fung 2001). The relative compactions, which are defined as the ratio between the dry density and the maximum dry density, of 100 samples taken from 68 pre-GCO fill slopes were measured and found to locate in the range between 65 and 85%. Hence, these filling slopes were considered to be in a very loose state (Fung 2001).

From the compaction tests, the maximum dry density of the CDG from Shau Kei Wun is 1650 kg/m^3, at an optimum water content of 21%, and the maximum dry density of the CDG taken from Beacon Hill is 1870 kg/m^3 at an optimum water content of 13% (Fung 2001). Yan (2003) also reported a similar CDG maximum dry density of 1.71 g/cm^3 at an optimum water content of 17.4%. Referring to Geotechnical Manual for Slopes (GEO 2004), a fill slope should be compacted to not less than 95% of the maximum dry density. Therefore, a relative compaction value of 0.85 can be used to represent the pre-GCO fill slopes and the natural loose slopes. A relative compaction value of 0.95 can be used to represent the standard fill slopes. In the laboratory experiments, soil dry density values are chosen with relative compaction values from 0.85 to 0.95.

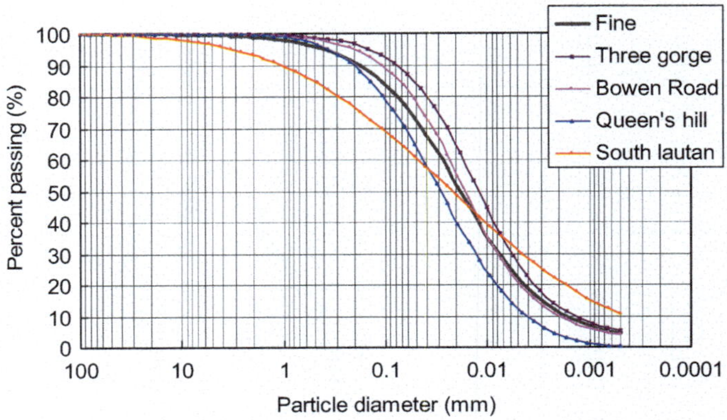

(a)

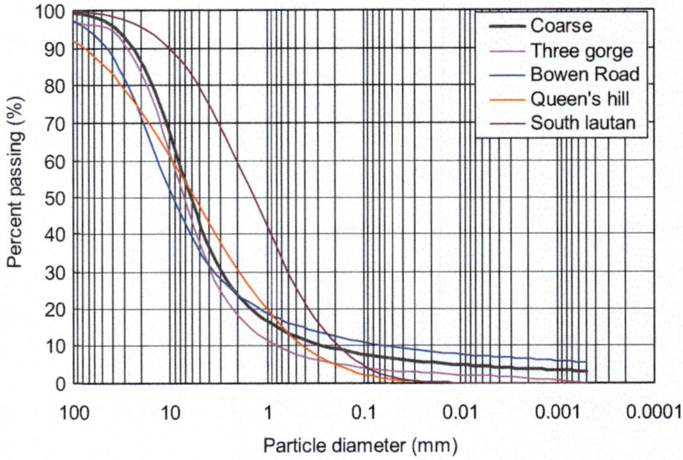

(b)

Fig. 2.14 GSDs of design fine soil and design coarse soil: **a** Fine parts of four natural bimodal soils. **b** Coarse parts of four natural bimodal soils

2.4 Soil Specimen Preparation

The soils used in the laboratory tests were prepared from natural completely decomposed granite at a construction site at Beacon Hill, Hong Kong. First, the natural CDG soil was sieved into 10 portions according to its particle sizes, namely <0.063 mm, 0.063–0.125 mm, 0.125–0.185 mm, 0.185–0.6 mm, 0.6–2 mm, 2–4.25 mm, 4.25–6 mm, 6–10 mm, 10–14 mm, and 14–20 mm. The grain-size distribution of the finest portion with a maximum particle diameter of 0.063 mm was measured using a hydrometer test (BS 1377, part 2 1990).

Second, a fine-grained soil was obtained by mixing the finest particle portion with particle diameters smaller than 0.063 mm and the particle portion with particle

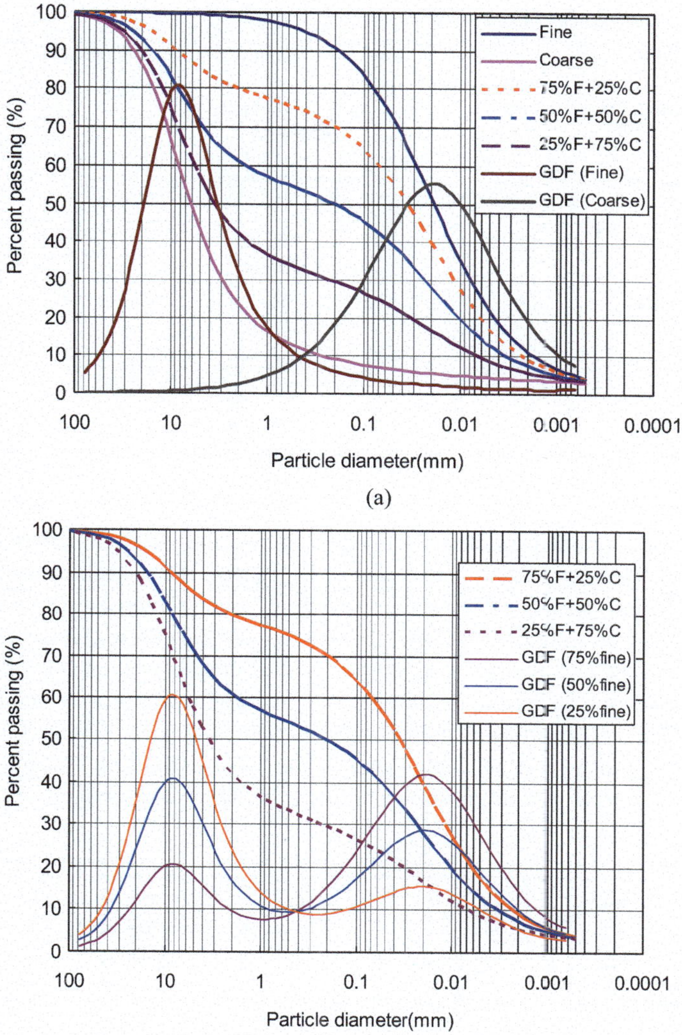

(a)

(b)

Fig. 2.15 GSDs of five design soils: **a** GSDs of five design soils. **b** Bimodal GSDs of the three widely-graded soils

diameters in the range of 0.063–0.125 mm by a weight proportion of 0.72: 0.28. Limited by the sieving technique, the grains with particle class smaller than 0.063 mm cannot be sieved into smaller particle classes. Hence, the GSD of the designed soil in this particle class must have a shape similar to that of the natural CDG soil. As shown in Fig. 2.16, the GSD of the designed fine soil in the particle class smaller than 0.063 mm is similar to that of the natural CDG (measured by hydrometer tests and normalized by the weight percentage of grains smaller than 0.063 mm).

Table 2.3 Soil properties of five design widely-graded soils

Group symbols (ASTM D2487)	Maximum dry density (kg/m^3)	Optimum water content (%)	Dry density (g/cm^3)	Saturated soil hydraulic conductivity (m/s)	Proportions		
					Fines (%)	Sand (%)	Gravel (%)
GW-GM with sand	1950	9	1.78	5E−3 to 3E−2	7	35	58
SM with gravel	1970	11	1.81	3E−5 to 5E−4	24	32	44
SC with gravel	1900	13	1.62	2E−7 to 4E−6	42	29	29
Sandy ML	1710	18	1.45	1E−7 to 3E−6	60	25	15
CL with sand	1550	21	1.32	3E−8 to 3E−7	80	20	0

Third, coarse-grained soil was obtained by mixing the different particle portions following the designed GSD (as shown in Fig. 2.16). The particles with diameters above 20 mm are cut off for practical reasons when the coarse soil was mixed according to the designed GSDs. After the fine soil and the coarse soil were prepared from the natural CDG, the fine-grained soil and the coarse-grained soil were mixed into three new soils according to three different weight proportions, namely 75% coarse-grained soil mixed with 25% fine-grained soil, 50% coarse-grained soil mixed with 50% fine-grained soil, and 25% coarse-grained soil mixed with 75% fine-grained soil. The GSDs of these soils are shown in Fig. 2.16. The coarse contents, f_c, of these

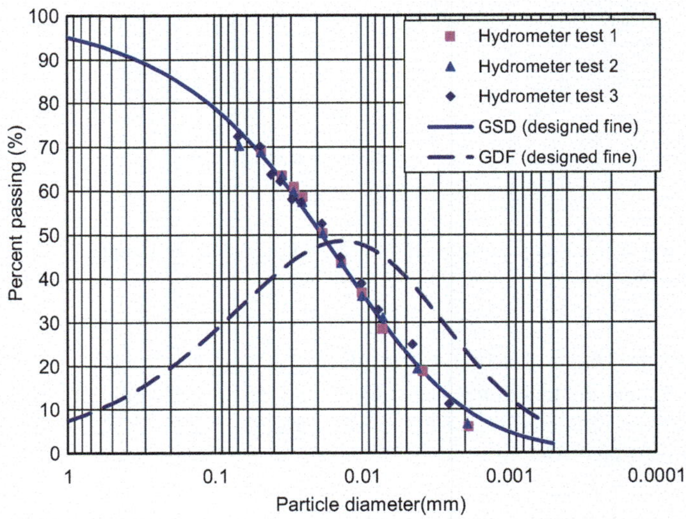

Fig. 2.16 GSD of fine fractions in natural CDG measured using hydrometer tests

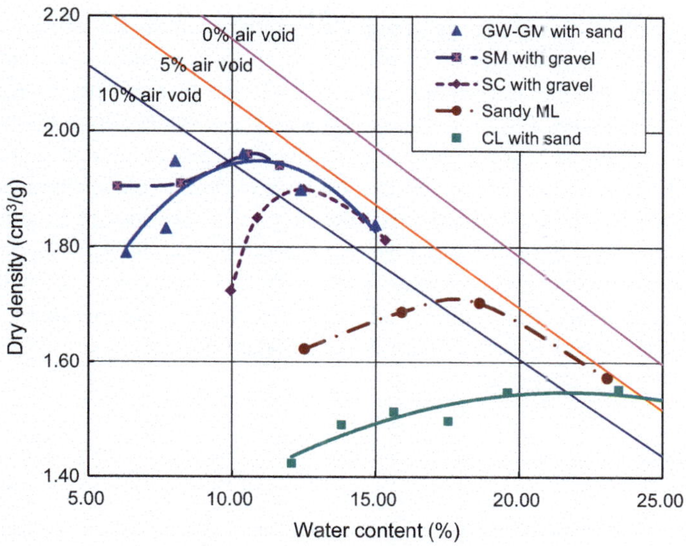

Fig. 2.17 Compaction curves of five colluvial soils used in the laboratory tests

soils are 20, 50, 68, 75, and 90%, respectively. The five types of soil are named as $f_c = 20\%, f_c = 50\%, f_c = 68\%, f_c = 75\%$, and $f_c = 90\%$.

2.5 Basic Soil Properties

The standard compaction test (BS 1377) was conducted and the compaction curves of the five design soils are shown in Fig. 2.17. The detailed parameters are shown in Table 2.3. The liquid limit, plastic limit, and shrinkage limit of CL with sand are 47, 32, and 8%, respectively. The specific gravity of the five soils is 2.64.

2.6 Summary

In this chapter, the characterization of natural widely-graded colluvial soils is reviewed. One fine soil is designed to represent the fine parts of natural colluvial soils and one coarse soil is designed to represent the coarse parts of natural colluvial soils. The mixed soils from the fine soil and the coarse soil will then represent different natural bimodal colluvial soils. Relative compaction values from 0.85 to 0.95 are appropriate to represent natural slopes.

Zhang et al. (2011a, b) conducted a field investigation on a giant fresh deposit of the Donghekou landslide, which was triggered by the Wenchuan earthquake.

Fascinating particle flow and segregation characteristics in the landslide process are found through field investigation of changes in the soil particle size, density, and fabric along the particle movement paths. They found that soil properties such as particle size, D_{60}, D_{50}, D_{30}, and D_{10} varied with the distance along the main flow line. So even when one soil deposit involving the same parent mineral is studied, the soils at different locations of the deposit still have different coarse contents.

Five soils were obtained by mixing different particle portions of a natural CDG. The five design soils have different coarser proportions (i.e., 90%, 75%, 68%, 50%, and 20%, respectively) and are classified as GW-GM with sand, SM with gravel, SC with gravel, sandy ML, and CL with sand, respectively. Among them, SM with gravel, SC with gravel and sandy ML show bimodal GSDs.

References

ASTM D2487 (2000) Standard practice for classification of soils for engineering purposes (unified soil classification system). In: 2000 annual book of ASTM standards. American Society for Testing and Materials, Philadelphia, PA

Anderson RS, Bunas KL (1993) Grain size segregation and stratigraphy in aeolian ripples modelled with a celluar automaton. Nature 365:740–743

BS 1377 (1990) Compaction-related tests. In: Methods of test for soils for civil engineering purposes. British Standards Institution, London

Casagli N, Ermini L, Rosati G (2003) Determining grain size distribution of the material composing landslide dams in the Northern Apennines: sampling and processing methods. Eng Geol 69:83–97

Costa JE, Schuster RL (1988) The formation and failure of natural dams. Geol Soc Am Bull 100:1054–1068

Dai FC, Xu C, Yao X, Xu L, Tu XB, Gong QM (2010) Spatial distribution of landslides triggered by the 2008 Ms 8.0 Wenchuan earthquake, China. J Asian Earth Sci 40(4):883–895

Deere DU, Patton FD (1971) Slope stability in residual soils. In: Proceedings of the 4th Pan-American conference on soil mechanics and foundation engineering, vol 1. San Juan, Puerto Rico, pp 87–170

Fugro (2001) Investigation of some selected landslides in 1998, vol 5. GEO report no. 112. Geotechnical Engineering Office, Civil Engineering Department, The Government of the Hong Kong Special Administrative Region

Fung WT (2001) Experimental study and centrifuge modeling of loose fill slope. Thesis. The Hong Kong University and Technology

GEO and Fugro (2001) Investigation of some selected landslides in 1999, vol 1. GEO no. 120. Geotechnical Engineering Office, Civil Engineering Department, The Government of Hong Kong Special Administrative Region

GEO and Halcrow China Ltd (2002) Investigation of some selected landslides in 2000, vol 1. GEO reports no. 129. Geotechnical engineering Office, Civil Engineering Department, The Government of the Hong Kong Special Administrative Region

GEO (1996) Report on the Shum Wan Road landslide of 19 August 1995, vol 2. Geotechnical Engineering Office, Civil Engineering Department, The Government of the Hong Kong Special Administrative Region

GEO (2004) Geotechnical manual for slopes, 2nd edn (1984)

Huang RQ (2008) Preliminary analysis of the development, distributions, and mechanisms of the geohazards triggered by the Great Wenchuan Earthquake, State Key Laboratory of Geohazards Prevention and Geological Environment Protection, Chengdu University of Technology, Chengdu, China

Iverson RM, Vallance JW (2001) New views of granular mass flow. Geology 29:115–118

Jullien R, Meakin P (1990) A mechanism for particle size segregation in three dimensions. Nature 344:425–427

Lacasse S, Nadim F, Hoeg K, Gregersen O (2004) Risk assessment in geotechnical engineering: the importance of engineering judgement. Advances in geotechnical engineering. In: The Skempton conference. Institution of Civil Engineers, London, pp 856–867

Legget RF (1976) Glacial till. Royal Society of Canada, Ottawa

Lee CT, Dong JJ, Lin ML (2009) Geological investigation on the catastrophic landslide in Siaolin Village, Southern Taiwan. Sino Geotech 122:87–94

Loganathan N, DeSilva S, Thurairajah A (1992) Strength correlation factor for residual soils. J Geotech Eng 118(4):593–600

MFJV (2007) Final report on the compilation of the enhanced natural terrain landslide inventory. In: Maunsell Fugro Joint Venture. Geotechnical Engineering Office, Hong Kong

Miyamoto H, Yano H, Scheeres DJ, Abe S, Barnouin-Jha O, Cheng AF, Demura H, Gaskel RW, Hirata N, Ishiguro M, Michikami T, Nakamura AM, Nakamura R, Saito J, Sasaki S (2007) Regolith migration and sorting on asteroid Ikokawa. Science 316:1011–1014

Morgenstern NR (2000) Report on the Kwun Lung Lau landslide of 23 July 1994. GEO no. 103. Geotechnical Engineering Office, Civil Engineering Department, The Government of the Hong Kong Special Administrative Region

Schuster RL, Logan RL, Pringle PT (1992) Prehistoric rock avalanches in the Olympic Mountains, Washington. Science 258:1620–1621

Tang WH, Zhang LM, Zheng YR (2006) Dealing with uncertainty in engineering design for large-scale gravel soil slopes in the Three Gorges Reservoir Zone. Keynote lecture, Geohazards-technical, economical and social risk evaluation. Berkeley Electronic Press, CD Rom

The Yangtze Water Conservancy Commission (1997) Engineering geology of the three-gorge project. Hubei Science and Technology Press, Wuhan

Vallejo LE (1979) An explanation for mudflows. Geotechnique 29(3):351–354

Vallejo LE (1989) An extension of the particulate model of slope stability for mudflows. Soils Found 29(3):1–13

Vallejo LE (1997) Surge wave development in debris flows. Soils Found 37(4):115–119

Wang Y, Hutter K (2001) Granular material theories revised. In: Geomorphological fluid mechanics. Springer, Berlin, pp 79–107

Wang FW, Cheng QG, Highland L, Miyajima M, Wang HB, Yan CG (2009) Preliminary investigation of some large landslides triggered by the 2008 Wenchuan earthquake, Sichuan Province, China. Landslides 6(1):47–54

Williams JC (1976) The segregation of particulate materials—a review. Powder Technol 15:245–251

Xie HP, Deng JH, Tai JJ, He CR, Wei JB, Chen JP, Li XY (2008) Wenchuan Earthquake and post-earthquake reconstruction-related geotechnical problems. Chin J Mech Eng 27:1781–1791 (in Chinese)

Yan WM (2003) Civil engineering experimental study and constitutive modelling of re-compacted completely decomposed granite. Ph.D. thesis. The Hong Kong University and Technology

Yin YP, Wang FW, Sun P (2009) Landslide hazards triggered by the 2008 Wenchuan earthquake, Sichuan, China. Landslides 6(2):139–152

Zhang LM (2009) Challenges in multi-hazard risk assessment and management: Geohazard chain in Beichuan Town caused by Great Wenchuan Earthquake. Geotechnical Risk and Safety, Taylor and Francis Group, London, pp 237–244

Zhang LM, Chen Q (2005) Predicting bimodal soil-water characteristic curves. J Geotech Geoenviron Eng 131(5):666–670

Zhang LL, Zhang J, Zhang LM, Tang WH (2011a) Stability analysis of rainfall-induced slope failure: a review. Geotech Eng 164(5):299–316

Zhang LM, Xu Y, Huang RQ, Chang DS (2011b) Particle flow and segregation in a grant landslide event triggered by the 2008 Wenchuan earthquake, Sichuan, China. Nat Hazard 11:1153–1162

Zhao HF, Zhang LM, Chang DS (2013) Behavior of Coarse Widely Graded Soils under Low Confining Pressures. J Geotech Geoenviron Eng 139(1): 35–48

Zhou GD (2010) The mechanisms of debris flow. Ph.D. thesis. Hong Kong University of Science and Technology, Hong Kong

Chapter 3
Microporosity Structures of Widely-Graded Soils

3.1 Introduction

The microporosity structure of soil provides important information on the shear strength, compressibility, hydraulic conductivity, and soil–water characteristics of the soil. The soil microporosity structure changes with stress state, transfer of water and air, temperature fluctuations, long-term gravimetric actions, and weathering. To investigate the formation of a microporosity structure during compaction, the evolution of microporosity structure after saturation, and the variation of microporosity structure during wetting–drying, scanning electron microscopy (SEM) was used to characterize the soil surface structure directly, and mercury intrusion porosimetry (MIP) was used to quantify the soil pore-size distribution (PSD).

A comprehensive study of microporosity structure was conducted for CL with sand. Based on the pore-size distribution curves obtained from the MIP tests, a mathematical model was developed to characterize the PSD and was used to relate the variations of PSD to the change of soil void ratio during a compaction process, a saturation process, or a drying process. Therefore, the PSD of a soil sample at any void ratio can be predicted. Comparison experiments of PSD were conducted for five colluvial soils (i.e., GW-GM with sand, SM with gravel, SC with gravel, sandy ML, and CL with sand) and used to investigate the influence of soil GSD on soil microporosity structure. For each soil, compacted samples, compacted and saturated samples, and compacted, saturated and oven-dried samples were tested to investigate the PSD variations induced by wetting–drying.

3.2 Experimental Techniques

The freeze-drying technique was used to dehydrate the wet soil samples for the MIP and SEM tests. The freeze-drier used was the Edwards Super Modulyo apparatus (Fig. 3.1a), and the freezing temperature was −40 °C.

X. Li et al., *Fundamental Behavior of Unsaturated Widely-Graded Soil*,
https://doi.org/10.1007/978-981-19-3402-5_3

(a) (b) (c) (d)

Fig. 3.1 Experimental devices used in the microporosity investigation: **a** Freezing dryer. **b** JSM-6300 scanning electron microscope. **c** Micromeritics AutoPore IV 9500 device. **d** Vacuum chamber for soil saturation

SEM tests were used to observe the soil surface. The dehydrated soil specimens used in the SEM tests were coated with gold before taking the SEM images. AJSM-6300 scanning electron microscope, as shown in Fig. 3.1b, was used in this study. The macro–microfabric at the s MIP is routinely and effectively used to measure the PSDs of powder and bulk materials with open and interconnected pore structures. In the MIP test, the assumption is made that cylindrical flow channels exist in the soil with a radius, r. The pores which are intruded by mercury under an applied pressure, p, can be calculated as (Diamond 1970):

$$r = \frac{2T_{s,m}\cos\alpha}{p} \tag{3.1}$$

where $T_{s,m}$ is the surface tension of the fluid (i.e., mercury); α is the contact angle of fluid–air interface to the solid (i.e., soil particles). The surface tension of mercury is 472 μN/m^2 at 20 °C, and the contact angle is taken as 135° at the mercury–air interface to the soil in this study.

The cumulative injected mercury volume indicates the cumulative pore volume. The mercury intrusion pressure indicates the pore radius according to Eq. (3.1). The pore-size distribution can be derived from:

$$f(r) = \frac{\mathrm{d}v(r)}{\mathrm{d}r} \tag{3.2}$$

where r is the pore radius; $v(r)$ is the volume of pores with radii larger than r in 1 g of dry soil; $f(r)$ is the pore-size distribution. When the pore radius approaches zero, $v(r)$ will be,

$$\lim_{x \to 0} v(r) = e/G_s \tag{3.3}$$

where e is the void ratio and G_s is the specific gravity.

The pores occupied by mercury as the mercury injection pressure increases from p_1 to p_2 is,

$$\int_{r_1}^{r_2} f(r)\mathrm{d}r = \int_{p_2}^{p_1} f(p)\mathrm{d}p \tag{3.4}$$

where r_1 and r_2 are the pore radius at p_1 and p_2, as respected by Eq. (3.1). The measurement range of pore radius is limited by the mercury injection pressure supplied by the MIP device. In this study, the measurement range of pore radius is from 0.003 to 72 μm as the intrusion pressure range supplied by Micromeritics AutoPore IV 9500V 1.04 MIP device as shown in Fig. 3.1c is from 9.3 kPa to 210 MPa.

3.3 Mathematical Description of Dual-Porosity Geometry

Bimodal pore-size distributions are characteristic of compacted soils (Alonso et al. 1987). Bimodal pore-size distributions are also frequently observed in structured soils such as undisturbed aggregated loams (Mallants et al. 1997), residual soils (Sudhakar and Revanasiddappab 2005), and sediments (Anandarajah and Lavoie 2002). Some porous media such as fractured rocks and cracked soils that exhibit significant secondary structures also have a bimodal pore-size distribution (Pruess et al. 1990; Zhang and Fredlund 2003).

The pore-size distribution of one compacted soil specimen is shown in Fig. 3.2. The curve exhibits a bimodal nature with one peak occurring at a pore radius of 10 μm and another peak occurring at 0.3 μm. Mathematically, a bimodal PSD, $f(r)$, can be obtained using the superposition of two overlapping unimodal pore-size distributions, (e.g., $f_1(r)$ and $f_2(r)$):

$$f(r) = f_1(r) + f_2(r) \tag{3.5}$$

An appropriate mathematical equation needs to be obtained that best fits the soil data. It is even difficult to find a commonly accepted mathematical equation for a soil with a unimodal pore-size distribution. Here, $f_1(r)$ and $f_2(r)$ are taken as simple log-normal functions. The bimodal pore-size distribution can be written as,

$$f(r) = a_1 f_1(r) + a_2 f_2(r) = \frac{a_1}{\sqrt{2\pi}\sigma_1 r} e^{-\frac{(\lg r - \mu_1)^2}{2\sigma_1^2}} + \frac{a_2}{\sqrt{2\pi}\sigma_2 r} e^{-\frac{(\lg r - \mu_2)^2}{2\sigma_2^2}} \tag{3.6}$$

where a_1 and a_2 are the parameters indicating the porosity values covered by distributions, $f_1(r)$ and $f_2(r)$, respectively; μ_1 and μ_2 are the corresponding mean radius values on a logarithmic scale; and σ_1 and σ_2 are the corresponding standard deviations on a logarithmic scale, respectively. The parameters a_1 and a_2, and the parameters defining $f_1(r)$ and $f_2(r)$, can be obtained by solving a set of moment equations:

$$\int r^i f(r) \mathrm{d}r = \int a_1 r^i f_1(r) \mathrm{d}r + \int a_2 r^i f_2(r) \mathrm{d}r \quad (i = 0, 1, 2, \ldots, n) \tag{3.7}$$

where n is equal to the number of the undetermined parameters minus 1.

The bimodal pore-size distribution can be artificially divided into two unimodal parts. Then a regression analysis can be applied to each unimodal part. The parameters can be obtained by solving,

$$\begin{cases} \int_R^\infty r^i f(r) \mathrm{d}r = a_1 \int r^i f_1(r) \mathrm{d}r \quad (i = 0, 1, 2, \ldots, n_1) \\ \int_0^R r^i f(r) \mathrm{d}r = a_2 \int r^i f_2(r) \mathrm{d}r \quad (i = 0, 1, 2, \ldots, n_2) \end{cases} \tag{3.8}$$

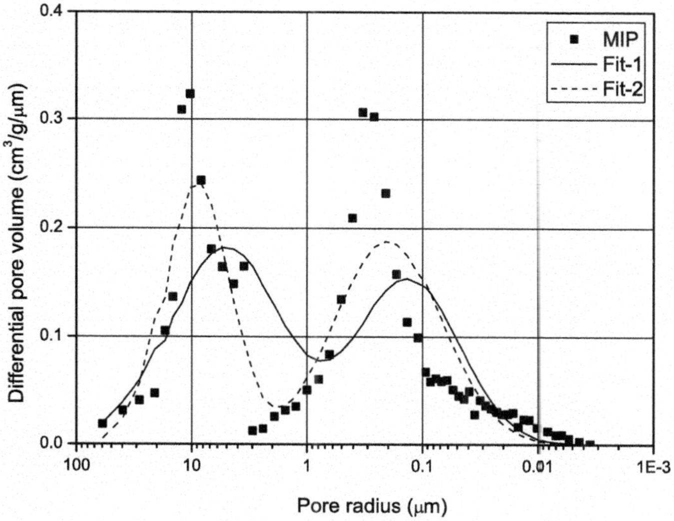

(a)

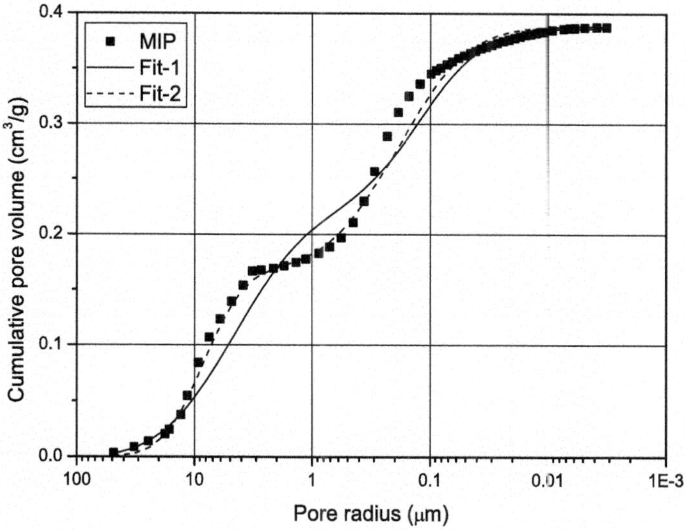

(b)

Fig. 3.2 Dual-porosity structure of one compacted soil sample: **a** Differential pore volumes. **b** Cumulative pore volume

where R is the boundary pore radius that is chosen to artificially split the bimodal PSD into two parts, namely $f_1(r)$, the normalized continuous distribution for describing the inter-aggregate pores (i.e., larger than R), and $f_2(r)$, the normalized continuous distribution for describing the intra-aggregate pores (i.e., smaller than R). The variables n_1 and n_2 are the numbers of parameters used to define $f_1(r)$ and $f_2(r)$ minus 1.

Six parameters need to be determined in the regression analysis for a bimodal pore-size distribution. Transforming the log-normal distribution into a normal distribution by using $x = \log r$ as an independent variable, Eq. (3.8) can be rewritten as (Lennart and Bertil 2004),

$$\begin{cases} \int_0^\infty f(x)\mathrm{d}x = a_1 + a_2 \\ \int_0^\infty x f(x)\mathrm{d}x = a_1\mu_1 + a_2\mu_2 \\ \int_0^\infty x^2 f(x)\mathrm{d}x = a_1(\mu_1^2 + \sigma_1^2) + a_2(\mu_2^2 + \sigma_2^2) \\ \int_0^\infty x^3 f(x)\mathrm{d}x = a_1\mu_1(\mu_1^2 + 3\sigma_1^2) + a_2\mu_2(\mu_2^2 + 3\sigma_2^2) \\ \int_0^\infty x^4 f(x)\mathrm{d}x = a_1(\mu_1^4 + 6\mu_1^2\sigma_1^2 + 3\sigma_1^4) + a_2(\mu_2^4 + 6\mu_2^2\sigma_2^2 + 3\sigma_2^4) \\ \int_0^\infty x^5 f(x)\mathrm{d}x = a_1\mu_1(\mu_1^4 + 6\mu_1^2\sigma_1^2 + 3\sigma_1^4 + 4\mu_1^2 + 12\sigma_1^2) \\ \quad + a_2\mu_2(\mu_2^4 + 6\mu_2^2\sigma_2^2 + 3\sigma_2^4 + 4\mu_2^2 + 12\sigma_2^2) \end{cases} \tag{3.9}$$

If the bimodal pore-size distribution is divided into two unimodal parts, $f1(x)$ and $f2(x)$, with a pore radius of R, Eq. (3.8) will be,

$$\begin{cases} \int_{\lg R}^\infty f(x)\mathrm{d}x = a_1 \\ \int_{\lg R}^\infty x f(x)\mathrm{d}x = a_1\mu_1 \\ \int_{\lg R}^\infty x^2 f(x)\mathrm{d}x = a_1(\mu_1^2 + \sigma_1^2) \\ \int_0^{\lg R} f(x)\mathrm{d}x = a_2 \\ \int_0^{\lg R} x f(x)\mathrm{d}x = a_2\mu_2 \\ \int_0^{\lg R} x^2 f(x)\mathrm{d}x = a_2(\mu_2^2 + \sigma_2^2) \end{cases} \tag{3.10}$$

As an example, if two log-normal distributions are used to simulate the PSD shown in Fig. 3.2, the solution to Eq. (3.9) gives six parameters: $a_1 = 0.22$, $\mu_1 = 0.66$, $\sigma_1 = 0.49$, $a_2 = 0.17$, $\mu_2 = -0.91$, and $\sigma_2 = 0.43$. The results are plotted as curve "Fit-1" in Fig. 3.2a. It should be mentioned that the integration range used is the measurable pore radius range in the MIP device, i.e., 0.003–72 μm. However, the probability distributions involving the solved parameters (mean and standard deviation) cover the full pore radius range, i.e., 0 to infinity, to describe the differential pore volumes. If the bimodal pore-size distribution is divided into two parts with a boundary pore radius equal to 2 μm and each part is simulated by a log-normal distribution, Eq. (3.10) gives $a_1 = 0.17$, $\mu_1 = 0.92$, $\sigma_1 = 0.27$, $a_2 = 0.22$, $\mu_2 = -0.72$, and $\sigma_2 = 0.48$. The results are plotted as curve "Fit-2" in Fig. 3.2a. For the inter-aggregate pores, the integration range used in Eq. (3.10) is 2–72 μm. For the intra-aggregate pores, the integration range used in Eq. (4.10) is 0.003–2 μm. The fitted cumulative pore volume curves are shown in Fig. 3.2b. Figure 3.2 shows that a better simulation is obtained using "Fit-2".

3.4 Comprehensive Study of Microporosity for LC with Sand

3.4.1 Experimental Technique

The experimental program involves a quantitative investigation of the microporosity structure, and the measurement of the soil pore-size distributions using MIP. Four unsaturated compacted samples at different void ratios were tested to study the soil microporosity structure formation due to compaction. Four saturated compacted samples at different void ratio values were tested to study the soil microporosity structure after saturation. Four additional saturated compacted samples at the same void ratio were dried to different degrees of saturation and tested to study the microporosity structure variation during the drying process. To qualitatively investigate the microporosity structure, the soil surface characteristics were also studied using SEM on a bulk soil sample, an unsaturated compacted sample, a saturated compacted sample, and an oven-dried saturated compacted sample. In total, 12 soil samples in three series were used for pore-size measurements and 4 soil samples were used for the surface characteristics study, as summarized in Table 3.1.

The unsaturated compacted samples were prepared in several steps. First, the soil was oven-dried in the laboratory. Then the dry soil was mixed with water corresponding to the optimum water content. Many large soil lumps formed and the mixed soil was sieved through a 2-mm sieve to break up the large soil lumps. Then the mixed soil was stored in a plastic bag for 7 days for moisture homogenization. The water content of the soil was measured as 20.5% (i.e., the optimum water content of CL with sand). The matured soil was statically compacted to a specified dry density value at a rate of 0.05 mm/min. Each soil sample was compacted inside a metal mold of 70 mm in diameter and 20 mm in height. The compaction was performed in two layers. Scratches were made on the soil between layers to avoid segregation.

Eight compacted samples were saturated for 10 days in a mold subject to vacuum, as shown in Fig. 3.1d. Five of them, with different void ratios, were tested using MIP directly (referring to Table 3.1), and three of them were subjected to various applied suction values (i.e., the difference between the pore-air pressure and the pore-water pressure). After the soil specimens were at equilibrium at suction of 100 kPa (i.e., test D10 as shown in Table 3.1), one soil specimen was taken from the SWCC device and

Table 3.1 Investigation of microporosity structure of CL with sand

Objective	Test no.	Void ratio at compaction	Void ratio at measurement	Samples preparation method	Technique
Investigation of microporosity structure formation due to compaction	C1	1.13	1.13	Compacted	MIP
	C2	1.05	1.05	Compacted	MIP
	C3	1.00	1.00	Compacted	MIP
	C4	0.76	0.76	Compacted	MIP
Investigation of microporosity structure after saturation	S5	1.17	1.17	Compacted and saturated	MIP
	S6	1.06	1.06	Compacted and saturated	MIP
	S7	0.91	0.91	Compacted and saturated	MIP
	S8	0.8	0.8	Compacted and saturated	MIP
Investigation of microporosity structure variation during drying process	S9	1.00	1.00	Compacted and saturated	MIP
	D10	1.00	0.96	Compacted, saturated, and drained to 100 kPa	MIP
	D11	1.00	0.84	Compacted, saturated, and drained to 500 kPa	MIP
	D12	1.00	0.65	Compacted, saturated, and oven-dried	MIP
Observation of microporosity structure on soil surfaces	C13	1.00	1.00	Compacted	SEM
	P14	–	–	Bulk, oven-dried	SEM
	S15	1.00	1.00	Compacted and saturated	SEM
	D16	1.00	0.65	Compacted, saturated, and oven-dried	SEM

used for MIP testing. The remaining soil specimens were further drained by applying a suction of 500 kPa (i.e., test D11 as shown in Table 3.1). Then one specimen was used for MIP testing, and another specimen was oven-dried and used for MIP and SEM testing (i.e., tests D12 and D16 as shown in Table 3.1).

3.4.2 *Dual-Porosity Structure of Compacted Soil*

Figure 3.3a shows the backscattered electron scanning image of one soil sample compacted at a water content of 20.5%. The void ratio of this soil sample was 1.00, and the corresponding dry density was 1317.5 kg/m^3. A dual-structure soil having both intra-aggregate pores and inter-aggregate pores can be seen in Fig. 3.3a. In Fig. 3.3a, the representative radius of the inter-aggregate pores is about 3 μm, and the representative radius of the intra-aggregate pores is about 0.6 μm.

For comparison purposes, the backscattered electron scanning image of the dry bulky soil sample is shown in Fig. 3.3b. The aggregates in the sample form a loose uniform microporosity structure. The pores in the dry bulky soil are more uniform compared to those in the compacted sample and have a typical pore radius around 1 μm. Thin leafage edges are present around the soil aggregates in the dry bulk sample. On the contrary, the edges around the soil aggregates are thicker in the compacted sample. The difference may be due to one of two main reasons. First, when mixing the soil particles and water, large soil aggregates form and appear to be associated with the water meniscus forces. Second, the compression stresses can reach 1200 kPa to produce a soil sample with a void ratio of 0.76 during the compaction process. The volume deformation of the bulk soil can reach 30% of the initial total volume after it is compacted to a void ratio of 0.76. The total pore volume in the bulk soil decreases to 50% after the sample is compacted. The distance between aggregates decreased because of the compression stresses and the stiff aggregates formed a strong skeleton to carry the compression force and formed large inter-aggregate pores. The distance between aggregates becomes closer as the compression stress is increased, and some aggregates are believed to break and merge with others during the compaction process. Hence, the inter-aggregate pores become smaller and the number of inter-aggregate pores becomes fewer as the compression stress is increased.

3.4.3 *Dual-Porosity Structure Change Due to Soil Compaction*

3.4.3.1 PSDs of Compacted Samples with Different Void Ratios

The MIP technique was used to characterize the changes in the soil dual-structure due to compaction. Four soil samples were compacted to void ratios of 1.13, 1.05, 1.00, and 0.76, respectively at optimum water content. The measured PSDs of the four samples are shown in Fig. 3.4a. The unsaturated compacted samples exhibit a distinct dual-porosity structure. In the intra-aggregate pore zone of the four distributions, the differential pore volume shows a peak at a radius of about 0.2 μm. The differential pore volumes of the four compacted samples are similar up to a radius of 2 μm. This

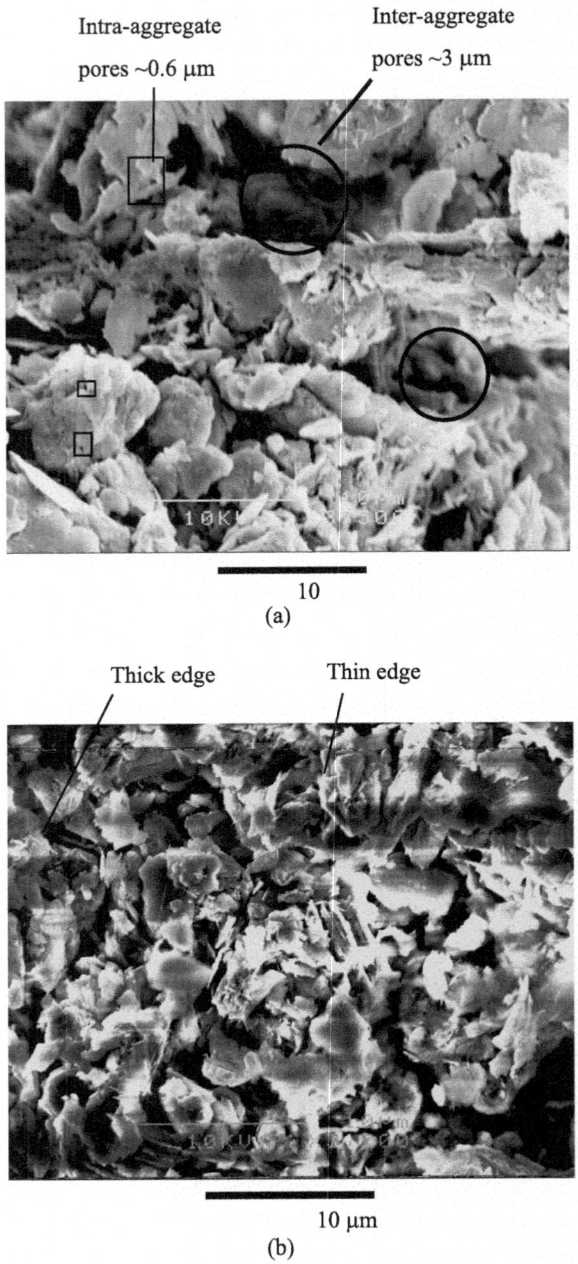

Fig. 3.3 Dual-porosity of CL with sand observed in SEM: **a** Unsaturated compacted soil sample, ×3500. **b** Dry bulky soil sample, ×3700

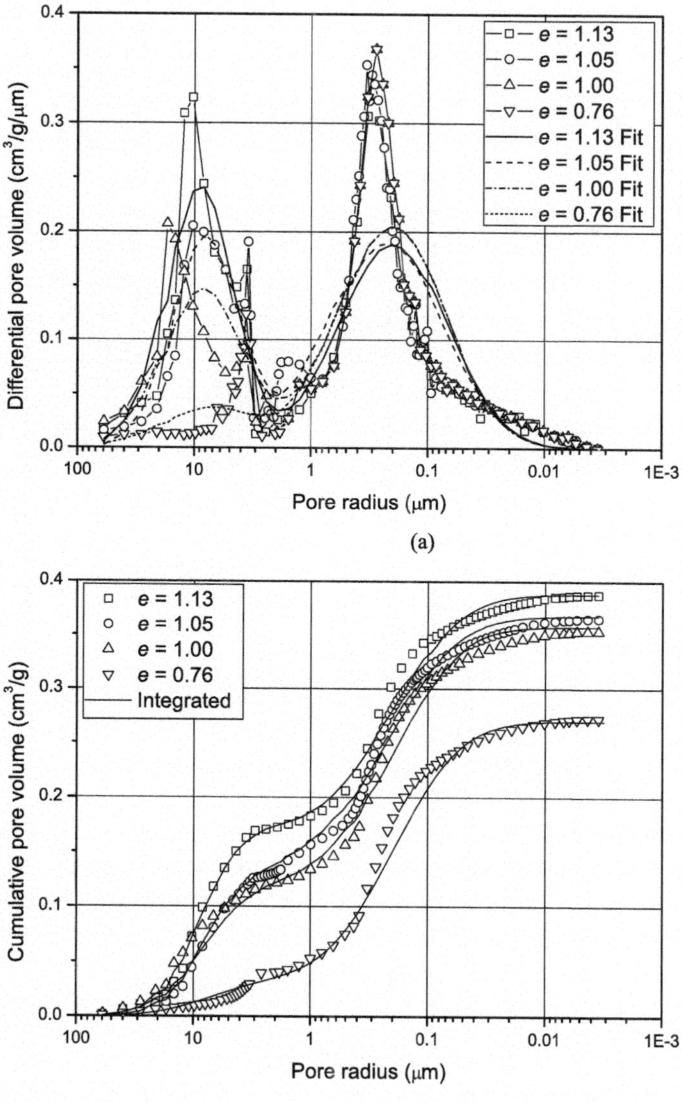

Fig. 3.4 Dual-porosity structure found in the unsaturated compacted samples (CL with sand): **a** Differential pore volumes. **b** Cumulative pore volumes

is considered to be the boundary between the intra-aggregate pores and the inter-aggregate pores. In the inter-aggregate pore zone of the four compacted samples, the differential pore volumes show another peak at around 7–8 μm. The differential pore volumes with pore radii larger than 2 μm decrease sharply as the void ratio decreases. The pore volumes in 1 g of dry soil are compared in Fig. 3.5 with respect to pore

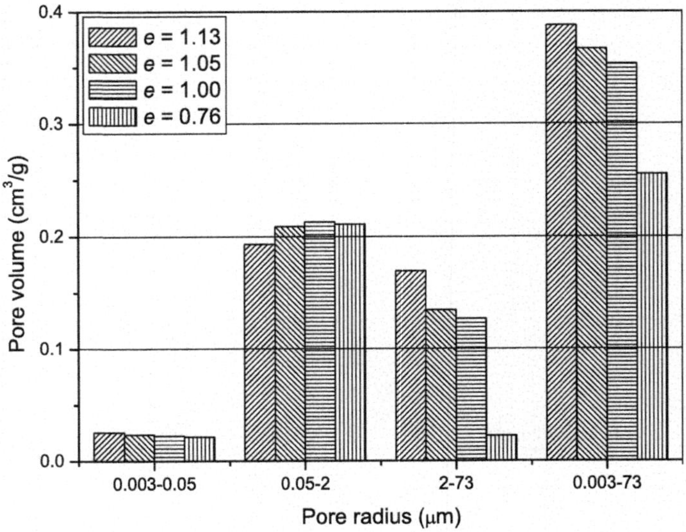

Fig. 3.5 Volume fractions of pore in different radii range in the unsaturated compacted samples (CL with sand)

radius. The pore volumes with radii smaller than 0.05 μm are similar in the four samples. The pore volumes with radii from 0.05 to 2 μm are also similar in the four samples. However, the pore volumes with radii larger than 2 μm (i.e., inter-aggregate pores) vary significantly, with a larger accumulated pore volume corresponding to a larger void ratio.

The measured pore-size distributions of the four compacted samples indicate that soil compaction changes the inter-aggregate structure of soil significantly but only changes the intra-aggregate structure slightly up to a relative compaction value of 0.97 ($e = 0.76$). The total soil volume and void ratio became smaller and smaller as the compaction stress was increased. Simultaneously, the distance between aggregates became smaller as the compaction stress was increased. Therefore, the total volume of the inter-aggregate pores and the differential pore volume in the inter-aggregate pore zone of the differential curve became smaller and smaller as the compaction stress was increased. The compaction stress applied in this test might not be sufficiently high to change the intra-aggregate pores.

Two log-normal distribution curves were used to characterize these bimodal pore-size distributions. The parameters for the log-normal distribution curves are listed in Table 3.2. The fitted PSD curves are shown in Fig. 3.4a. The measured cumulative pore volume curve and the cumulative pore volume curves integrated from the fitted PSD curves are shown in Fig. 3.4b. The measured pore volume curves and the integrated curves agree well.

Parameters a_1 and μ_1 shown in Table 3.2 indicate the porosity and log mean radius of the inter-aggregate pores. These values are shown to decrease significantly with

Table 3.2 Best-fit analysis results for PSDs of samples compacted at optimum water content

Soil parameters	PSD of inter-aggregate pores			PSD of intra-aggregate pores		
Void ratio	a_1 (cm^3/g)	μ_1	σ_1	a_2 (cm^3/g)	μ_2	σ_2
1.13	0.17	0.91	0.28	0.22	−0.74	0.46
1.05	0.13	0.87	0.28	0.23	−0.66	0.49
1.00	0.12	0.88	0.33	0.24	−0.72	0.46
0.76	0.035	0.84	0.38	0.24	−0.73	0.46

the void ratio, whereas parameters a_2 and μ_2, which indicate that the porosity and log mean radius of the intra-aggregate pores vary little with void ratio.

One noticeable phenomenon is that the sum of a_1 and a_2 is about 5–10% smaller than e/G_s [referring to Eq. (3.3)]. The difference occurs because there are some pores with radii outside the measurement range (i.e., from 0.003 to 60 μm for Micromeritics AutoPore IV 9500). These pores cannot be measured and have not been included in the total pore volume.

Through regression analysis using the data in Table 3.2, parameter a_1, which indicates the pore volume with radius larger than 2 μm, linearly decreases with the void ratio (as shown in Fig. 3.6a). Parameter μ_1 also changes linearly with the void ratio (as shown in Fig. 3.6b). However, parameters a_2 and μ_2 show insignificant change with the void ratio.

3.4.3.2 Formulation of Dual-Structure Porosity Change Due to Compaction

An attempt has been made to relate the pore-size distribution changes to changes in void ratio. Referring to Eq. (3.3), the total soil volume change is the sum of the changes in the volumes of the inter-aggregate pores and the intra-aggregate pores:

$$\Delta a_1 + \Delta a_2 = \frac{\Delta e}{G_s} \tag{3.11}$$

Introducing a parameter k_1, which indicates the percentage of inter-aggregate pore volume change over the total soil volume change, Eq. (3.11) can be rewritten as,

$$\Delta a_1 = k_1 \frac{\Delta e}{G_s} \tag{3.12}$$

$$\Delta a_2 = (1 - k_1) \frac{\Delta e}{G_s} \tag{3.13}$$

Changes in the mean logarithmic radii of inter-aggregate pores and intra-aggregate pores are considered to be linearly related to changes in void ratio,

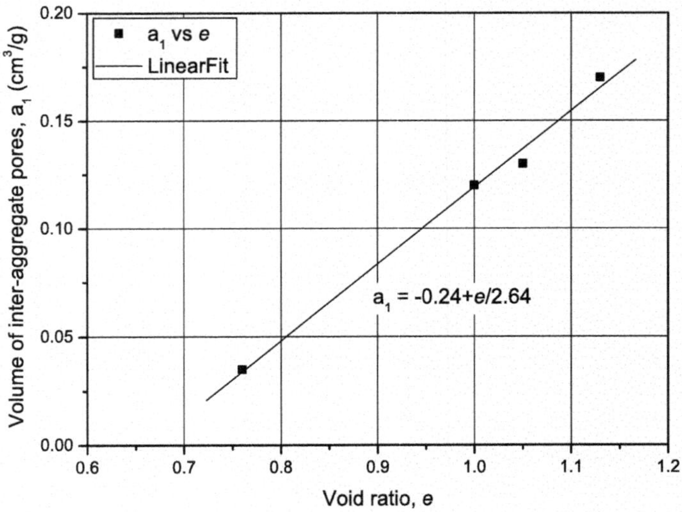

(a)

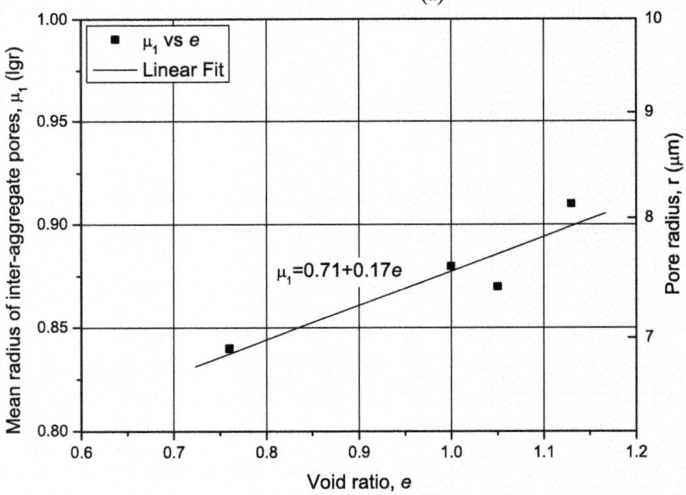

(b)

Fig. 3.6 Regression analysis of parameters indicating PSDs of the unsaturated compacted samples (CL with sand): **a** The pore volume of inter-aggregate pores in soil of 1 g solid, a_1. **b** The arithmetic mean of the log-normal inter-aggregate PSD, μ_1

$$\Delta\mu_1 = \lambda_1 \Delta e \tag{3.14}$$

$$\Delta\mu_2 = \lambda_2 \Delta e \tag{3.15}$$

where λ_1 and λ_2 are empirical parameters.

The standard deviations, σ_1 and σ_2, indicate the shape of the inter-aggregate PSD and the shape of the intra-aggregate PSD, respectively. Assuming R is the boundary radius between the inter-aggregate pores and the intra-aggregate pores, and only a certain percentage of the inter-aggregate pores is smaller than the boundary pore radius, then σ_1 is,

$$\sigma_1 = \lambda_3(\mu_1 - \lg R) \tag{3.16}$$

where λ_3 is related to the specified percentage through the standard normal distribution function. If $\lambda_3 = 0.5$, Eq. (3.16) can be transformed as $[(\mu_1 - \lg R)/\sigma_1 = 2]$. That is to say, only 2.28% of inter-aggregate pores are smaller than R. If $\lambda_3 = 1$, then 15.87% of inter-aggregate pores are smaller than R. Similarly, assuming a certain percentage of the intra-aggregate pores is larger than the boundary pore radius,

$$\sigma_2 = \lambda_4(\lg R - \mu_2) \tag{3.17}$$

where λ_4 is, similar to λ_3, related to the specified percentage through the standard normal distribution function. The empirical parameters, k_1, λ_1, λ_2 λ_3, and λ_4, are calculated from the fitting parameters, a_1, a_2, μ_1, μ_2, σ_1, and σ_2, at different void ratios as shown in Fig. 3.5 and are considered dependent on soil type and soil state.

Parameters used to describe the PSDs of inter-aggregate pores and intra-aggregate pores can be related to its void ratio through the application of Eqs. (3.12)–(3.17). Parameters k_1, λ_1, λ_2, λ_3, and λ_4 are all related to soil type and boundary conditions and need to be calibrated to experimental data. The regression analysis based on the data in Table 3.2 gives $k_1 = 1$, $\lambda_1 = 0.17$, $\lambda_2 = 0$, $\lambda_3 = 0.6$, and $\lambda_4 = 0.5$. Note that $k_1 = 1$ indicates the soil volume change due to soil compaction in this study is entirely based on compression of the inter-aggregate pores. After k_1, λ_1, λ_2, λ_3, and λ_4 have been determined the pore-size distribution of a compacted sample at any void ratio can be predicted based on one known pore-size distribution. The pore-size distributions at four different void ratios can be predicted and are shown in Fig. 3.7. As the fitted pore-size distributions in Fig. 3.4 are smoother than the measured curves and show lower peaks, the predicted pore-size distributions are also smoother than the measured curves shown in Fig. 3.4 and show lower peaks. This result indicates that the accuracy of prediction is highly related to the fitting curves used to calculate the empirical parameters (i.e., k_1, λ_1, λ_2, λ_3, and λ_4) and the equation used to fit the measured pore-size distributions. A better fitting equation can result in better fitting curves and better predictions.

The formulation for the dual-porosity structure of a compacted soil can be coupled with a model describing void ratio changes during soil compaction. For instance, if soil compaction is assessed using a linear relationship between void ratio e and the logarithm of the compression stress p, then Eqs. (3.12)–(3.17) can be transformed by substituting p for e.

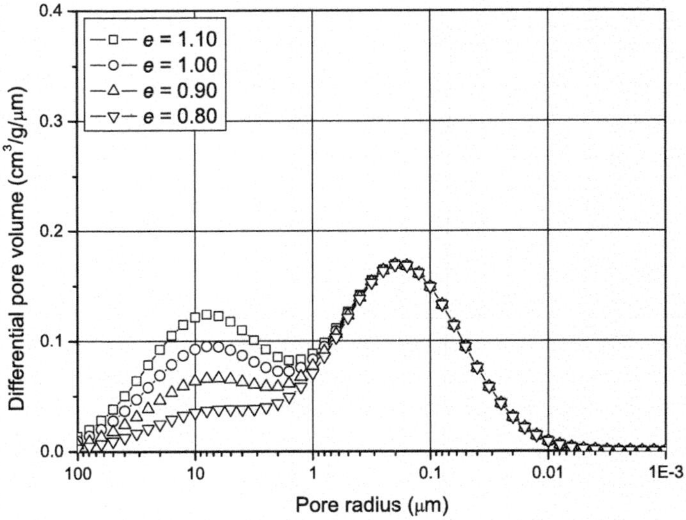

Fig. 3.7 Predicted PSDs of samples compacted at optimum water content with different void ratios

3.4.4 Dual-Porosity Structure Evolution During Soil Saturation

3.4.4.1 PSDs of Saturated Samples with Different Void Ratios

The PSDs of five saturated compacted samples with different void ratios (i.e., 1.17, 1.06, 1.00, 0.91, and 0.80) were measured using the MIP technique and are shown in Fig. 3.8. These differential pore volume curves indicate that the porosity of the saturated samples essentially consists of the intra-aggregate pores with little inter-aggregate pores. In the intra-aggregate pore zone of these five saturated soil samples, the differential pore volumes show peaks at pore radii around 0.3–0.7 μm. The peak value and the pore radius at peak value are smaller for samples with lower void ratios among these five saturated samples. The pore volumes in 1 g of dry soil are compared in Fig. 3.9. The accumulated pore volumes in the inter-aggregate pore zone (i.e., pore radii larger than 2 μm) of the five saturated soil samples only consist of 8% of the total void volume.

The PSDs of one unsaturated and one saturated compacted sample at the same void ratio are compared in Fig. 3.10. Although the unsaturated compacted sample and saturated compacted sample have the same void ratios, their PSDs are quite different. Distinct bimodal PSD features can be seen for the unsaturated compacted sample while only weak bimodal PSD features are present for the saturated compacted sample. In the intra-aggregate pore zone of the differential distributions (i.e., pore radii smaller than 2 μm), the pore radius at the peak value in the unsaturated compacted soil sample is 0.27 μm, while the pore radius at the peak value in the saturated soil sample is 0.64 μm. The pore volumes in 1 g of dry solid of the compacted

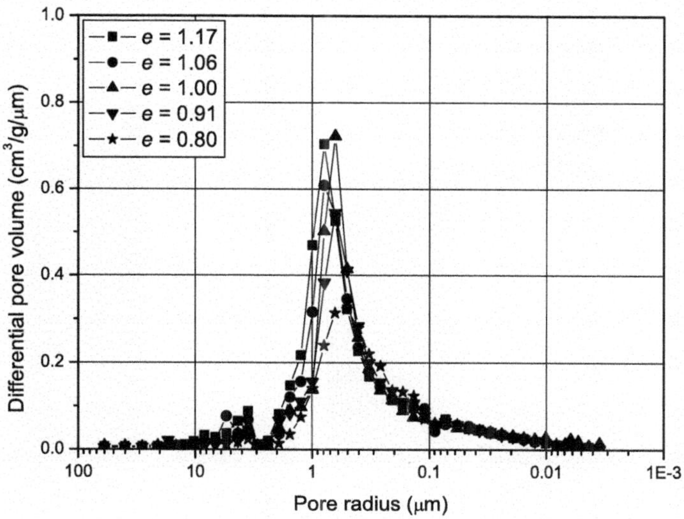

(a)

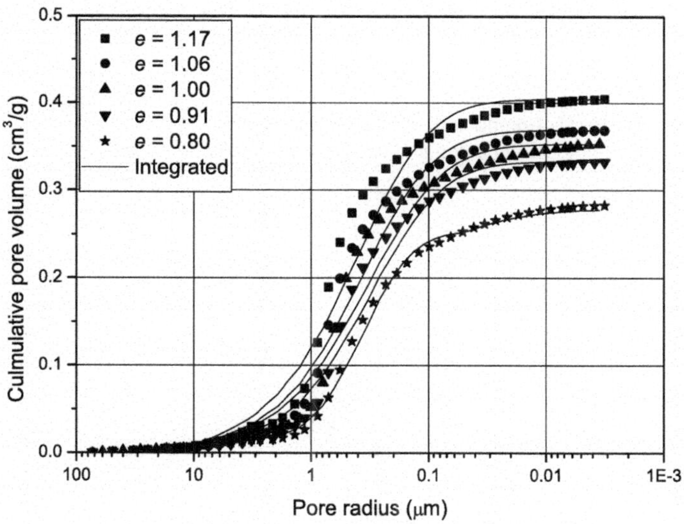

(b)

Fig. 3.8 Dual-porosity structure of the compacted and saturated samples (CL with sand): **a** Differential pore volume. **b** Cumulative pore volume

samples and saturated samples are compared in Fig. 3.11. In the intra-aggregate pores zone, the accumulated pore volume in the compacted soil sample is smaller than that in the saturated soil sample. In the inter-aggregate pores zone, the accumulated pore volume in the compacted soil sample is much larger than that in the saturated soil sample.

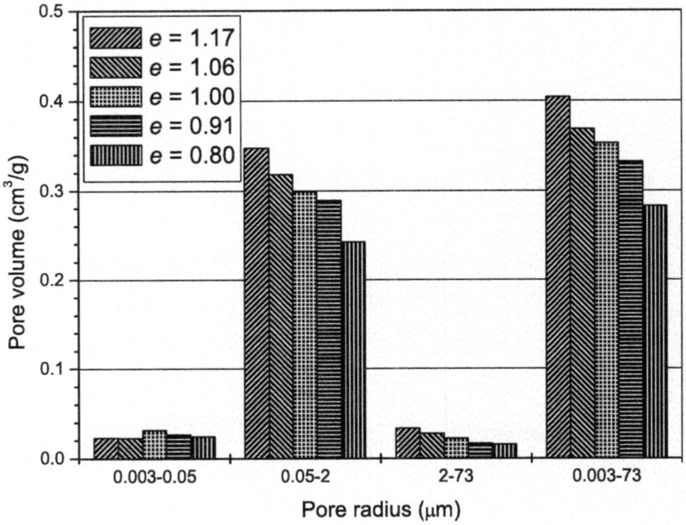

Fig. 3.9 Volume fractions of pore in different radii range in the compacted and saturated samples (CL with sand)

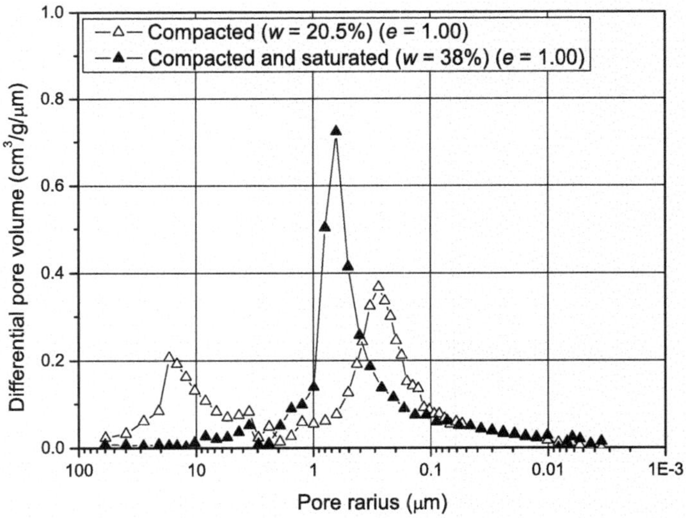

Fig. 3.10 PSD change after saturation (CL with sand)

The cyclic swelling and shrinkage of clays have been studied by numerous investigators (e.g., Basma et al. 1996). The lean clay with sand used in this study was sieved from completely decomposed granite in Hong Kong. A mineral analysis by Zhao et al. (2005) showed that the fines of completely decomposed granite contained

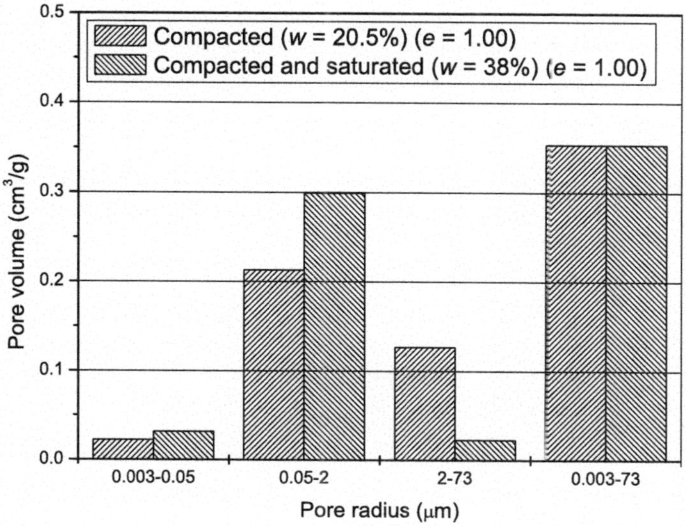

Fig. 3.11 Change of volume fractions of pore in different radii range after saturation (CL with sand)

approximately 60% of kaolinite and 40% of ledikite. As the soil used in the laboratory tests contains 80% fines and has a shrinkage limit of 8%, it is anticipated to swell upon wetting when its water content is larger than 8% (i.e., shrinkage limit). When the soil sample is being saturated under free boundary conditions, the soil sample will swell and the void ratio of the soil sample will increase. As part of the saturation method used in this study, the soil sample was placed inside a metal mold (Fig. 3.1d), and therefore, the total volume and void ratio of the soil sample could not change. As the soil aggregates swelled during the saturation process, the intra-aggregate pores would become larger and the distances between soil aggregates would become smaller because the sample was confined by the rigid mold. Consequently, the inter-aggregate pores became smaller as water content increased. The accumulated volume of inter-aggregate pores (i.e., the pores above 2 μm) decreased after saturation and the accumulated volume of intra-aggregate pores (i.e., the pores smaller than 2 μm) increased after soil saturation, as shown in Fig. 3.11. However, the total volume of soil remained approximately the same.

3.4.4.2 Formulations for Dual-Porosity Structure of Saturated Soil Samples

Two log-normal distributions were again combined to describe the pore-size distributions of the five saturated soil samples. One log-normal distribution was used to characterize the inter-aggregate pores (larger than 2 μm) and another was used to characterize the intra-aggregate pores with radii smaller than 2 μm. The parameters obtained from the regression analyses are listed in Table 3.3. The measured cumulative pore volume curves and the integrated pore volume curves based on the fitted PSDs are shown in Fig. 3.8b.

The inter-aggregate pore-size distributions and the intra-aggregate pore-size distribution are related to the void ratio through the application of Eqs. (3.12)–(3.17). Regression analyses were carried out based on the data listed in Table 3.3. The parameters obtained are $\lambda_1 = -0.54$, $\lambda_2 = 0.5$, $\lambda_3 = 0.9$, $\lambda_4 = 0.6$, and $k_1 = 0.75$. Note that $k_1 = 0.75$ means that about 75% of the volume change occurs in the intra-aggregate pores and only about 25% of the volume change occurs in the inter-aggregate pores for saturated soil samples. The pore-size distribution of a soil sample at any void ratio can be predicted based on its pore-size distribution associated with one void ratio (i.e., 1.17 in this case as shown in Fig. 3.8a). The pore-size distributions at four different void ratios are predicted and shown in Fig. 3.12.

Table 3.3 Best-fit analysis results for PSDs of the compacted and saturated samples

Soil parameters	PSD of inter-aggregate pores			PSD of intra-aggregate pores		
Void ratio	a_1 (cm^3/g)	μ_1	σ_1	a_2 (cm^3/g)	μ_2	σ_2
1.17	0.040	0.67	0.36	0.36	−0.43	0.49
1.06	0.031	0.71	0.34	0.34	−0.47	0.49
1.00	0.027	0.74	0.41	0.33	−0.52	0.52
0.91	0.022	0.79	0.42	0.31	−0.53	0.49
0.8	0.017	0.87	0.40	0.27	−0.62	0.51

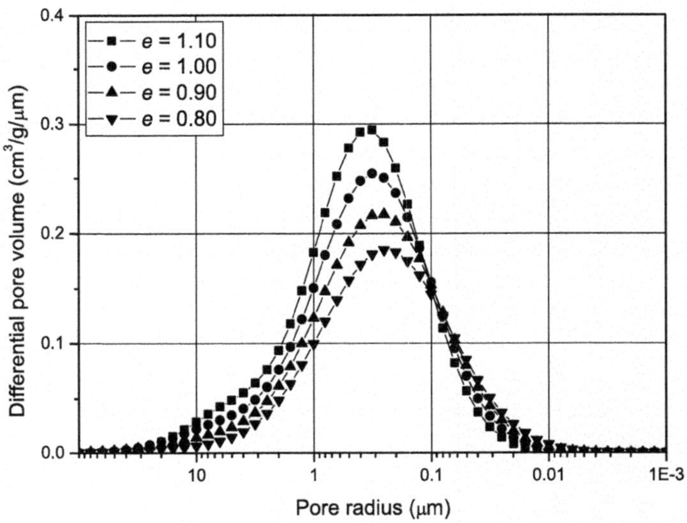

Fig. 3.12 Predicted PSDs of the compacted and saturated samples with different void ratios

3.4.5 Dual-Porosity Structure Changes During Drying

The backscattered electron scanning images of one saturated compacted sample and one compacted sample that was first saturated and then oven-dried are shown in Fig. 3.13a, b, respectively. The aggregates in the oven-dried sample are much closer and larger than those in the saturated sample. A dual-porosity structure can be found in the saturated soil sample similar to that of the unsaturated compacted sample as shown in Fig. 3.3. Some inter-aggregate pores with radii between 1 μm and 5 μm are present in the saturated soil sample. On the contrary, few inter-aggregate pores are present in the oven-dried sample. A large quantity of intra-aggregate pores with radii smaller than 1 μm is present in both the saturated and oven-dried samples.

The measured PSDs of four soil samples at different degrees of saturation are shown in Fig. 3.14a. Generally speaking, the pores after the drying process are essentially intra-aggregate pores with few inter-aggregate pores. The pore volumes in 1 g of dry soil are compared in Fig. 3.15 with respect to pore radius. Both the volume of inter-aggregate pores and the volume of intra-aggregate pores decrease as the degree of saturation decreases. The total pore volume also decreases as the degree of saturation decreases, which indicates that the soil shrinks during the drying process.

As soil suction is increased to 100 kPa ($S_r = 0.8$), the differential pore volume of the pores above 0.5 μm decreases comparing to that in the saturated soil sample. At the same time, the differential pore volume of the pores smaller than 0.5 μm is similar with that in the saturated soil sample. The soil sample shrinks freely during the drying process. The pores are expected to shrink due to the water meniscus progressing from larger pores to smaller pores as suction increases. Referring to Eq. (3.1) and choosing the surface tension of water T_s at a temperature of 25 °C as 72 μm/m^2 and a contact angle of 0°, the maximum pore radius filled with water at suction of 100 kPa is 1.44 μm. As suction is increased to 100 kPa the pores with radii larger than 1.44 μm will drain and are expected to shrink. Limited influence is expected on the pores with radii smaller than 1.44 μm as the suction is increased to 100 kPa.

As soil suction is increased to 500 kPa ($S_r = 0.45$), the pore densities between 0.1 μm and 2 μm form a trough. A decrease in differential pore volume with radii larger than 0.3 μm and an increase of differential pore volume with radii smaller than 0.3 μm were found in the sample drained from a suction of 100–500 kPa. As suction is increased to 500 kPa, the pores with radii larger than 0.29 μm will drain and are expected to shrink. The differential pore volumes with radii smaller than 0.29 μm will decrease. As many pores larger than 0.29 μm shrink into smaller pores, the quantity and differential pore volumes of the pores smaller than 0.29 μm will increase.

The differential pore volumes with radii larger than 0.4 μm decreased significantly comparing to that in the soil sample at a suction of 500 kPa ($S_r = 0.45$) as the soil sample was oven-dried ($S_r \sim 0$). Dineen (1997) and Cafaro (2002) discussed a noticeable void ratio decrease during soil drying and suggested that the particle system minimizes its potential energy during drying through an immediate change of

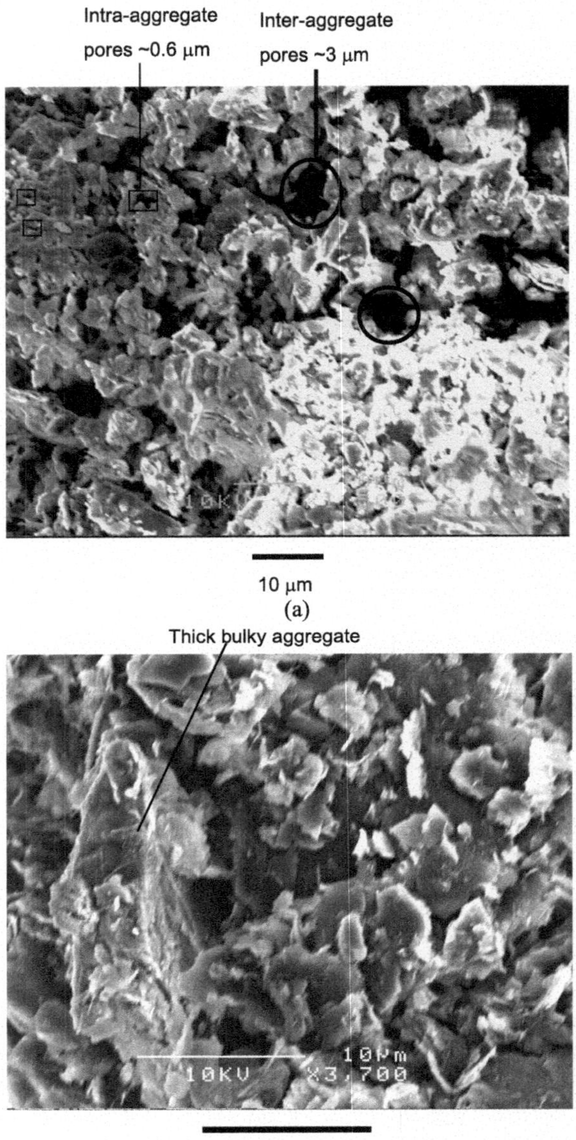

Fig. 3.13 Dual-porosity changes after drying observed in SEM (CL with sand): **a** Saturated compacted soil sample, ×1500. **b** Oven-dried saturated compacted sample, ×3700

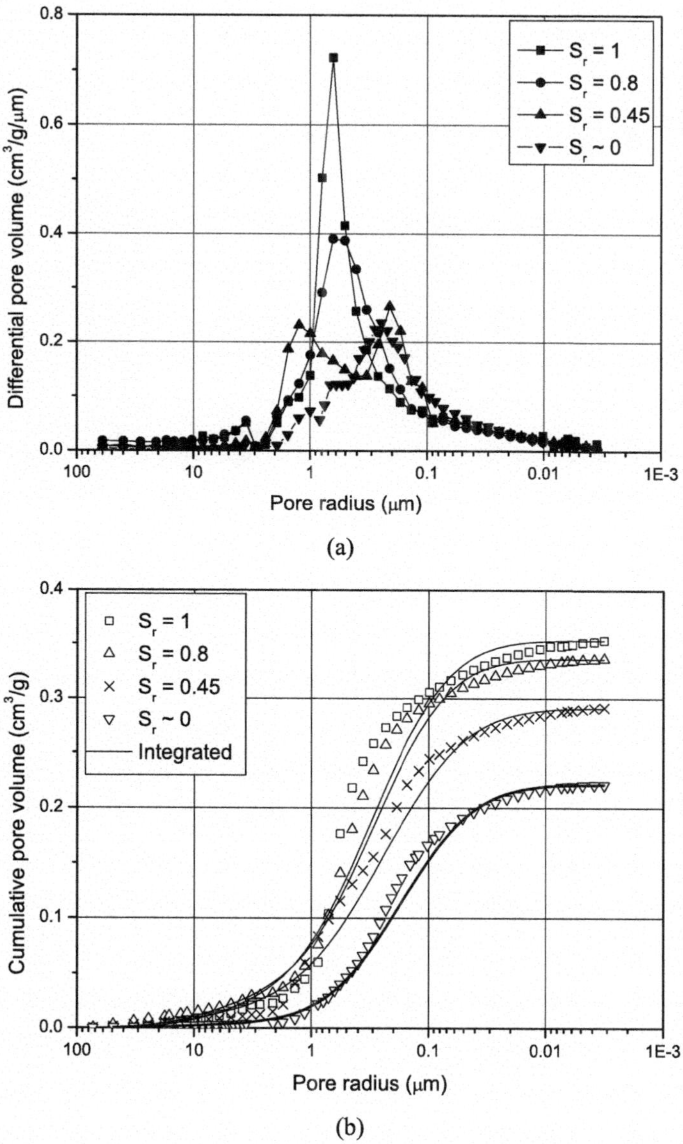

Fig. 3.14 Pore shrinkage during the drying process (CL with sand): **a** Differential pore volume. **b** Cumulative pore volume

the pore-size distribution taking place when suction reaches a threshold value (i.e., 1400 kPa suggested by Dineen). The observed decrease in differential pore volume with radii larger than 0.4 μm in this study may occur at the threshold suction. The pores larger than 0.4 μm might collapse at the threshold suction. The loss of water at the boundary due to evaporation may induce arching or bridges that bond the soil

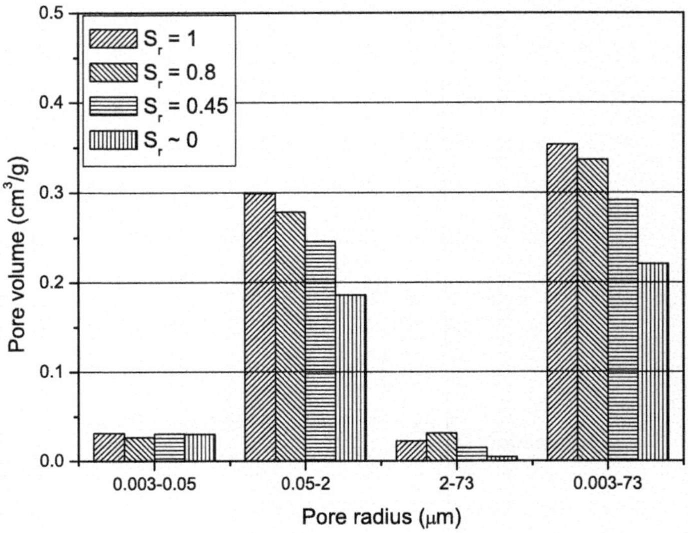

Fig. 3.15 Changes of volume fractions of pore in different radii range during drying process (CL with sand)

particles together. It should also be considered that an increase in pore water salinity may take place during drying associated with water evaporation (Cafaro 2002).

Two log-normal distributions can be combined to describe the pore-size distributions of the samples obtained during the soil drying process similar to the cases of soil compaction and saturation. The parameters obtained from the regression analysis are listed in Table 3.4. The measured cumulative pore volume curves and the integrated pore volume curves based on the fitted PSDs are shown in Fig. 3.14b.

Changes in the volume of inter-aggregate pores and intra-aggregate pores are equal to the total soil volume change. With the application of Eq. (3.11), the void ratio values of the soil samples at 100 kPa and 500 kPa suction values, and the oven-dried soil sample are calculated (as shown in Table 3.1). Furthermore, the inter-aggregate pore-size distributions and the intra-aggregate pore-size distribution are related to the void ratio. Regression analyses based on the data listed in Table 3.4 give $\lambda_1 = -0.46$, $\lambda_2 = 0.69$, $\lambda_3 = 0.9$, $\lambda_4 = 0.6$, and $k_1 = 0.75$. Again $k_1 = 0.75$ means that about 75% of volume change occurs in the intra-aggregate pores and only 25% of volume change occurs in the inter-aggregate pores for the saturated soil samples.

Table 3.4 Best-fit analysis results for PSDs of the compacted, saturated, and oven-dried samples

Suction (kPa)	Water content (%)	Degree of saturation	Void ratio	PSD of inter-aggregate pores			PSD of intra-aggregate pores		
				a_1 (cm^3/g)	μ_1	σ_1	a_2 (cm^3/g)	μ_2	σ_2
<1	37.8	1	1.00	0.027	0.74	0.41	0.33	−0.52	0.52
100	29.2	0.80	0.96	0.036	0.72	0.47	0.30	−0.54	0.50
500	14.2	0.45	0.84	0.021	0.86	0.50	0.27	−0.61	0.57
>10^5	~0	~0	0.65	0.006	0.88	0.52	0.22	−0.76	0.50

3.5 Investigation of Microporosity Structure for Coarse Soils

3.5.1 Experimental Program

The largest soil sample was limited to 15 cm^3 in the MIP tests. Therefore, the soil particles with diameters above 10 mm, which presents in sandy ML, SC with gravel, SM with gravel, and GW-GM with sand, will introduce non-homogeneity to soil samples. To maintain the homogeneity of soil samples, soil particles with diameters above 10 mm were taken out from the soil when preparing the soil samples for the MIP tests. The particles with diameters above 10 mm are assumed to be a cubic solid without any pores and the remaining part retains the void ratio and PSD of the soil, the density of the soil sample used in the MIP tests was corrected as,

$$\rho_c = \frac{(1-p)\rho}{1-\frac{p\rho}{G_s}} \tag{3.18}$$

where ρ is the initial soil density; ρ_c is the corrected soil density for preparation soil sample used in MIP test; p is the weight proportion of the soil particles with diameters larger than 10 mm; G_s is the gravimetric density of the soil solid.

To investigate the microporosity structure changes during wetting and drying for coarse soils, the pore-size distributions of unsaturated compacted sample, compacted and saturated sample, saturated and oven-dried sample are measured. For each soil type, three samples at the same dry density, such as one compacted sample at optimum water content, one compacted and saturated sample, and one compacted, saturated and oven-dried sample, are tested.

The unsaturated compacted samples were prepared in several steps. First, the soil was oven-dried in the laboratory. Then the dry soil was mixed with water corresponding to the optimum water content. Then the mixed soil was stored in a plastic bag for 7 days for moisture homogenization, and the final water content of the soil was measured. The matured soil was statically compacted to a specified dry density value at a rate of 0.05 mm/min. Each soil sample was compacted inside a metal mold of 70 mm in diameter and 40 mm in height. The compaction was performed in two layers. Scratches were made on the soil between layers to avoid segregation.

Some compacted samples were saturated for 10 days in a mold while being subjecting to a vacuum (as shown in Fig. 3.1d). Some saturated samples were oven-dried as saturated and oven-dried soil sample.

3.5.2 Estimation of Volume of Pores with Radii Out of Measurement Range

In this study, the measurement range of pore radius is from 0.003 to 72 μm as the intrusion pressure range supplied by Micromeritics AutoPore IV 9500V 1.04 MIP device is from 9.3 kPa to 210 mPa [referring to Eq. (3.4)]. The volumes of pores with radius out of the measurement range, i.e., volume of pores with radius smaller than 3 nm and pore volume with radius larger than 72 μm, are needed to be estimated.

The pore volume with radius smaller than 3 nm was estimated as the water content of soil sample exposed to laboratory atmosphere over weeks (referring to Table 3.5). The soil sample exposed to laboratory atmosphere over weeks was considered to equilibrium to an atmosphere humidity of 62% (an average value measured in the laboratory, which corresponds to a suction of 72 mPa), and the pores with radius smaller than ~2 nm in the sample were considered to be filled by water. Therefore, in the soil sample exposed to laboratory atmosphere over weeks, the water content should be equal to pores volume with radius smaller than ~2 nm. The volume of pores with radius between 2 and 3 nm was ignored, and the water content was taken as the pore volume with radius smaller than 3 nm.

As the pore volume with radius smaller 3 nm was estimated and the pore volume with radius in range of (3 nm–72 μm) could be measured, the pore volume with radius larger than 72 μm was calculated as the difference between the total void (determined by the soil dry density) and the pore volume with radius smaller than 72 μm. For the unsaturated compacted soil sample, the soil dry density was known. For the compacted and saturated soil sample, the soil dry density is remained as the value at compaction (no volume change is permitted during the confined saturation, as shown in Fig. 3.14b). For oven-dried soil sample, the soil dry density was unknown because shrinkage might occur during the drying process and induce sample volume change. Therefore, for oven-dried soil sample, the pore volume with radius larger than 72 μm was estimated from the volume of pores with radius in range of (2–72 μm) by using the same proportion in the compacted and saturated soil sample, which is calculated as the ratio between the pore volume with radius larger than 72 μm and the pore volume with radius in range of (2–72 μm).

Table 3.5 Widely-graded soils used in the investigation of microporosity experiments

Group symbols (ASTM D2487)	Dry density (kg/m^3)	Water content of samples exposed to atmosphere (%)	proportions		
			Fines (%)	Sand (%)	Gravel (%)
GW-GM with sand	1849	0.7	7	35	58
SM with gravel	1802	1	24	32	44
SC with gravel	1734	1.5	42	29	29
Sandy ML	1659	1.8	60	25	15
CL with sand	1320	2.3	80	20	0

3.5.3 Evolution of Dual-Porosity Structure Upon Wetting

The unsaturated compacted soil samples and the compacted and saturated soil samples were analyzed using MIP. The PSDs for the four widely-graded colluvial soils, i.e., GW-GM with sand, SM with gravel, SC with gravel, and sandy ML, are shown in Figs. 3.16, 3.17, 3.18 and 3.19.

The PSDs for sandy ML are shown in Fig. 3.16a. The PSD of compacted sandy ML sample shows dual-porosity, i.e., macro-pores (radius above 2 μm) and micro-pores (radius smaller than 2 μm). The PSD of saturated sandy ML sample shows a unimodal, micropores porosity, where the PSD shows one peak at 0.4 μm. In the samples with 1 g of solid, the pore volumes with radius in different ranges are shown in Fig. 3.16b. After wetting, the pore volume with radius larger than 2 μm decreases distinctly and the pore volume with radius in range of (0.3–2 μm) increases a lot.

The soils used in this study were sieved from completely decomposed granite in Hong Kong. A mineral analysis by Zhao et al. (2005) showed that the fines of completely decomposed granite contained approximately 60% of kaolinite and 40% of ledikite. The fines (i.e., clay aggregates) are anticipated to swell upon wetting. Because the intra-aggregate pores swell into larger pores, the PSD shows a larger average radius a larger differential volume of intra-aggregate pores after wetting (i.e., from 0.2 to 0.4 μm as shown in Fig. 3.16). When the soil sample is saturated under free boundary conditions, the soil sample may swell and show some increase of void ratio. In the saturation method used in this study, however, the soil sample was placed inside a metal mold (Fig. 3.1d) and therefore the total volume and void ratio of the soil sample could not change. Consequently, the volume of inter-aggregate pores (i.e., the pores with radius larger than 2 μm in Fig. 3.16) decreases after wetting.

Similar phenomena, where the decrease of macro-pores (radius above 2 μm) and the increase of micropores occur upon wetting, are found for SC with gravel (Fig. 3.17), SM with gravel (Fig. 3.18), and GW-GM with sand (Fig. 3.19).

3.5.4 Micropore Shrinkage After Drying

The compacted and saturated soil samples and the compacted, saturated and oven-dried soil samples were analyzed using MIP. The PSDs of different soils are shown in Figs. 3.16, 3.17, 3.18, and 3.19. During the drying process, some intra-aggregate pores (i.e., pores radius among 2–0.5 μm) formed by fines, shrink to smaller pores. Therefore, the differential pore volumes with radius in range of (0.5–2 μm) decrease the differential pore volumes with radius in range of (0.1–0.5 μm) increase (Figs. 3.16, 3.17, 3.18 and 3.19). For sandy ML and SC with gravel, the differential pore volumes with radius above 2 μm decrease a lot after drying. On the contrary, for SM with gravel and GW-GM with sand, the differential pore volumes with radius above 2 μm increase after drying. The total pore volumes of sandy ML and SC with

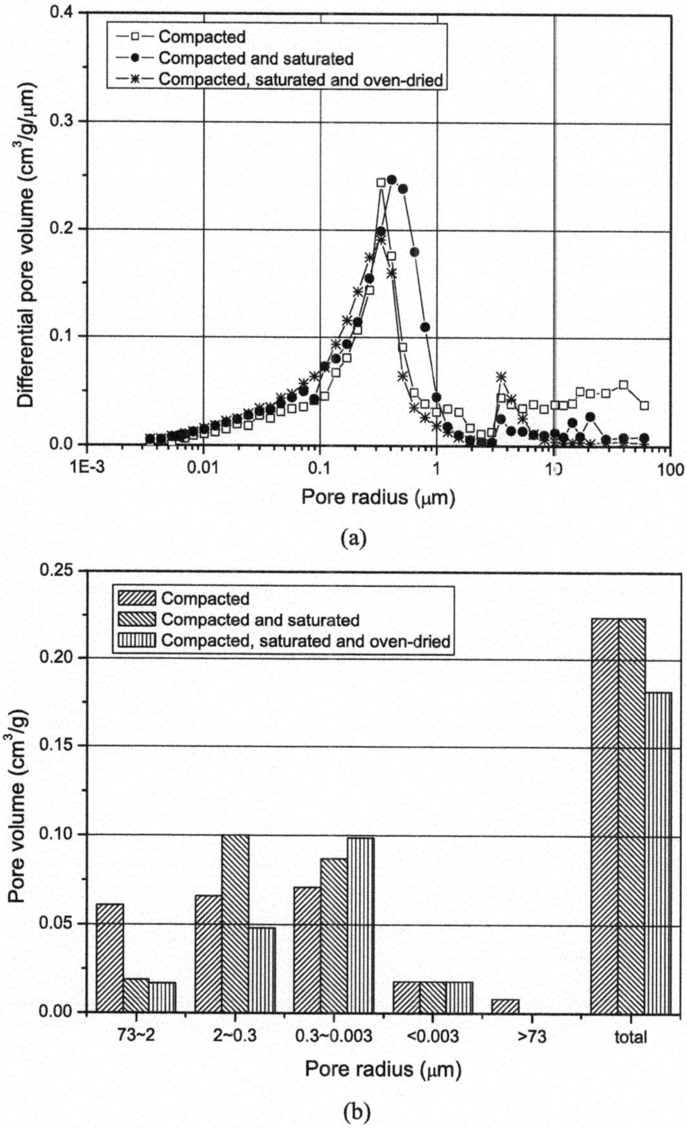

Fig. 3.16 PSD changes during wetting and drying for Sandy ML: **a** Differential pore volume. **b** Volume fractions of pore in different radii range

gravel decrease significantly after drying, whereas the total pore volumes of SM with gravel, and GW-GM with sand show little changes.

For sandy ML and SC with gravel, the soils, which have high proportion of fine contents, form structures substantially controlled by fines (i.e., supported by skeleton composed of clayey aggregates). During the drying process, the intra-aggregate pores

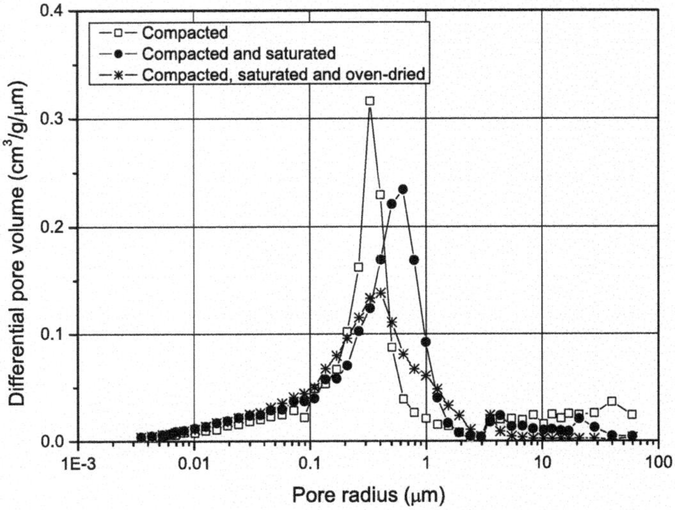

(a)

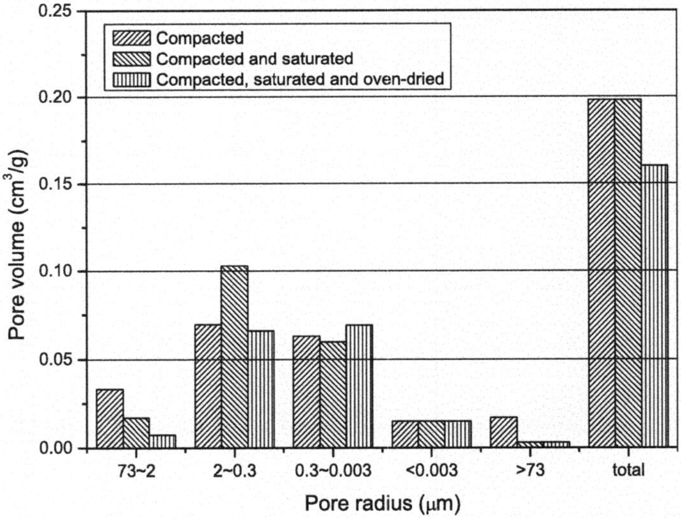

(b)

Fig. 3.17 PSD changes during wetting and drying for SC with gravel: **a** Differential pore volume. **b** Volume fractions of pore in different radii range

(pores with radius smaller than 2 μm) shrink because of the contraction of soil aggregates. As a result, the total soil volume and the volume of shrinkable pores [pores with radius in range of (0.5–2 μm)] decrease after drying.

For SM with gravel and GW-GM with sand, the soils form structures supported by skeleton composed of coarse particles. Because the coarse particles are much more

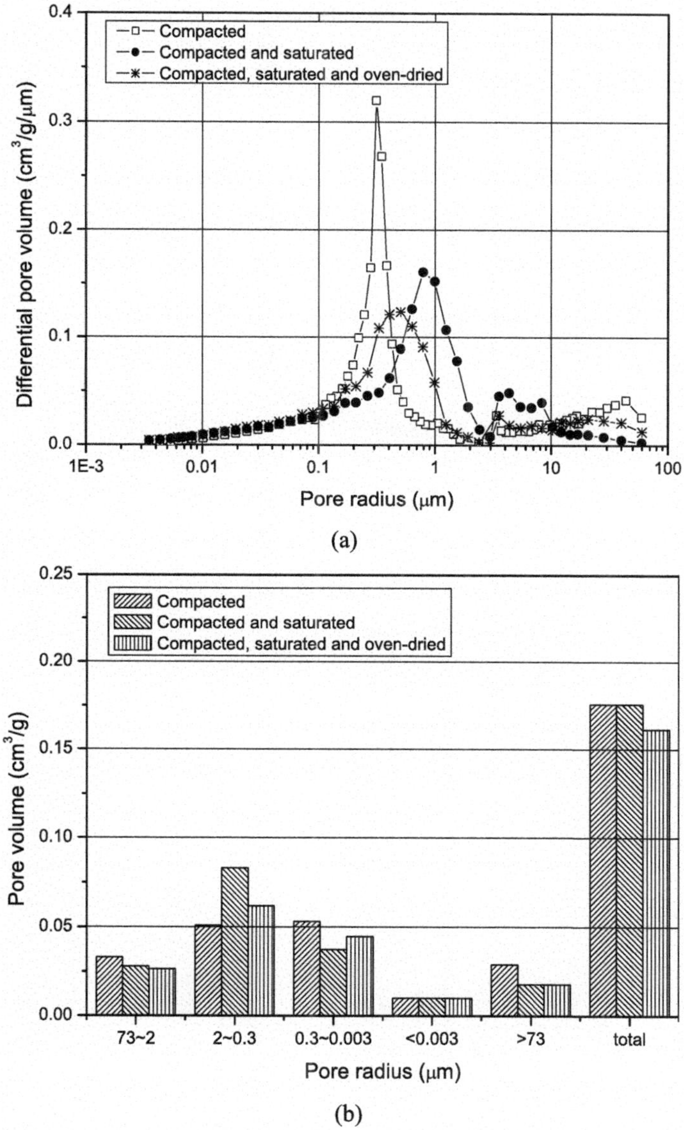

Fig. 3.18 PSD changes during wetting and drying for SM with gravel: **a** Differential pore volume. **b** Volume fractions of pore in different radii range

stable comparing to the clayey aggregates upon water actions, the soil structures are considered to remain upon drying. Therefore, the total pore volumes show few changes after drying for SM with gravel and GW-GM with sand. The soil plasticity and shrink–swell potential increase with increasing the content of amorphous clay-size materials in the soil (Wan et al. 2002). The three finer soils used in the laboratory

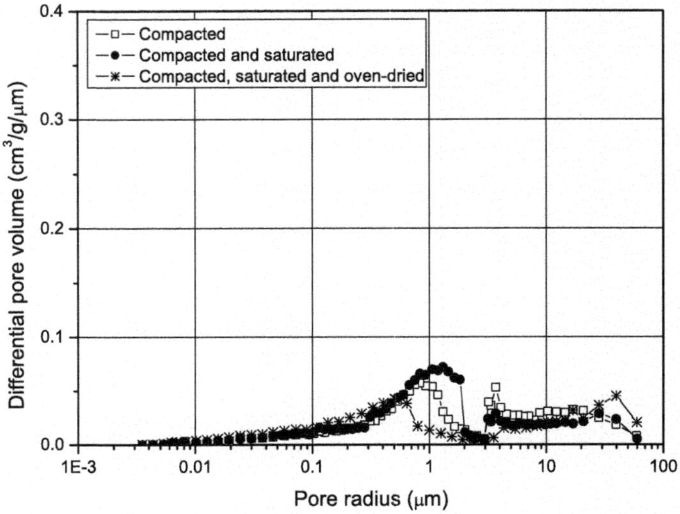

(a)

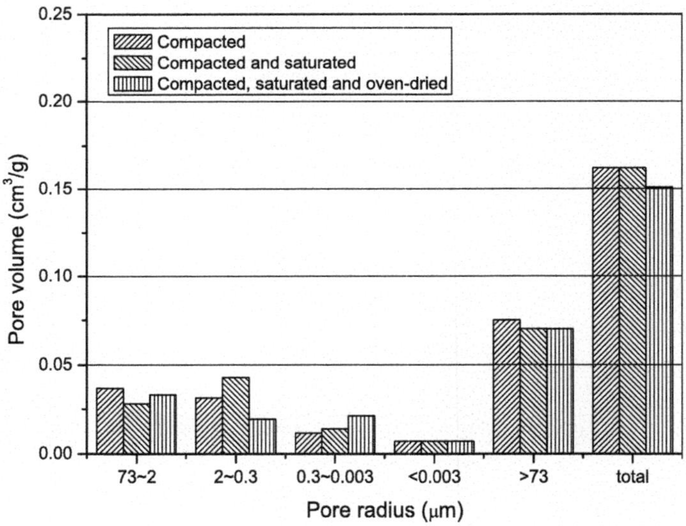

(b)

Fig. 3.19 PSD changes during wetting and drying for GW-GM with sand: **a** Differential pore volume. **b** Volume fractions of pore in different radii range

test, which have fractions of fines ranging from 80 to 42, show high shrinkage ability. The two coarser soils, which have fractions of fins ranging from 7 to 24, show weak shrinkage ability.

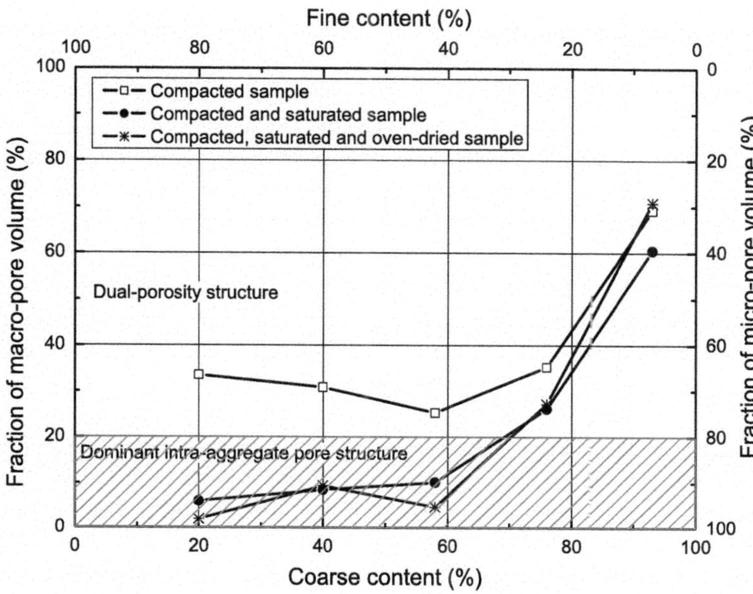

Fig. 3.20 Microporosity structure changes upon wetting–drying

3.5.5 Effect of the Coarse Fraction on Soil Porosity Structure

The proportions of macro-pores (with radius above 2 μm) of the five different soils are compared (in Fig. 3.20). All the unsaturated compacted soil samples of the five different soils have considerable fractions of macro-pores (pores with radius larger than 2 μm) volume above 25% (Fig. 3.20) and show bimodal PSDs (Figs. 3.4, 3.16, 3.17, 3.18, and 3.19). This indicates that a dual-porosity structure is common for compacted colluvial soil samples, no matter the soil has a unimodal GSD or bimodal GSD.

As the soil becomes coarser, the fine content in the soil decreases (i.e., 80% in CL with sand, 60% in sandy ML, 42% in SC with gravel, 24% in SM with gravel, and 7% in GW-GM with sand.) As the soil becomes coarser, the difference between the PSD of the compacted sample and the saturated soil sample becomes smaller (referring to Figs. 3.16, 3.17, 3.18, 3.19, and 3.20). For compacted and saturated soil samples, the fractions of macro-pores volume are smaller than 10% for CL with sand, sandy ML, and SC with gravel. Also, no distinct bimodal PSD feature is found in the PSD of CL with sand, sandy ML, and SC with gravel (referring to Figs. 3.4, 3.16, and 3.17). That is to say, the dual-porosity structures formed during the compaction evolve into an intra-aggregate pores structure after saturation for CL with sand, sandy ML, and SC with gravel. Compacted and saturated soil samples of SM with gravel and GW-GM with sand show different behaviors. The fractions of macro-pores volume in the compacted and saturated soil samples are larger than 25% for SM with gravel and GW-GM with sand. The dual-porosity structures remain after saturation for SM with gravel and GW-GM with sand.

For a same soil, the oven-dried soil sample shows the same feature as the saturated soil sample. The dual-porosity structure is found in the oven-dried soil sample for SM with gravel and GW-GM with sand. And an intra-aggregate pores structure is found in the oven-dried soil sample for CL with sand, sandy ML, and SC with gravel. That is to say, the change of soil structure is not significant during drying for these colluvial soils.

As the soil becomes coarser, the soil structure gradually changes from a fines-controlled structure (supported by the fine aggregate skeleton) to a coarse-controlled structure (supported by the coarse particles). The fines-controlled structure is easily influenced by the water wetting or drying process. In the fines-controlled structure, the fines aggregates are anticipated to swell upon wetting and are anticipated to shrink upon drying, and the fines aggregate is easily reorganized or destroyed during the wetting–drying process. Therefore, in the cases for CL with sand, sandy ML, and SC with gravel, the dual-porosity structure formed due to soil compaction is fine-controlled and evolves into an intra-aggregate pores structure during the confined saturation, and the intra-aggregate pores structure shows distinct shrinkage during the drying process. In the cases for GW-GM with sand and SM with gravel, a dual-porosity, coarse-controlled structure is formed during the soil compaction and shows limit changes after wetting and drying.

A critical value of the fraction of macro-pore volume is used to identify a dual-porosity structure, or an intra-aggregate pore structure. The critical value of the fraction of macro-pore volume is recommended as 20% (referring to Fig. 3.20), which is approximately the average value between the macro-pore volume in compacted sample and that in saturated sample for these fines-controlled soils. Corresponding to this critical value, the soils with fines contents smaller than 30% are considered as coarse-controlled soils and show constant dual-porosity structures. The soils with fines contents above 30% are considered as fines-controlled soils and show dominant intra-aggregate pore structures after wetting. The variations of the microporosity structure are summarized in Figs. 3.21 and 3.22.

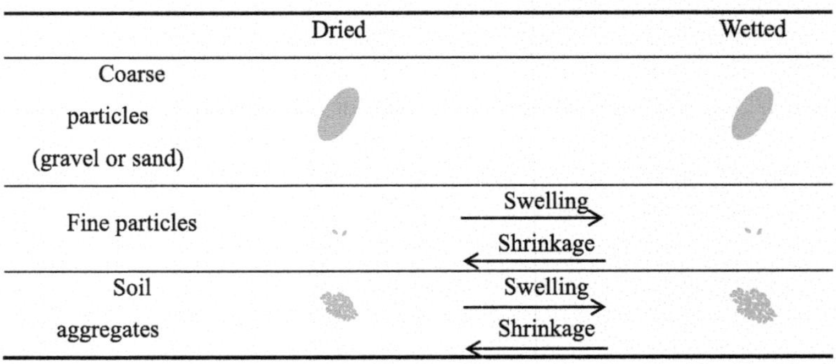

Fig. 3.21 Three basic elements in widely-graded soils

State	Sketch	PSD change	Porosity structure
At compaction	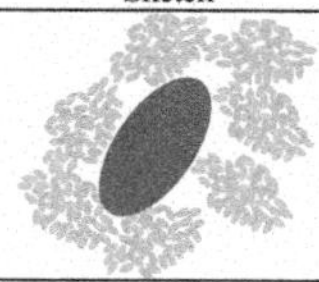		Dual-porosity structure
After saturation (confined boundary)	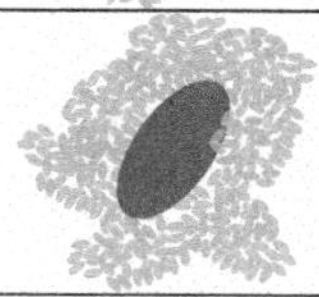	Soil aggregate swelling → intra-aggregate pores increase; inter-aggregate pores decrease; no volume change	Dominant intra-aggregate pore structure
Oven-dried	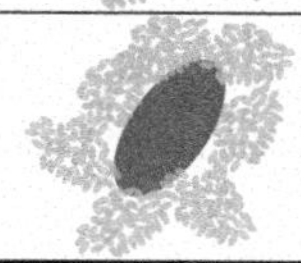	Soil aggregate shrinkage → intra-aggregate pores decrease; volume decrease	Dominant intra-aggregate pore structure

(a)

State	Sketch	PSD change	Porosity structure
At compaction	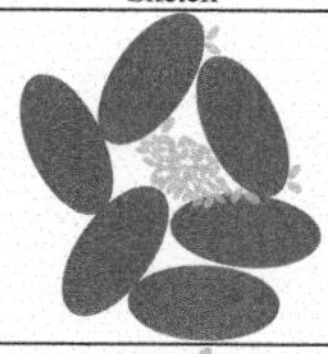		Dual-porosity structure
After saturation (confined boundary)	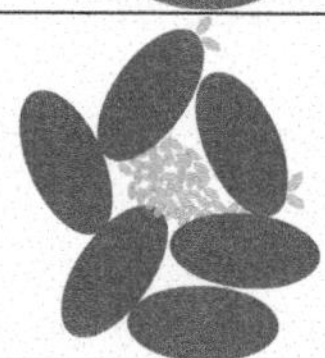	Soil aggregate swelling → intra-aggregate pores increase; inter-particle pores decrease; no volume change	Dual-porosity structure
Oven-dried		Soil aggregate shrinkage → intra-aggregate pores decrease; inter-particle pores increase; limited volume change	Dual-porosity structure

(b)

Fig. 3.22 Soil structure of widely-graded soils: **a** Fines-controlled structure. **b** Coarse-controlled structure

Similar critical fines contents had also been found in the investigation of the effect of fines content on soil porosity and shear strength. Yang et al. (2006) studied the porosity of sands with varied fine contents and reported that a minimum porosity was reached at a fines content of 30% (70% coarse content) under the same compaction effort. Vallejo and Mawby (2000) reviewed the studies on the shear strength of soils with different coarse contents and summarized that: (1) the shear strength of soils with coarse contents lower than 50% was controlled by the fines, (2) the shear strength of soils with coarse content larger than 76% was controlled by the coarse particles, and (3) the shear strength of soils with coarse content among 50–76% was controlled by both the fines and coarse particles.

3.6 Summary

For CL with sand, an unsaturated sample, a bulky soil sample, a saturated sample, and an oven-dried sample were tested using SEM to observe the microporosity structures on soil surfaces. A dual-porosity structure is found on the surfaces of these samples. The soil aggregates become thicker and closer after the saturated soil sample is oven-dried.

The pore-size distributions of a series of compacted samples, saturated samples, and dried samples were measured using MIP to quantitatively characterize the characteristics and variations of soil microporosity structure during compaction, saturation, and drying. The following conclusions can be drawn based on the results of these tests:

(1) A dual-porosity structure can be represented by inter-aggregate pores (macro-pores) and intra-aggregate pores (micropores). A critical value of the fraction of macro-pores volume is used to identify a dual-porosity structure or a substantial intra-aggregate pores structure and is recommended as 20% (referring to Fig. 3.20).
(2) For colluvial soils, a dual-porosity structure may form during the soil compaction process.
(3) For colluvial soils, as the soil becomes coarser, the soil structure changes from a fines-controlled structure (supported by fine aggregate skeleton) to a coarse-controlled structure (supported by the coarse particles).
(4) For the colluvial soils with a fines fraction above 30%, such as CL with sand, sandy ML, and SC with gravel, a dual-porosity fines-controlled structure is formed during soil compaction and evolves into an intra-aggregate pores structure after the confined saturation, and in the further drying process, the intra-aggregate pores structure shows substantial shrinkage. The clayey aggregates in the fines-controlled structure are anticipated to swell upon wetting and are anticipated to shrink upon drying. Therefore, the soil structure supported by the skeleton composed of soil aggregates is easily influenced by the wetting or drying process and the sample volume decreases with the shrinkage of intra-aggregate pores during the drying process.
(5) For the colluvial soils with a fines fraction smaller than 30%, such as GW-GM with sand and SM with gravel, a coarse-controlled, dual-porosity structure is formed during the soil compaction and shows limit changes after wetting and drying. A possible interpretation is that inside the loading-collapse surface, the soil structure supported by skeleton composed of substantial coarse particles is stable upon wetting and drying.

(6) The compaction effort can cause a significant reduction in the volume of the inter-aggregate pores, but can only cause slight influences on the intra-aggregate pores during compaction.
(7) On the condition of a rigid boundary during soil saturation, the intra-aggregate pores show remarkable increase accompanying with some decrease of inter-aggregate pores.
(8) During the drying process, finer soils show higher shrinkage ability. For fine soils, both the intra-aggregate pores and the inter-aggregate pores decrease after drying. For coarse soils, the intra-aggregate pores (or inter-particle pores) show some decrease, whereas the inter-particle pores increase. As a result, the total volume of coarse soil changes little after drying.
(9) A mathematical model was developed to describe the bimodal PSDs and to relate the soil PSDs with void ratio. The PSD at an arbitrary void ratio subjected to different deformation histories can be predicted using the empirical parameters calibrated through the MIP tests.

References

Alonso EE, Gens A, Hight DW (1987) Special problem soils: General report. In: Proceedings of the 9th European conference of soil mechanics and foundation engineering, vol 30. Dublin, pp 1087–1146

Anandarajah A, Lavoie D (2002) Numerical simulation of the microstructure and compression behavior of Eckernforde Bay sediments. Mar Geol 182:3–27

Basma AA, Al-Homoud AS, Husein M, Ai-Bashabsheh MA (1996) Swelling—shrinkage behavior of natural expansive clays. Appl Clay Sci 11(2):211–277

Cafaro F (2002) Metastable states of silty clays during drying. Can Geotech J 39(4):992–999

Diamond S (1970) Pore size distributions in clays. Clays Clay Miner 18:7–23

Dineen K (1997) The influence of soil suction on compressibility and swelling. Ph.D. thesis. University of London, UK

Lennart R, Bertil W (2004) Mathematics handbook for science and engineering, 5th edn. Springer, New York

Mallants D, Mohanty P, Vervoort A, Feyen J (1997) Spatial analysis of saturated hydraulic conductivity in a soil with macropores. Soil Technol 10:115–131

Pruess K, Wang Y, Tsang W (1990) On thermo hydrologic conditions near high-level nuclear wastes emplaced in partially saturated fractured tuff, effective continuum approximation. Water Resour Res 26:1249–1261

Sudhakar M, Revanasiddappab K (2005) Role of microfabric in matrix suction of residual soils. Eng Geol 80:60–70

Vallejo LE, Mawby R (2000) Porosity influence on the shear strength of granular material-clay mixtures. Eng Geol 58(2):125–136

Wan Y, Kwong J, Brandes et al (2002) Influence of amorphous clay-size materials on soil plasticity and shrink-swell behavior. J Geotech Geoenviron Eng 128(12):1026-1031

Yang SL, Lacasse S, Sandven RF (2006) Determination of the transitional fines content of mixtures of sand and non-plastic fines. Geotech Test J 29(2):102–107

Zhang LM, Fredlund DG (2003) Characteristics of water retention curves for unsaturated fractured rocks. In: proceedings of the 2nd Asian conference on unsaturated soils. Osaka, Japan, pp 425–428
Zhao JJ, Wang SJ, Shang YJ, Yue ZQ (2005) Consolidation characteristics of completely decomposed granite in Hong Kong. J Hohai Univ Nat Sci 33(1):85–88

Chapter 4
Techniques for Measuring Hydraulic Properties for Unsaturated Soils

4.1 Introduction

The soil water characteristic curve (SWCC) and the soil hydraulic conductivity function (SHCF) are two important soil parameter functions required when performing seepage analysis for saturated/unsaturated soil systems. Examples of engineering applications where a soil hydraulic conductivity function is involved include the triggering of landslides due to rainfall infiltration, and the modeling of flow and volume change in collapsing soils, compacted soils, and expansive clays. The SHCF is also required when modeling contaminant migration in vadose zones and the design of capillary barriers and cover systems. In this chapter, the techniques for measuring the hydraulic properties of unsaturated coarse granular soils (i.e., SWCCs and SHCFs) were introduced.

The Widely-graded soils with dual-porosity structure (as shown in Fig. 3.20) may have a remarkable volume of pores with radius above 72 μm [such as GW-GM with sand (Fig. 3.19) and SM with gravel (Fig. 3.18)]. Such soils are considered to have a significant change of water content at suctions smaller than 2 kPa. A new SWCC device was developed for measuring the SWCC at extremely low suctions (smaller than 2 kPa). At the extremely low suctions, the suction difference along the sample height is large, as 0.2–1 kPa, and may cause distinct errors. A new interpretation equation is proposed to reduce that error for measuring SWCC at extremely low suctions.

The measurement of SHCF includes the measurement of saturated soil hydraulic conductivity and the measurement of unsaturated soil hydraulic conductivity. The saturated coefficients of soil hydraulic conductivity of five colluvial soils were measured using traditional constant-head and falling head tests. Existing unsaturated SHCF measurement methods are often time-consuming, limited in the available measurement ranges of coefficient of soil hydraulic conductivity or are dependent on the selected SHCF models for analyzing experimental data. It is desirable to develop a direct method for the measurement of the SHCF that is time-saving and applicable to a wide range of coefficients of soil hydraulic conductivity and a wider range of soil

X. Li et al., *Fundamental Behavior of Unsaturated Widely-Graded Soil*,
https://doi.org/10.1007/978-981-19-3402-5_4

suction. To achieve this objective, a new experimental method called the "wetting front advancing method" (WFAM) was developed. In this method, a large-scale soil column test is used to simulate the flow of water into an unsaturated soil column. The monitored data are used to calculate the SHCF with the help of a newly developed analytical interpretation procedure.

In the last part, a digital WFAM is proposed and a thorough comparison of IPM and WFAM is also carried out. Both the data of soil column tests reported by Cui et al. (2008) and Ng and Leung (2012) and those from numerical tests are analyzed using these two methods. Then based on the results, the accuracy and the applicable conditions of the two methods are discussed. Finally, a combination of IPM and digital WFAM is proposed to measure SHCF over a full suction range and its applicability is also examined by numerical tests.

4.2 Development of a SWCC Device for Testing Gravelly Soils at Very Low Suctions

4.2.1 Difficulties in Measuring SWCC of Gravel Soils at Very Low Suctions

One of the difficulties encountered in the measuring SWCC of colluvial soils is the lack of a proper device for the measurement of SWCCs at very low suctions (i.e., smaller than 2 kPa). There are at least two methods for measuring SWCCs. One method is to measure both the water content of a soil specimen and the suction (total suction or matric suction) in the soil specimen, or the humidity around the soil specimen. This method includes the techniques using filter papers (Marinho and Oliveira 2006), psychrometers (Leong et al. 2003), traditional tensiometers (Fredlund and Rahardjo 1993), suction probes (i.e., tensiometers for measuring high soil suctions) (Ridley and Burland 1996; Take and Bolton 2003; Cui et al. 2008), and thermal conductivity sensors (Feng and Fredlund 2003; Nichol et al. 2003). The suction measurement ranges of different sensors are shown in Fig. 4.1a. These sensors may not be suitable for use at very low suctions. Another method is to control the suction and measure the water content in the soil specimen. This method includes the axis-translation technique (Fredlund and Rahardjo 1993), the osmotic technique (Delage and Cui 2008; Monroy et al. 2007), and the humidity control technique (Likos and Lu 2003). The axis-translation technique is the most popular SWCC measurement method because of its accuracy and validity in the most concerned suction range. But the traditional axis-translation devices always use the air pressure to control the soil suction and show low accuracy in the low suction range. Water head control can help solve this problem in the axis-translation technique. The four different methods for controlling soil suction, as well as their suction ranges, are summarized in Fig. 4.1b. The water head control axis-translation technique seems to be the most suitable method for measuring SWCCs at very low suctions.

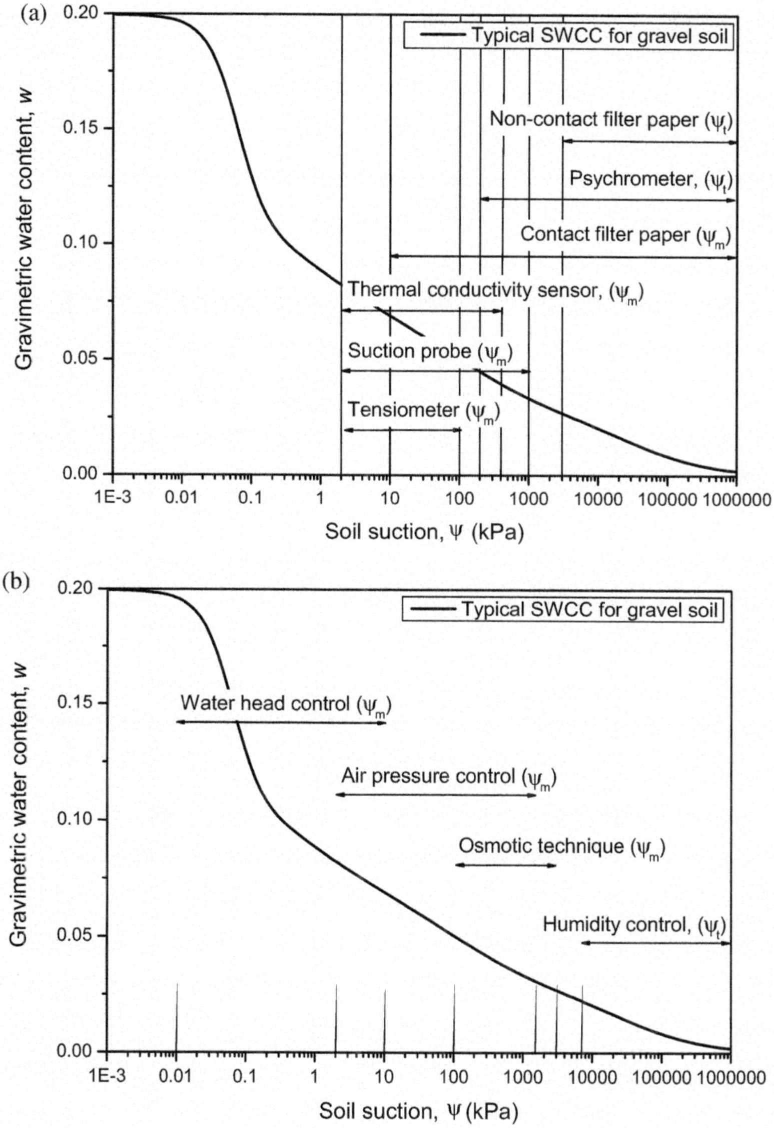

Fig. 4.1 Applicable suction range of different SWCC measurement techniques (ψ_m is matric suction; ψ_t is total suction): **a** Suction measurement methods. **b** Suction controlling methods

Another difficulty encountered in measuring the SWCCs for gravel soils at very low suctions is the inaccuracy of suction associated with the sample size. For example, the suction difference due to different hydrostatic pore water pressures at the top and bottom of a 50 mm height sample is 0.5 kPa. Such a suction difference is unacceptable for measuring a SWCC at very low suctions. An interpretation procedure that can reduce the inaccuracy associated with sample size at very low suctions is needed.

To overcome this limitation, a modified axis-translation device was developed for measuring the SWCCs for gravel soils at very low suctions and a new data interpretation method was proposed to reduce the inaccuracy associated with sample size at very low suctions.

4.2.2 *A New SWCC Device for Using at Very Low Suctions*

A sketch of a device designed for SWCC measurement at very low suctions is shown in Fig. 4.2a. The low suction SWCC device includes a SWCC cell, a water supply and pressure control system, and an air pressure control system. A photograph of the test device is shown in Fig. 4.2b.

4.2.2.1 SWCC Cell

A SWCC cell, as shown in Fig. 4.3, is used to hold the soil specimen and to maintain soil suction. The SWCC cell includes a cell pedestal, a cylindrical cell, and a cell cover. O-rings are used to seal the joints. The cylindrical cell is 300 mm in inner diameter and 150 mm in height and used as a soil specimen container. The cylindrical cell is connected with the cell pedestal and the cell cover with screw poles. A 0.5-bar air entry value ceramic disk, which is 276.23 mm in diameter and 10.31 mm in height, is fixed on the pedestal. The pedestal is machined with 5 mm deep grooves for water supply and diffused air bubble flushing.

The soil matric suction is defined as $u_a - u_w$, where u_a is the pore-air pressure and u_w is the pore water pressure in the soil specimen. When a water supply and pressure control system is connected to the cell pedestal, the water pressure controlled by the system will come to equilibrium with the water pressure in the soil specimen through the ceramic disk. When an air pressure control system is connected to the cell cover, the air pressure controlled by the system will be equal to the cell air pressure, i.e., the air pressure in the soil specimen, u_a. Therefore, the soil matric suction in the SWCC device can be controlled.

4.2.2.2 Water Supply and Pressure Control System

A water supply and pressure control system, as shown in Fig. 4.4, is used to adjust the water head, record the water volume, and flush air bubbles. A horizontal glass tube is fixed on an active board, which can be adjusted up and down along the screw poles and is used to control the water head. The water level adjustment precision is 0.5 mm. Therefore, the pore water pressure or suction can be changed at a precision of 0.005 kPa. During an experiment, the water tail is kept inside the horizontal glass

(a)

(b)

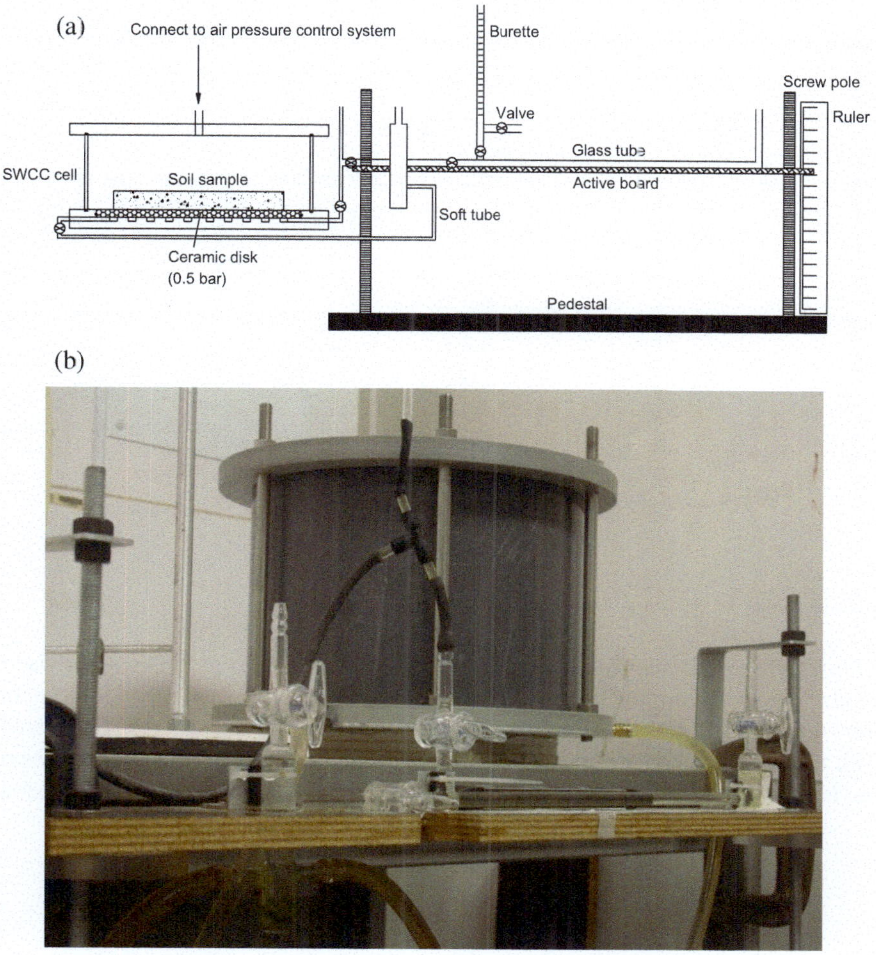

Fig. 4.2 A new SWCC device for use at very low suctions: **a** Sketch of the new low suction SWCC device. **b** Photograph of the new low suction SWCC device

tube, as shown in Fig. 4.5. Therefore, the water level will maintain at the height of the horizontal glass tube (i.e., the active board). The volume of outflow (drained from the soil specimen) or inflow (absorbed by the soil specimen) is measured as the difference between two measurements (i.e., the recorded scales on the burette in Fig. 4.5a, d).

During the experiment, diffused air bubbles may accumulate below the ceramic disk. One function of the water supply and pressure control system is to flush the diffused air bubbles. As a blow ball forces water to flow from tube A to tube B (as shown in Fig. 4.6a), the air bubbles below the ceramic disk will be flushed. After removing the blow ball, water will flow from tube B to tube A and the air bubbles below the ceramic disk will be flushed again, as shown in Fig. 4.6b.

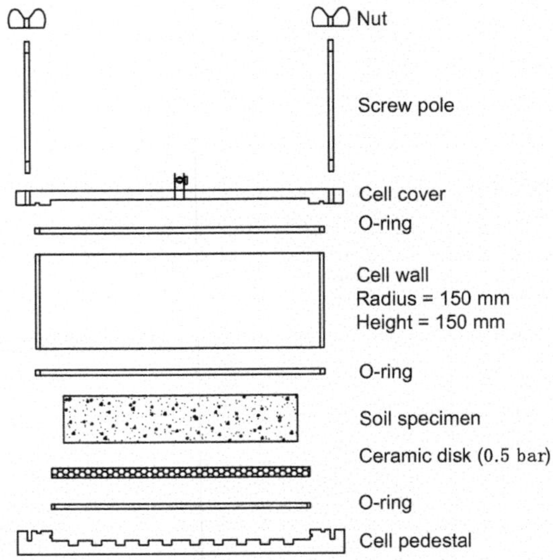

Fig. 4.3 Sketch of the SWCC cell

4.2.2.3 Air Pressure Control System

Although the water supply and pressure control system can be used to control water heads up to −1000 mm (i.e., a suction of 10 kPa), an air pressure control system shown in Fig. 4.7 is more convenient to control suctions above 2.5 kPa. Compared with the water supply and pressure control system, the air pressure control system can be operated more easily but has lower precision (e.g., 0.1 kPa). Such precision

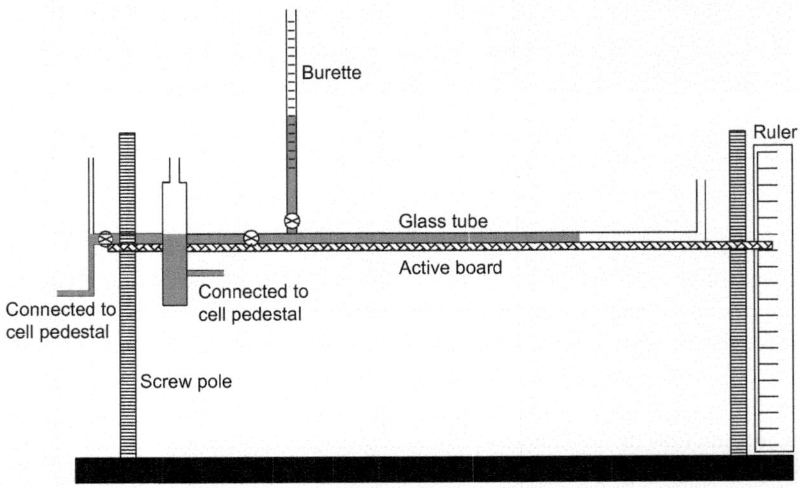

Fig. 4.4 Sketch of the water supply and pressure control system

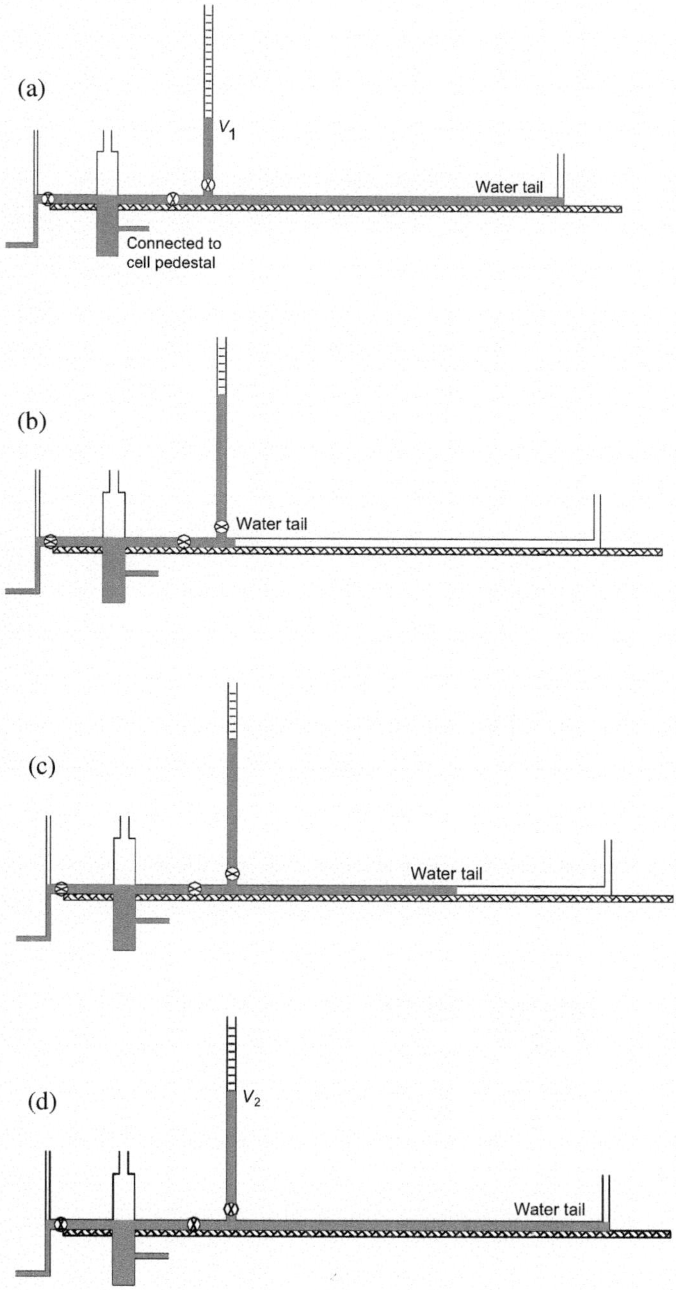

Fig. 4.5 Measurement of drained water volume ($\Delta V = V_2 - V_1$): **a** Recording before a drying step. **b** Start of a drying step. **c** End of a drying step. **d** Recording after a drying step

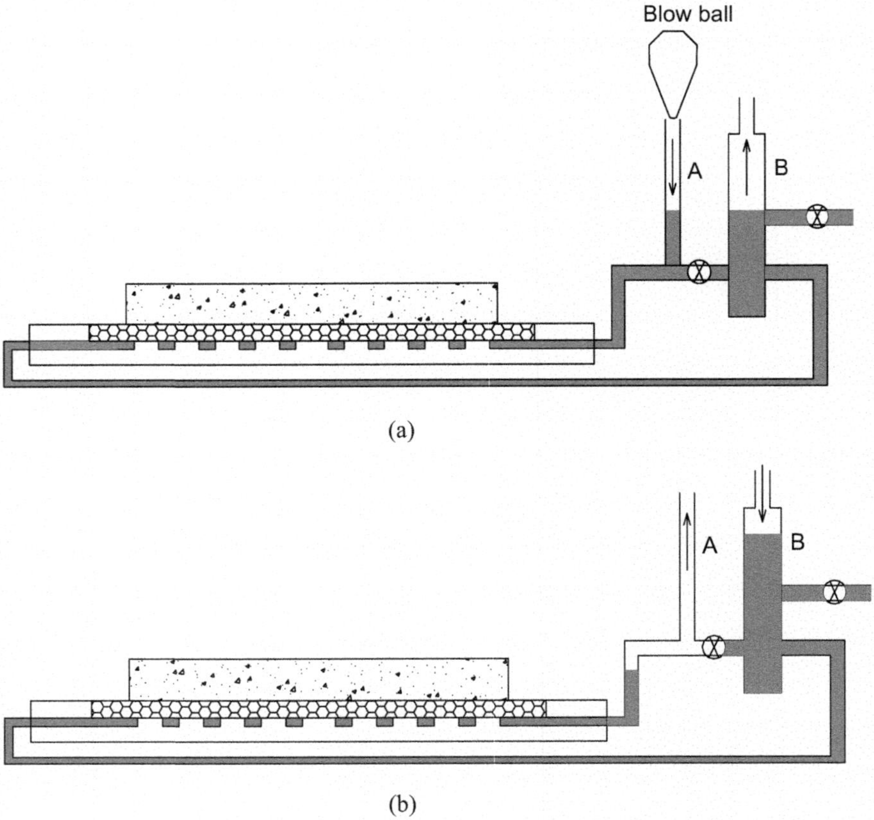

Fig. 4.6 Air bubble flushing: **a** Inducing water flow by using a blow ball. **b** Recovering water level due to gravity equilibrium

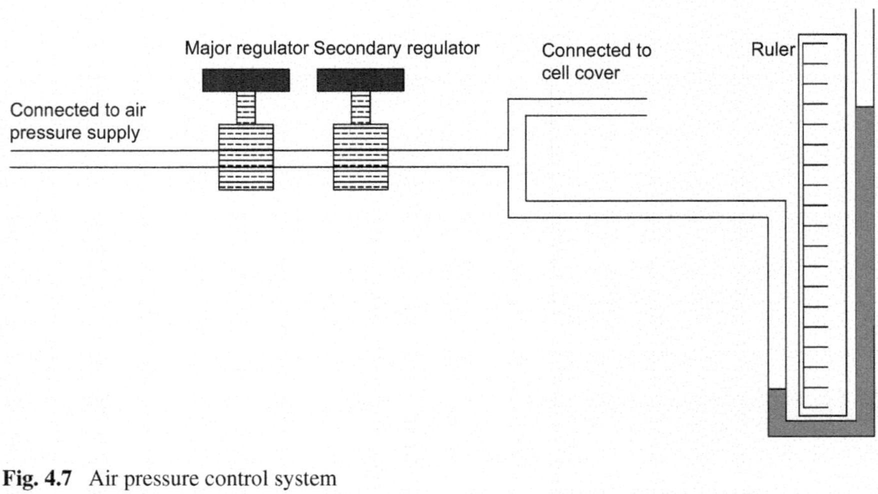

Fig. 4.7 Air pressure control system

is sufficient for measuring suctions above 2.5 kPa (i.e., error < 4%) but insufficient for use at suctions less than 1 kPa (i.e., error > 10%).

When the suction in the system is smaller than 2.5 kPa, the air pressure control system is open to the atmosphere. The soil suction is solely controlled by the water supply and pressure system. When the suction in the system is larger than 2.5 kPa, the air pressure control system is connected to an air pressure supply in the laboratory. A major pressure regulator and a secondary pressure regulator are used together to adjust the air pressure at high accuracy. The air pressure is indicated by the water head difference in the transparent tubes (Fig. 4.7).

4.2.3 Experimental Procedure

The following steps were used in preparing test samples:

(1) The oven-dried soils and water were mixed to the optimum water content.
(2) The mixed soil was sealed in a plastic bag and matured for 7 days. Then, the soil in the sealed plastic bag was considered as uniform soil and used for soil compaction.
(3) During soil compaction, a steel mold, 50 mm in height, 232 mm in inner diameter and 252 mm in outer diameter, was used. The soil was compacted in two layers. Each layer was 25 mm thick. For each layer, one bag of matured soil was used. The soil was statically compacted to a specified dry density value at a rate of 0.05 mm/min.
(4) After compaction, the soil specimen was put into the SWCC cell for saturation. During soil specimen saturation, water was introduced into the SWCC device to submerge the soil specimen. A heavy metal plate with several holes was placed on the surface of the metal mold for preventing soil swelling. Vacuum was then applied through the top tube and hold for 24 h. The soil specimen was then considered saturated.

The following steps were used in measuring the SWCC:

(1) After the soil sample was saturated, the water level in the water supply and pressure control system was adjusted to the ceramic disk surface and maintained until no more water drained. Then, the experiment was started.
(2) The drying SWCC measurement was conducted first and followed by the wetting SWCC measurement. During the drying/wetting SWCC measurement, the soil suction was decreased/increased in steps. In each step, the suction, ψ, was maintained until the drainage of water in 8 h was smaller than 0.01% of the total dry soil weight (e.g., about 0.4 g in this study).
(3) Then, the drained water volume from the soil specimen was recorded and used to calculate the gravimetric water content, w. Finally, the equilibrium suctions and water contents at different steps were plotted as a drying/wetting SWCC.

(4) At the end of the experiment, the soil suction would be adjusted to a relatively high suction (e.g., $\psi = 5$ kPa) and reached the equilibrium state. This would help prevent drainage of water from the soil specimen when the soil specimen is moved out from the device before weighting, which was possible for a gravel soil specimen at the end of a wetting SWCC test (i.e., $\psi \sim 0$ kPa). Then, the SWCC device was uninstalled and the final water content of the soil specimen was measured by the oven-dry method.

4.2.4 Error Evaluation

At the start of the experiment, the mass of the SWCC cell together with the soil specimen was measured and noted as M_s. At the end of the experiment, the mass was measured again and noted as M_e. The cumulative outflow/inflow water mass change (or water volume change) was calculated and noted as M_w. Theoretically speaking, $(M_s - M_e - M_w)$ should be 0. Some errors would inevitably be generated during the experiment. Denote the total dry soil mass as M_d and use a ratio, $\eta = (M_s - M_e - M_w)/M_d$, to describe the test error. If η is smaller than 0.01, the error of measured water contents would be smaller than 1%. In this study, η varied in the range of −0.003 to 0.008, so the experimental data are considered acceptable.

4.2.5 Data Interpretation Procedure

Assume the sample height is L and the equilibrium water level is H below the bottom of soil specimen (i.e., the surface of ceramic disk). The suction head at the bottom of the soil specimen will be H, and the suction head at the top of the soil specimen will be $H + L$. The suction head difference between the top and bottom of soil specimen is L. If the soil has unique dying/wetting SWCCs, an average suction head (i.e., $H + L/2$) inside the specimen can be used. So the water content can be calculated as,

$$w(H + L/2) = w(H + L/2 + \Delta h) - \Delta M/M_d \tag{4.1}$$

where M_d is the total mass of dry soil in the specimen; ΔM is the drained water mass in one step; and Δh is the water head change in one step. Then, the gravimetric water content at the corresponding equilibrium suctions can be back calculated from the final gravimetric water content and the cumulative water volume change (i.e., drained water volume or inflow water volume) at each step.

The total water mass in the soil specimen, M, can be calculated as,

$$M = \int_{H}^{H+L} w(h)\rho_d A dh \tag{4.2}$$

where ρ_d is the soil dry density; A is the cross-sectional area of soil specimen; $w(h)$ is the soil gravimetric water content corresponding to a water head of h (i.e., a suction of $h\gamma_w$); h is the water head in the soil specimen; and γ_w is the unit weight of water.

The soil is assumed to be homogenous with unique dying/wetting SWCCs. Consider a water level change of Δh (i.e., a suction step), after equilibrium the drained water mass in this step, ΔM, can be calculated as

$$\begin{aligned}\Delta M &= \int_{H}^{H+L} w(h)\rho_d A dh - \int_{H+\Delta h}^{H+L+\Delta h} w(h)\rho_d A dh \\ &= \int_{H}^{H+\Delta h} w(h)\rho_d A dh - \int_{H+L}^{H+L+\Delta h} w(h)\rho_d A dh \\ &= \rho_d A\Delta h[w(H+0.5\Delta h) - w(H+L+0.5\Delta h)]\end{aligned} \quad (4.3)$$

Substituting $M_d = \rho_d AL$ into Eq. (4.3), the equation can be rewritten as,

$$w(H+0.5\Delta h) = w(H+L+0.5\Delta h) + \frac{L\Delta M}{M_d \Delta h} \quad (4.4)$$

Since Δh and ΔM are recorded during the experiment and L and M_d are two constants, the second part in Eq. (4.4) is known. Then, the water content at a lower suction, $H+0.5\Delta h$, can be back calculated from the water content at a higher suction, $H+L+0.5\Delta h$.

If the water level is higher than the surface of the ceramic disk, water will flow into the chamber and the measured water mass will not reflect the water mass change inside the soil specimen. Therefore, the water level must be lower than the ceramic disk (i.e., $H \geq 0$). If H is adjusted to 0, the minimum measurable water content using Eq. (4.1) will be $w(L/2)$ and the minimum measurable water content using Eq. (4.4) will be $w(\Delta h/2)$. The minimum measurable suction head is decreased from $L/2$ (i.e., 25 mm in this study) to $\Delta h/2$ (i.e., 2.5 mm in this study) by using Eq. (4.4) instead of Eq. (5.1). So the minimum measurable suction head is extended from centimeters (i.e., 0.1 kPa) to millimeters (i.e., 0.01 kPa).

Theoretically speaking, Eq. (4.1) gives a water content at a suction head precision of $L/2$ (i.e., average water content in a suction head range of H to $H+L$) and Eq. (4.4) gives a water content at a suction head precision of $\Delta h/2$ (i.e., average water content in a suction head range of H to $H+\Delta h$). When $\Delta h > L$, Eq. (4.1) is more accurate, and when $\Delta h < L$, Eq. (4.4) is more accurate. In this study, Eq. (4.1) was used for water content calculation when $H > 250$ mm, where Δh was larger than L (i.e., 50 mm), and Eq. (4.4) was used for water content calculation when $H < 250$ mm, where Δh was smaller than L.

Please note that the accuracy of Δh highly influences the value of the second part in Eq. (4.4) and the accuracy of calculated water contents. Because Δh is at a precision of 0.5 mm and the accuracy of Δh is $0.5/\Delta h$, Δh should be much larger than 0.5 mm. In this study, Δh was chosen as 5 mm.

4.3 Development of a Device for Measuring SHCF Using Wetting Front Advancing Column Tests

A new experimental method called the "wetting front advancing method" (WFAM) was developed for the measurement of the SHCF directly. This method is time-saving and applicable to a wide range of coefficients of soil hydraulic conductivity and a wider range of soil suction. To realize this objective, a soil column device was developed to measure coefficients of soil hydraulic conductivity for the colluvial soils over a wide range of suction values. The soil column device is shown in Fig. 4.8 consists of a column cell, a system for suction and water content monitoring, and a water flow control and measurement system.

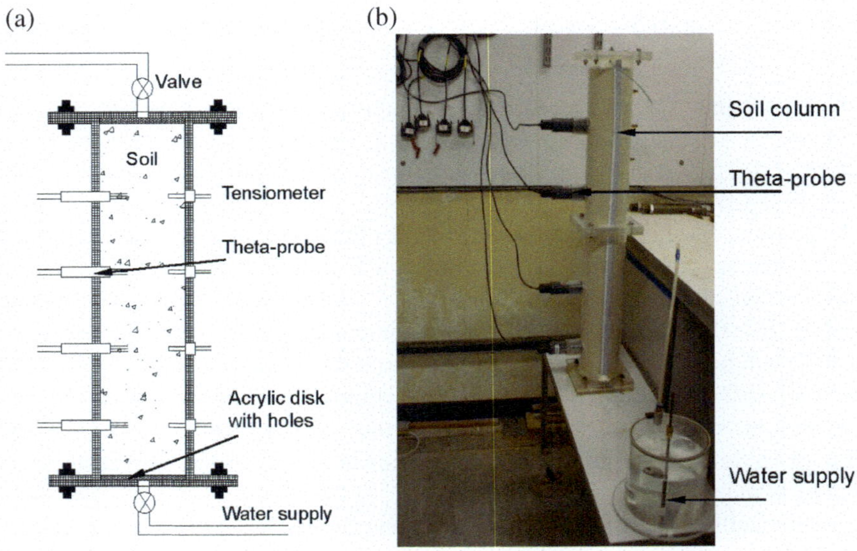

Fig. 4.8 Soil column device developed in this research program: **a** Sketch, **b** Photograph

4.3.1 A Large-Scale Column Device for Measuring SHCF

4.3.1.1 Soil Column Cell

A 120 mm diameter, 1000 mm high acrylic column cell was used to hold the soil sample and support soil suction and water content monitoring sensors. A top cap with connections to a water supply tube and a water level indicator tube was designed to prevent evaporation. A porous-disk pedestal with many drilled holes of 2 mm in diameter was connected to another water supply tube and a water level indicator tube. A number of openings were cut along the cell for mounting the measurement sensors. O-rings were used to seal the joints between the soil column and the top cap, as well as between the sockets and the sensors, and the soil column and the pedestal.

4.3.1.2 Suction and Water Content Monitoring System

Four Theta-probes were installed for water content measurement as well as four tensiometers for suction measurement. The sensors were installed along the column cell at heights of 100, 300, 600, and 800 mm above the pedestal. Miniature-tip tensiometers were used to monitor soil suction. Each tensiometer consisted of a high air entry, porous ceramic cup connected to a pressure-measuring device through a small plastic tube. The ceramic cup was saturated and installed in the soil. The pore-water pressure in the ceramic cup came to equilibrium with the pore-water pressure in the soil. The pore-water pressure in the soil was translated from the ceramic cup to the pressure-measuring device. The working suction range of the tensiometers was limited to between 0 and 90 kPa.

ML2x Theta-probe soil moisture sensors from Delta-T Devices Ltd., UK (Miller and Gaskin 1997), were used to monitor water content. Based on calibration curves for the test soil, the voltage signal from the sensor was transformed to soil volumetric water content.

Wet soil showed as a different color than the dry soil as shown in Fig. 4.9a. During the wetting process, the dry soil in the soil column turned darker as shown in Fig. 4.9b. A clear edge appeared between the wetted soil and the initially drier soil. This demarcation is called the wetting front and can be clearly identified and measured during the wetting process. The location of the wetting front was taken from two graduation labels placed on the outer surface of the column cell (see Fig. 4.9b).

4.3.1.3 Water Flow Controlling and Measuring System

A special water tank as shown in Fig. 4.10a provided a constant-head water supply at a high flow rate (i.e., larger than 30 ml/h). The working principle of the water supply is same as that of a Mariotte bottle (McCarthy 1934). The hydraulic head remained at the bottom of a copper tube, which was open to the atmosphere. The water tank

(a) (b)

Fig. 4.9 Observed wetting front in column test: **a** Wetted soil and drier soil. **b** Observed wetting front in column test

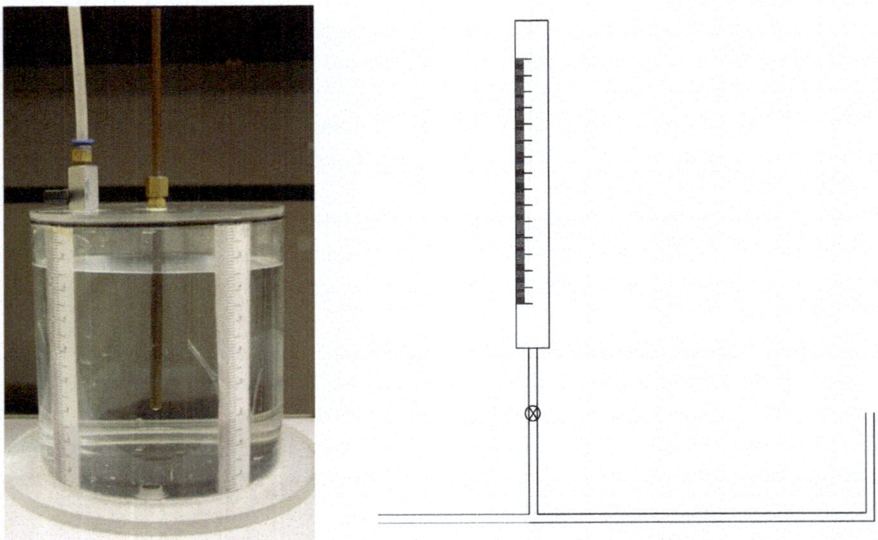

(a) (b)

Fig. 4.10 Constant water supplies: **a** Mariotte bottle (McCarthy 1934). **b** Hysteresis attachment (Soil Moisture Equipment Inc. 2005) was marked with scales that indicated the volume of water remaining in the tank. The water volume remaining in the tank was recorded over time and used to calculate the water inflow rate.

A hysteretic attachment shown in Fig. 4.10b was used to provide a constant water head supply at low flow rates (i.e., smaller than 30 ml/h). The working principle and method of water volume measurement can be referred to Soil Moisture Equipment Inc. (2005).

4.3.2 Experimental Procedure and Boundary Conditions

Each soil specimen was prepared using the following steps. First, oven-dried soil was exposed in the air for two days and the water content was measured. Second, the soil was compacted into the column in layers, each layer being 40 mm thick. The soil mass was calculated based on the cell volume, the soil water content, and the required dry density. Four Theta-probes for water content measurement were installed into the soil during soil compaction.

After the compaction of the soil specimen was completed, a flux boundary condition was applied. The laboratory tests can be divided into two types in accordance with the imposed flux boundary condition. One test is called a capillary-rise test, and another test is called an infiltration test. In the capillary-rise test, a constant water head about 5 mm higher than the bottom of the soil was applied at the bottom of the soil column. The column device at the beginning of a capillary-rise test is shown in Fig. 4.10b. After the rate of wetting front rise (as shown in Fig. 4.9b) reached a slow value (e.g., 1 mm/day), the capillary-rise test was terminated.

In the infiltration test, a constant water head (e.g., 100 mm higher than the top of the soil column) was applied at the top of the soil column. The infiltration test was terminated when steady-state conditions were achieved (i.e., when the water flowed out from the bottom of the soil column at a constant flow rate). In both types of tests, when soil suction at a monitoring section was estimated to be equal to or smaller than about 80 kPa (based on the monitored water content and the SWCC), a miniature-cup tensiometer was installed in the soil column through a predrilled hole along the column. The tensiometers were then used to monitor soil suction.

4.3.3 Interpretation of SHCF Based on the WFAM

Figure 4.11a shows the case of capillary rise from the bottom of the column. The top of the soil column is at section A. The water content and suction measurement sensors were installed at section B. The observed wetting front rises with time and is recorded as shown in Fig. 4.11b during the capillary-rise process. Figure 4.11a shows that at time t_1, the wetting front reaches section F1, and at time t_2, the wetting front reaches section F2. The flux passing through section B between two time intervals

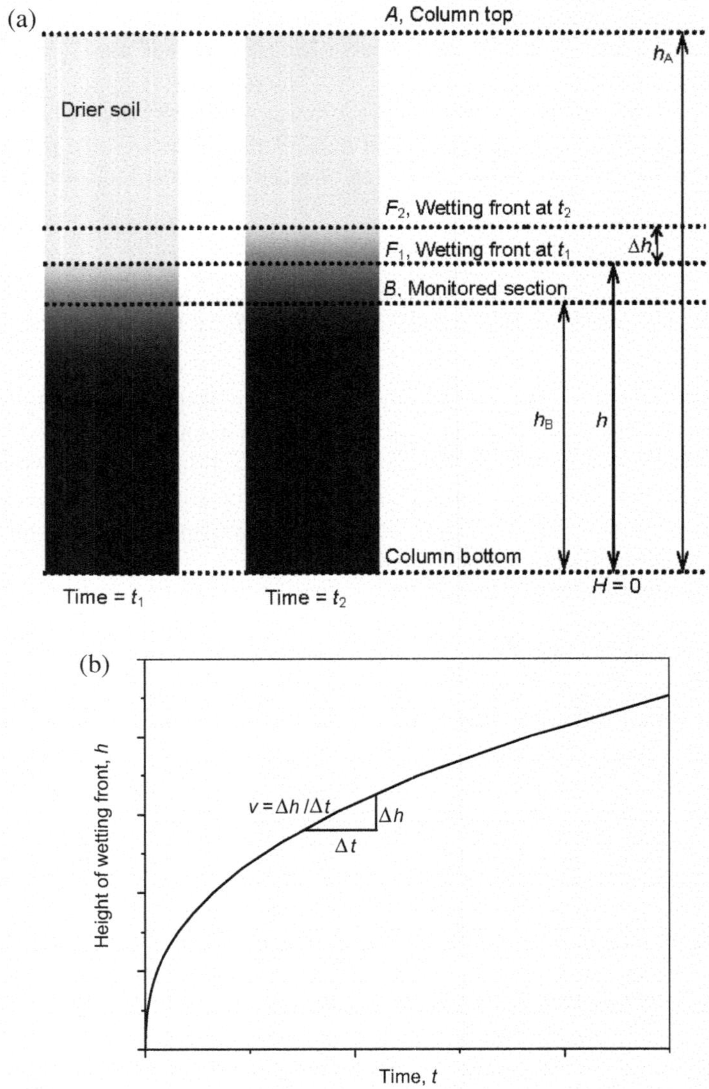

Fig. 4.11 Water capillary-rise process in an initially air-dried soil column: **a** Soil column at two moments during capillary-rise process. **b** Height of wetting fronts recorded over time

can be calculated as,

$$Q_B = \Delta W_{B-A} + Q_A \tag{4.5}$$

where Q_B is the flux passing through section B from t_1 to t_2; Q_A is the flux passing through section A from t_1 to t_2, which is assumed to be zero during the capillary-rise process; ΔW_{B-A} is the water storage change between section B and section A in this period and can be calculated as,

$$\Delta W_{B-A} = \int_{h_B}^{h_A} \theta(h, t_2) C dh - \int_{h_B}^{h_A} \theta(h, t_1) C dh \tag{4.6}$$

where $\theta(h, t)$ is the water content distribution function; C is the cross-sectional area of the soil column; t is time; h is the section height from the bottom of the column; h_B is the height at section B; and h_A is the height at section A.

During the wetting process, the wetted zone is assumed to advance uniformly. That is to say, the soil crossing a horizontal section has the same water content at any time. Hence, the wetting front observed on the column surface can indicate the wetted zone in the soil column. Figure 4.10a shows that in a short time period of Δt (i.e., $t_2 - t_1$), the wetting front advances a distance of Δh (i.e., height difference between F_1 and F_2). If the water content contour is assumed to advance smoothly during the time interval, then the water content distribution function obeys

$$\theta(h, t + \Delta t) = \theta(h - \Delta h, t) \tag{4.7}$$

and the suction distribution function obeys

$$\psi(h, t + \Delta t) = \psi(h - \Delta h, t) \tag{4.8}$$

Then, the total flow volume passing through section B in Eq. (4.5) can be written as,

$$\begin{aligned} Q_B &= \int_{h_B}^{h_A} \theta(h - \Delta h, t_1) C dh - \int_{h_B}^{h_A} \theta(h, t_1) C dh \\ &= \int_{h_B - \Delta h}^{h_B} \theta(h, t_1) C dh - \int_{h_A - \Delta h}^{h_A} \theta(h, t_1) C dh \end{aligned} \tag{4.9}$$

The first part of Eq. (4.9) can be calculated as,

$$\begin{aligned} \int_{h_B - \Delta h}^{h_B} \theta(h, t_1) C dh &\approx 0.5[\theta(h_B, t_1) + \theta(h_B - \Delta h, t_1)] C \Delta h \\ &\approx 0.5[\theta(h_B, t_1) + \theta(h_B, t_2)] C \Delta h \end{aligned} \tag{4.10}$$

If the initial water content is very low and therefore the coefficient of soil hydraulic conductivity is extremely small, the water content change in the initially air-dried soil zone can be ignored during the limited wetting period. Therefore, the water contents in this zone remain at the initial water content, θ_i, during the entire wetting process. The second part of Eq. (4.9) can be calculated as follows,

$$\int_{h_A-\Delta h}^{h_A} \theta(h,t_1)C\mathrm{d}h \approx 0.5[\theta(h_A,t_1)+\theta(h_A-\Delta h,t_1)]C\Delta h \approx \theta_i C\Delta h \tag{4.11}$$

Q_B can be calculated by substituting Eqs. (4.10) and (4.11) into Eq. (4.9),

$$Q_B \approx 0.5[\theta(h_B,t_1)+\theta(h_B,t_2)-2\theta_i]C\Delta h \tag{4.12}$$

The flow quantity can be written using Darcy's law,

$$Q = kiC\mathrm{d}t \tag{4.13}$$

where k is the instantaneous coefficient of soil hydraulic conductivity at a specific section along the soil column; i is the hydraulic gradient at that section, which is calculated as

$$i = \frac{-\mathrm{d}h + \mathrm{d}\psi/\gamma_w}{\mathrm{d}h} = \frac{\mathrm{d}\psi}{\gamma_w \mathrm{d}h} - 1 \tag{4.14}$$

The hydraulic gradient at t_2 at section B can be written by substituting Eq. (4.8) into Eq. (4.14):

$$\begin{aligned} i &= \frac{-\psi(h_B,t_2)+\psi(h_B+\Delta h,t_2)}{\gamma_w\Delta h} - 1 \\ &= \frac{\psi(h_B,t_1)-\psi(h_B,t_2)}{\gamma_w\Delta h} - 1 \end{aligned} \tag{4.15}$$

The average coefficient of soil hydraulic conductivity between t_1 to t_2 at section B can be calculated by substituting Eqs. (4.12) and (4.15) into Eq. (4.13):

$$\begin{aligned} k_{\text{ave}} &= \frac{Q_B}{iC(t_2-t_1)} \\ &= \frac{[\theta(h_B,t_1)+\theta(h_B,t_2)-2\theta_i]\gamma_w\Delta h^2}{2\big[\psi(h_B,t_1)-\psi(h_B,t_2)-\gamma_w\Delta h\big](t_2-t_1)} \end{aligned} \tag{4.16}$$

The wetting front advancing velocity (referring to Fig. 4.11b) is defined as,

$$v = \frac{\Delta h}{\Delta t} \tag{4.17}$$

Using v instead of Δh, Eq. (4.16) becomes,

$$k_{\text{ave}} = \frac{[\theta(h_B, t_2) + \theta(h_B, t_1) - 2\theta_i]\gamma_w v^2(t_2 - t_1)}{2\left[\psi(h_B, t_1) - \psi(h_B, t_2) - \gamma_w v(t_2 - t_1)\right]} \tag{4.18}$$

The calculated coefficient of soil hydraulic conductivity corresponds to a suction of $0.5[\psi(h_B, t_1) + \psi(h_B, t_2)]$.

4.3.4 Limitations of the Application of WFAM

The proposed WFAM is able to measure the SHCF in a few days resulting in considerable time-saving as compared to traditional steady-state methods and the instantaneous profile method. The proposed method is applicable to a wide range of soil suctions and soil types. In the wetting front advancing method, an initially air-dried soil column is subjected to a wetting event. During the wetting process, a wetting front advances with time and gives a wetting front advancing velocity function and the soil suction and water content are monitored.

The instantaneous profile method (IPM) is an unsteady-state method that can be used either in the laboratory or in situ to calculate the SHCF (Richards and Weeks 1953; Bruce and Klute 1956; Watson 1966; Klute 1972; Hamilton et al. 1981; Daniel 1983; Roth et al. 1992; Meerdink et al. 1996).

However, according to Li et al. (2009a, b), there are at least three limitations that hinder the application of WFAM:

(1) The original design of WFAM depends on a wetting front that is visible to human eyes.
(2) The accuracy of WFAM lacks a thorough verification. As the "true" soil hydraulic conductivity function (SHCF) of soil is always unknown, the error of SHCF measured in physical tests is difficult to evaluate.
(3) Only the hydraulic conductivities of relatively dry soil (e.g., for soil with a degree of saturation lower than 0.9 and has a relative higher suction) can be measured in the original WFAM.

Because of the above limitations, WFAM is still not widely applied in the geotechnical community and the time-consuming IPM with severe soil disturbance is still the most popular method for SHCF measurement (Plagge et al. 2006; Cui et al. 2008; Krisdani et al. 2009; Scheffler and Plagge 2011; Askarinejad et al. 2012; Leung et al. 2016; Su et al. 2018; Ni et al. 2019).

4.4 Measuring SHCF Over a Wide Suction Range

4.4.1 Combining WFAM and IPM for the Measurement of Hydraulic Conductivity

In this part, a digital WFAM is proposed to overcome the aforementioned limitations, targeting at:

(1) Removing the requirement of a visible wetting front. In WFAM, there exist two assumptions: wetting front advances smoothly during the infiltration process and soil remains at the initial water content before wetting front approaches. These two assumptions require that the soil is as dry as possible so that a visible wetting front can be formed and captured. In this improved WFAM, a digital method is proposed to identify the wetting front, in which the monitored data is used to infer the wetting front. In other words, a digital wetting front is used instead of a visible wetting front. Both numerical tests and laboratory tests are used to verify the proposed digital method.

(2) Examining the accuracy and major error sources of WFAM. A series of numerical tests are carried out. In each numerical test, the input SHCF can be regarded as the correct one and used to examine the accuracy of WFAM. Based on the results, the applicability conditions for WFAM can be further concluded.

(3) Enlarging the measurement range of WFAM. A new test program is designed to measure SHCF in a full suction range.

Besides the above contributions, a thorough comparison of IPM and WFAM is also carried out. Both the data of soil column tests reported by Cui et al. (2008) and Ng and Leung (2012) and those from numerical tests are analyzed using these two methods. Then based on the results, the accuracy and the applicable conditions of the two methods are discussed. Finally, a combination of IPM and digital WFAM is proposed to measure SHCF over a full suction range and its applicability is also examined by numerical tests.

4.4.1.1 A New Digital Determination Method for Wetting Front Velocity Based on Sensor Reading

In early section, the wetting front advancing velocity $v(t)$ is extracted from a clear wetting front that can be observed by human eyes (as shown in Fig. 4.9b). In fact, there is another option, which is to calculate the wetting front advancing velocity from the monitoring data. In this way, the dependence on the forward process of wetting front observed by the naked eye can be avoided. The detailed procedure is as follows:

First, a water content θ_d can be selected as the digital indicator of wetting front. It can be achieved after the wetting front has passed a monitoring section and will be applied to all monitoring sections. Obviously, the volumetric water content θ_d that used as the digital indicator of wetting front should be larger than the initial water content and smaller than the volumetric water content at the quasi-steady state in the infiltration process.

Second, the wetting front advancing process $h(t)$ can be determined using θ_d. As shown in Fig. 4.12, there are three monitoring sections *A*, *B*, and *C*. In a laboratory test, the infiltration process should be well controlled, either applying a constant water head boundary, or using a proper infiltration rate that is smaller than the saturated hydraulic conductivity and can avoid ponding phenomenon on the surface. In the process of infiltration, when the water content of certain monitoring sections rises to θ_d, it can be regarded as the wetting front advances to this position at this instant time. The distances between the infiltration boundary and sections *A*, *B*, and *C* are h_A, h_B, and h_C, respectively. The time for the wetting front to reach the three cross-sections is t_A, t_B, and t_C. The relation between distance and time is plotted in Fig. 4.13 and fitted by an exponential function.

Third, by deriving the fitted wetting front advancing curve, a digital wetting front advancing velocity v_d can be solved and used in Eq. (4.18) to calculate SHCF.

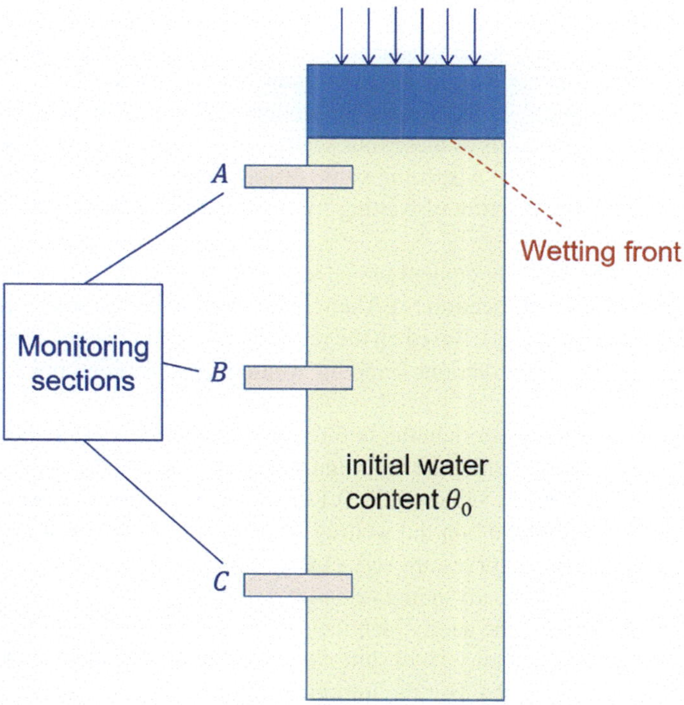

Fig. 4.12 Wetting front in soil column test

Fig. 4.13 Fitted wetting front advancing curve $h(t)$

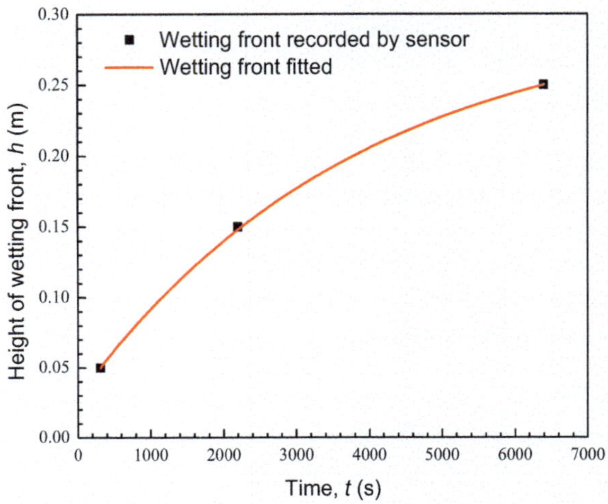

4.4.1.2 Comparison of Digital and Visual Observation Methods

The data from Li et al. (2009a, b) are used to verify the digital wetting front advancing velocity. A capillary rising column test with sandy silt reported in Li et al. (2009a, b) is used as an example. Four monitoring sections were used in Li et al. (2009a, b). The water content records at these four sections (named Section *A*, *B*, *C*, and *D*) are illustrated in Fig. 4.14a. A specific value of the volumetric water content θ_d is selected to indicate the location of wetting front and named as the digital indicator of wetting front.

Once the monitored water content in one section (e.g., section *A*) reaches θ_d, the time will be recorded (e.g., section t_A). Then, four points can be obtained as (h_A, t_A), (h_B, t_B), (h_C, t_C), and (h_D, t_D). Based on the result in Fig. 4.14a, the relation between the locations of monitoring sections *h* and the wetting front arriving time *t* is plotted in Fig. 4.14b.

θ_d is supposed to be larger than the initial water content θ_0 and smaller than the maximum water content in the primary rising stage, as shown in the available range for θ_d in Fig. 4.14a. Four θ_d values, 0.06, 0.1, 0.15, and 0.26, are used to investigate the influence of the digital θ_d on the wetting front advancing curve. It can be seen from Fig. 4.14b that these curves are very close.

Further, the wetting front advancing velocities are derived from the wetting curves in Fig. 4.14b and found to be nearly identical, as shown in Fig. 4.14c. Obviously, if a larger θ_d is used, the $h(t)$ curve will shift a little on the time domain (a shift to the right as shown in Fig. 4.14b). But the important thing is that the derivation of $h(t)$ curve is almost constant and is nearly independent with the selection of θ_d. That is to say, ν (Eq. 4.17) and κ (Eq. 4.18) are not sensitive to the selection of the digital indicator of wetting front. When the digital indicator of wetting front is selected in a reasonable range, the wetting advancing velocity and the calculated SHCF are nearly

SHCF are irrelevant to the digital indicator.

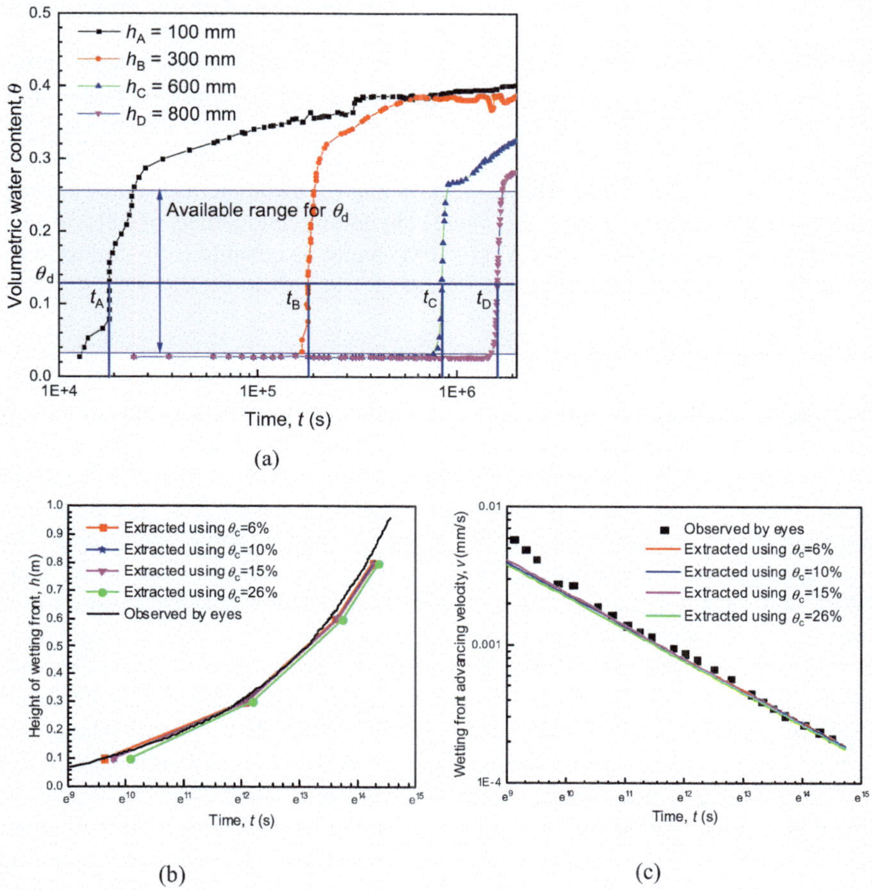

Fig. 4.14 An example of the digital wetting front in a capillary rising soil column test: **a** Identify the time when the wetting front arrives four monitoring sections. **b** Wetting front advancing curve $h(t)$. **c** Wetting front advancing velocity $v(t)$

identical. That is to say, the wetting advancing velocity and the calculated SHCF are irrelevant to the digital indicator.

When the soil initial water content is relatively high, the wetting front may be difficult to distinguish with the naked eyes, but can be digitally identified. By using the digital indicator of wetting front, WFAM is applicable to such conditions. It can be concluded that a digital wetting front can be used instead of the visually observed wetting front in WFAM. The removal of this limitation can greatly expand the applicability of WFAM.

4.4.2 Numerical Tests

In this study, numerical tests are used to evaluate the accuracy and applicability of WFAM and examine the digital wetting front identification method proposed in Sect. 4.4.1. In a numerical test, the input hydraulic conductivity can be regarded as the "true" value and used to evaluate the accuracy of the hydraulic conductivity calculated by WFAM or IPM. This is the advantage of the numerical test comparing to the laboratory test where the "true" value of hydraulic conductivity of soil is always unknown and the accuracy of WFAM or IPM cannot be quantitatively evaluated.

4.4.2.1 Numerical Model

In the numerical tests, the inflow process in a column filled with homogenous soil is considered. The unsaturated flow process is simulated by a commercially available code "Comsol 4.3." As shown in Fig. 4.12, a soil column test with a height of 30 cm is considered. The soil column is vertically oriented. The soil column is homogeneous in the numerical model and the soil has the same initial conditions (e.g., same matric suction and hydraulic conductivity) and properties (e.g., same SWCC and SHCC). The left and right boundaries are impermeable. During the inflow process, the boundary condition maintains and a constant inflow flux is applied at the top surface of the soil column. The mesh of model is unified as quadrilateral mesh in fine size. The effect of soil properties, initial soil suction, and the boundary condition on the accuracy and applicability of WFAM or IPM is to be studied.

In the numerical tests, no real sensor is needed. Three virtual volumetric water content sensors and three matric suction probes are installed evenly in the height direction of soil column, i.e., at the location of 5, 15, and 25 cm from the bottom of soil column. In the practical test, three sections are enough to record the wetting front advancing process. While in the numerical test, due to the convenience of data extraction, data from more sections may be output to obtain the details of wetting front advancing curve. The duration curves of volumetric water content and soil matric suction at these sections are exported (as the data in the supplemental files) from the numerical test results and further used to calculate SHCF by IPM or WFAM.

The data process analysis is as follows: (1) The arriving time at a virtual monitoring section is determined by θ_d and three points of $h(t)$ can be obtained; (2) Based on $h(t)$, the wetting front advancing velocity can be determined as, $v(t) = \partial h/\partial t$; (3) Calculate the hydraulic conductivity using Eq. (4.18) at a time of t; (4) The hydraulic conductivity at time t is plotted versus the suction at time t, forming a SHCC.

4.4.2.2 Input Soil Hydraulic Properties and Numerical Test Plan

The input hydraulic properties of the soil used in the numerical tests are plotted in Fig. 4.15, including the soil water characteristic curve and hydraulic conductivity

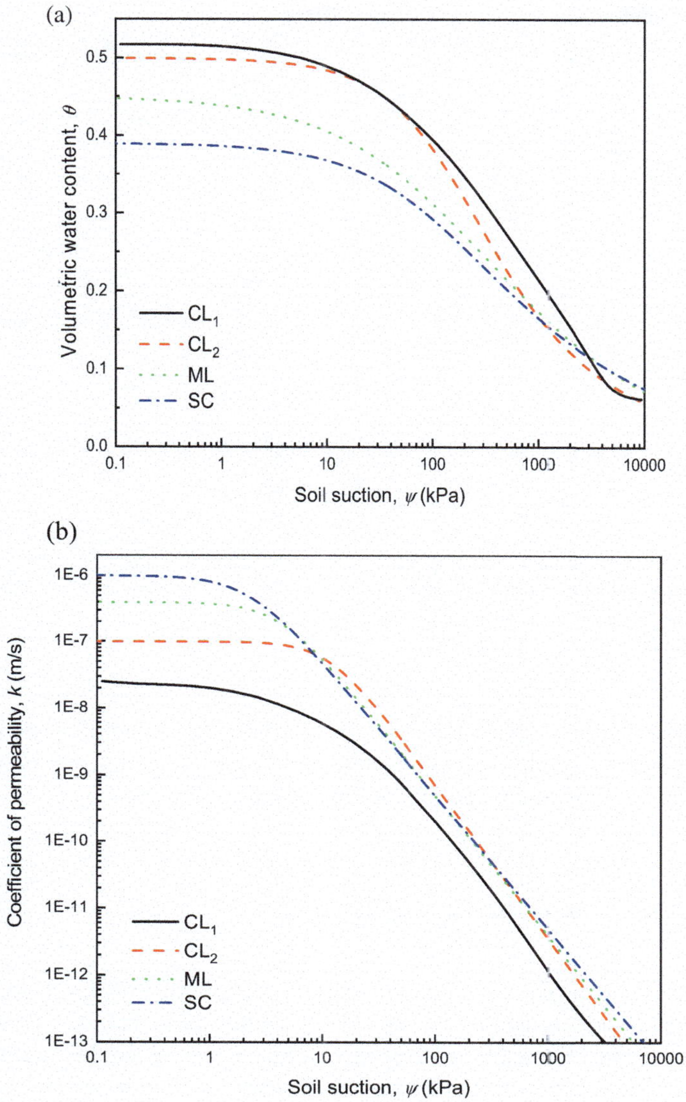

Fig. 4.15 The input soil hydraulic properties: **a** The soil–water characteristic curves. **b** The hydraulic conductivity functions

function. van Genuchten model (van Genuchten 1980) and Mualem model (Mualem 1976) are used to describe the SWCC and the SHCF, respectively. The numerical test plan is listed in Table 4.1. To check the applicability of WFAM, four types of soil are used in this study. According to previous research, the boundary conditions and initial conditions are two major influence factors on the SHCC measurement. In this study, the initial water content is set from 7.5 to 23.5% and the soil is under a

Table 4.1 Numerical test plan

Test No.	Soil type	θ_0 (%)①	Boundary condition②	κ_s (m/s)③	Analysis method
C1, C2, C3	CL_1	7.5, 12.5, 21.5	q_{top} = 2e−8 m/s	2.5e−8	WFAM
C4, C5	CL_1	7.5	q_{top} = 1e−9 m/s, 5e−9 m/s		
C6	CL_1	14.5	H_{top} = 50 kPa		
C7	CL_2	7.5	q_{top} = 9e−8 m/s	1e−7	IPM
M8	ML	7.5	q_{top} = 3e−7 m/s	4e−7	
S9	SC	7.5	q_{top} = 2e−7 m/s	1e−6	IPM and WFAM
S10, S11, S12	SC	7.5	q_{top} = 1e−7 m/s, 1e−8 m/s, 1e−9 m/s	1e−6	
C13	CL_2	9	H_{top} = 0 kPa	1e−7	
M14	ML	17.5	H_{top} = 0 kPa	4e−7	
S15	SC	23.5	H_{top} = 0 kPa	1e−6	

Notes; H_{top} is total water head at the upper boundary in the infiltration test
①θ_0 is the initial volumetric water content
②q_{top} is inflow flux from the top boundary in the infiltration test
③k_s is the saturated hydraulic conductivity

relatively low degree of saturation. Two types of boundary conditions are used: (1) 0 kPa water head is applied on the top boundary; and (2) flow rate that smaller than ks, which normally will not cause the surface ponding problem.

4.4.3 Validation of the Digital WFAM

The results of numerical test C1 are used to check the validity of WFAM. To simulate the work conditions in laboratory WFAM test (referring to Fig. 4.12), only three virtual monitored sections are used to extract data in numerical test. In each monitored section, the time sequence records of water content and suction are monitored and used to calculate the hydraulic conductivity by Eq. (4.18).

4.4.3.1 Monitored Time Sequence Records of Water Content and Suction

The time sequence records of water content and suction at three virtual monitored sections in numerical test C1 are plotted in Fig. 4.16. In this test, the infiltration process from the top boundary is started at time of 0 s. The wetting front moves to the bottom boundary at a time around 4 × 106 s, and this time is named as tout

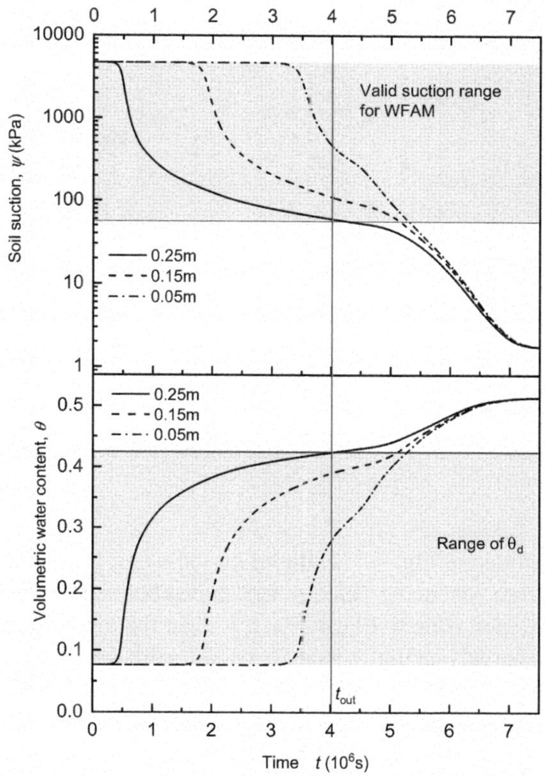

Fig. 4.16 Time sequence records of water content and suction (Test No. C1)

and marked as a red line in Fig. 4.16. After the wetting front moves to the end, the infiltration process is still continued until the whole soil column is fully saturated.

Referring to Fig. 4.16, there are three stages in wetting process. Before the wetting front arrives in a monitor section, the soil water content changes are negligible, i.e., a non-change range. After wetting front arrives a monitor section, a rapid increase of soil water content happens in a very short time, i.e., a rapid increase range, and follows with a slow increase of soil water content, i.e., a slow increase range. This phenomenon is similar to that observed in the laboratory test (Fig. 4.14a).

4.4.3.2 Selection of Digital Wetting Front θ_d

Based on the results in Fig. 4.16, the effect of the selection of θ_d is examined. As the initial water content is 7.5% and the maximum water content monitored at the time t_{out} is 42%, θ_d that used to indicate the wetting front should be selected in the range of (7.5%, 42%).

θ_d is taken in the range of 7.5–42% in 12 levels. By using 7 virtual monitoring sections, the wetting front advancing curves can be determined by various θ_d, as

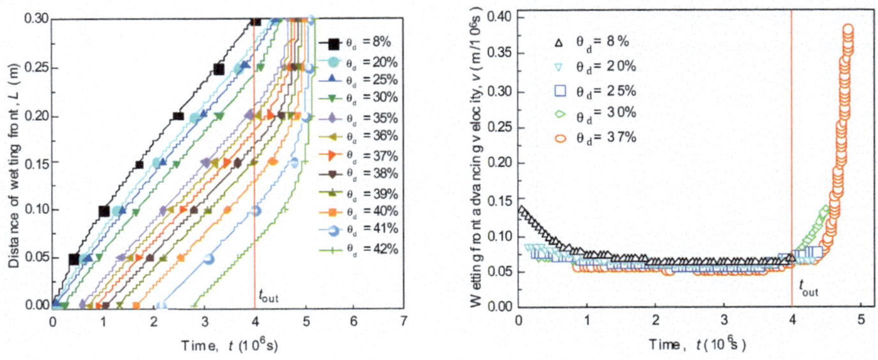

(a) (b)

Fig. 4.17 Digital wetting front advancing process (Test No. C1): **a** $h(t)$ affected by θ_d, **b** $v(t)$ affected by θ_d

plotted in Fig. 4.17a. Based on the wetting front advancing curves $h(t)$, the wetting front advancing velocity v_d can be obtained by Eq. (4.17), as plotted in Fig. 4.17b.

Referring to Fig. 4.17a, it is clear that before time t is larger than t_{out}, the wetting front advancing processes, i.e., $h(t)$ relation, are parallel. When $t < t_{out}$, the wetting front advancing velocity v_d values (i.e., $\partial h/\partial t$) are nearly identical and not sensitive to the selection of θ_d, as shown in Fig. 4.17b.

Referring to Fig. 4.17, several hints can be drawn for the selection of θ_d:

(1) if t is larger than t_{out}, the wetting front advancing v_d has no physical meaning and becomes inaccurate.

(2) θ_d should be smaller than the maximum water content achieved in the monitored section at time t_{out}. Referring to Fig. 4.17b, if θ_d is selected from the water content range in the rapid increase stage (Fig. 4.16), the calculated v_d is nearly identical.

(3) if θ_d is larger, the time required for the water content to achieve θ_d in the first monitor section is larger. Hence, the missing time range of v_d is larger.

(4) if θ_d is low, the wetting front advancing velocity will be large at the initial stage of infiltration. For example, $\theta_d = 8\%$ shown in Fig. 4.16 is close to the initial water content 7.5%; there is a much faster wetting front advancing velocity comparing to use a θ_d of a higher value. This phenomenon is owing to that an extremely large hydraulic gradient at the wetting/drying boundary will occur at the beginning and it will decrease to a reasonable value after the initial stage of infiltration. Therefore, θ_d should be selected in the median range, such as $\theta_d = 20$–30% in Fig. 4.17b.

4.4.3.3 Valid Range of WFAM

Given the digital wetting front advancing velocity (Fig. 4.17b), the hydraulic conductivity function can be calculated, as plotted in Fig. 4.18. The results indicate that:

(1) when θ_d is within the range (7.5%, 25%), the hydraulic conductivity functions are nearly identical, showing that the selection of θ_d in the rapid increase range of soil water content (Fig. 4.16) is proper and will not cause severe errors to the results.
(2) the calculated hydraulic conductivity is consistent with the input hydraulic conductivity in the range of (60, 4500) kPa. However, when the soil suction is lower than 60 kPa, a significant difference is found between the calculated hydraulic conductivity and the input value.

Such a phenomenon is owing to the wetting front advancing process. Referring to Fig. 4.16, when the soil suction is lower than 60 kPa, the data is located in the slow increase stage ($t > t_{out}$). In the slow increase stage ($t > t_{out}$), the assumption that the water content contour advances smoothly during the wetting process is not valid and Eqs. (4.7) and (4.8) is not applicable. When $t < t_{out}$, the assumption of WFAM is valid and the valid ranges for WFAM are consistent in suction records (Fig. 4.16) and calculated SHCF (Fig. 4.18).

In short, only the data in the rapid increase range ($t < t_{out}$) is valid for WFAM and can lead to accurate hydraulic conductivity.

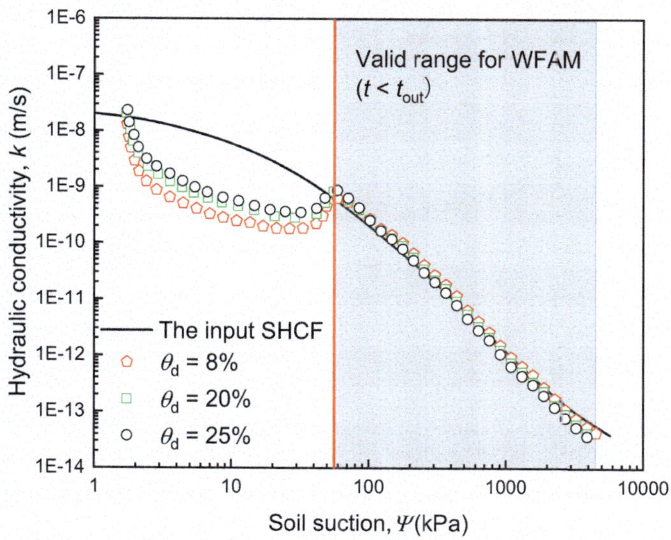

Fig. 4.18 SHCF obtained by WFAM in the numerical test (Test No. C1)

4.4.3.4 Effect of Sensor Spacing

If there is a clear interface between wetted soil and dry soil as that in Li et al. (2009a, b), one monitor section is enough to calculate the hydraulic conductivity function. When a digital wetting front is used, at least 2–3 monitor sections are needed to obtain the wetting front advancing velocity. Obviously, if the space of monitor section, i.e., the sensor space, is closer, the wetting front advancing velocity is more accurate.

The results from various sensor spaces are compared in Fig. 4.19. The data analysis process is as follows:

(1) First, the arriving time t_i for the i monitor section is taken as the time when the monitored water content $\theta_i\ (t)$ reaches θ_d, i.e., $\theta_i\ (t) = \theta_d$.
(2) Second, for the i monitor section, the point (t_i, h_i) is obtained and will be plotted on the wetting front advancing distance curve $h(t)$.
(3) Third, exponential functions are used to fit $h(t)$ shown in Fig. 4.19a.
(4) Forth, the wetting front advancing velocity $v(t)$ is calculated by deriving the wetting front advancing distance curve (Eq. 4.17).
(5) Further, Eq. (4.18) is used to calculate the hydraulic conductivities as shown in Fig. 4.19b.

The results in Fig. 4.19 indicate that the sensor space has limited influences on $v(t)$ and the calculated SHCF. Considering the fitting requirement of $h(t)$, three monitor sections are recommended and can assure the accuracy of hydraulic conductivity solved by digital WFAM.

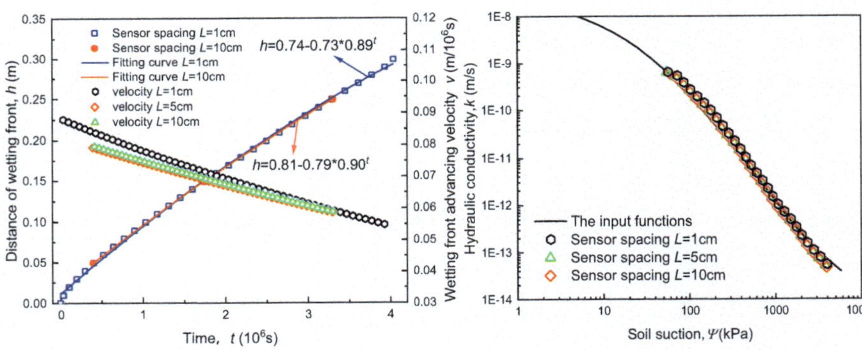

(a) (b)

Fig. 4.19 Unsaturated SHCF affected by different spacing of sensors (Test No. C1): **a** $v(t)$ affected by sensor spacing. **b** SHCF solved by digital WFAM

4.4.4 Comparison of WFAM and the Traditional IPM

4.4.4.1 Performance of WFAM

In the WFAM, the initial conditions and boundary conditions can affect the wetting process and the solved hydraulic conductivities. A series of soil column tests listed in Table 4.1 are used to investigate the influence of these two factors.

Three values for initial soil water content are used in the numerical tests, and the hydraulic conductivities are solved by WFAM. In these tests, silty clay is used. For numerical test C1, C2, and C3, the initial water content is 7.5%, 12.5%, and 21.5%, respectively, and corresponding to an initial suction of 4500 kPa, 2500 kPa, and 990 kPa, respectively. It is noteworthy that for initially wet soil, the candidate range for θ_d is much narrower compared with the initially dry soil. In a laboratory test, to reduce the possible error, it is an applicable and convenient choice to select θ_d as half of the initial water content θ_0 and the water content θ_1 monitored at time t_{out}. The results are plotted in Fig. 4.20a and demonstrate that a wider range can be obtained in a test with a drier initial condition.

Using an initial θ_0 of 7.5% and varying the infiltration flux, the effect of infiltration flux is studied and the results are illustrated in Fig. 4.20b (numerical Test No. C1, C4, and C5). Clearly, when the initial water content is constant (7.5% in C1, C4, and C5), a larger infiltration flux is used, a wider range of solved SHCF can be obtained.

Based on Fig. 4.20, a drier soil or a larger infiltration flux used in the test can lead to a wider range of hydraulic conductivities. With this consideration, a constant head is applied on the top of soil column to enlarge the infiltration flux as faster as possible and it is supposed to have a SHCF with a wider suction range. 50 kPa water head is applied on the top boundary of the soil column. The results are plotted in Fig. 4.21. In this numerical test, a correct SHCF in the full range can be obtained. However, in laboratory tests, such infiltration speed is too fast for the sensor to respond. For

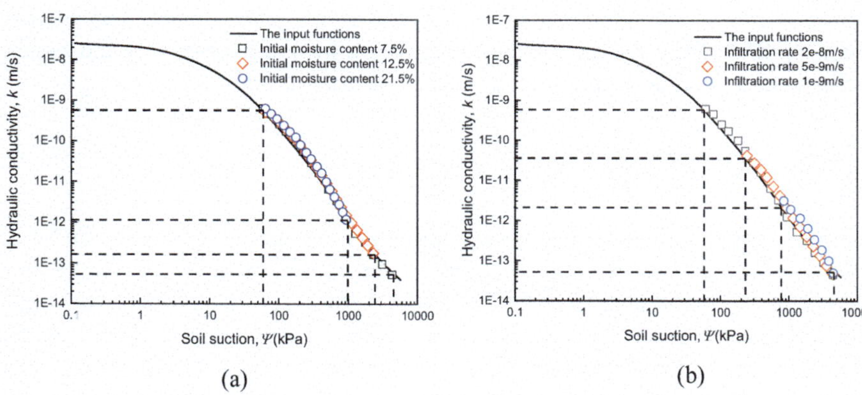

Fig. 4.20 Performance of digital WFAM (Numerical Tests No. C1–C5): **a** Effect of initial water content. **b** Effect of infiltration flux

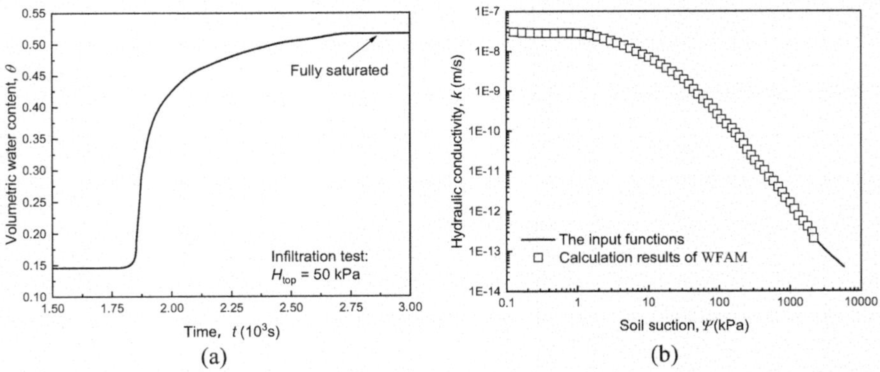

Fig. 4.21 The solved SHCF with applying 50 kPa water head at the boundary (Test No. C6): **a** The monitored water content duration curve. **b** Solved SHCF by WFAM

example, the widely used TDR or FDR sensor is designed to measure the volumetric water content in a soil zone of several centimeters. If the infiltration speed is too fast, the measured data of TDR of FDR will be an average value, but not an accurate value at the monitoring section. Hence, this test can only numerically verify the assumption of WFAM, but is not applicable in practice.

In WFAM, to obtain a wetting front in infiltration test, the ratio of applied infiltration rate q_{top} to the initial soil hydraulic conductivity k_0 should be at least larger than 100. Otherwise, a smooth advancing process of wetting front cannot be guaranteed. The higher the ratio of q_{top}/k_0, the wider range of solved SHCF can be obtained.

4.4.4.2 Performance of Instantaneous Profile Method (IPM)

Theory of IPM

IPM uses a series of monitored sections to obtain the water content and suction profile. In IPM (Meerdink et al. 1996), an average hydraulic gradient is calculated by the suction monitored at two adjacent sections, as follows,

$$i* = \frac{\psi S1 - \psi S2}{\gamma wL} + a \tag{4.19}$$

where ψ_{s1} and ψ_{s2} are the suction at two adjacent monitoring sections $S1$ and $S2$; L is the distance between two adjacent sections. Obviously, there are two choices for calculating the hydraulic gradient $i*$, i.e., one is $i*_-$ that is calculated in the backward direction and another is $i*_+$ that is calculated in the forward direction. For example, to calculate the hydraulic gradient $i*_+$ at Section B in Fig. 4.12, $i*_-$ is that between section A and section B and $i*_+$ is that between section B and section C.

In IPM, the flux is calculated by water balance. Similar to the hydraulic gradient i^*, there are also two choices for calculating flux q^*, as

$$q_{+}^{*} = \sum_{n=j}^{m} \Delta\theta_n S L_n + q_{\text{out}} \tag{4.20}$$

or

$$q_{-}^{*} = q_{\text{in}} - \sum_{0}^{n=j} \Delta\theta_n S L_n \tag{4.21}$$

where m is the total number of monitoring sections; n is the number of the monitoring section that is counting from 0 to m along the following direction; j is the number of the monitoring section used for calculation; q^* is the flow flux passing the j monitoring section in a time interval Δt; $\Delta\theta_n$ is the average change value for water content in Δt at the n monitoring section; L_{n} is the average sensor spacing at the n monitoring section; S is the cross area of soil column; q_{in} and q_{out} are the inflow and outflow flux at the boundary of soil column in Δt. With using q^* and i^*, the hydraulic conductivity k^* in IPM can be calculated as:

$$k^* = \frac{q^*}{i^* S \Delta t} \tag{4.22}$$

where q^* can be used as the average value of q^*_+ and q^*;_ i^* can be used as $\sqrt{i^* - i_+^*}$.

Referring to Eqs. (4.19)–(4.22), a linear assumption between two adjacent sections is adopted in the IPM. Hence, the inflow flux and the spacing cf the monitored section are two essential parameters that affect the degree of nonlinearity for the infiltration process and determine the accuracy of solved SHCF.

Effect of Sensor Spacing on IPM

For numerical tests No. C7, M8, and S9, IPM is used to calculate the SHCF. As shown in Fig. 4.22, the results from various section spaces are plotted. The results demonstrate that the solved SHCF oscillates near the input values and the larger the monitoring section spacing is, the greater the oscillation amplitude is. That is to say, with the decrease of monitoring section spacing, the accuracy of SHCF increases. However, even a space of 1 cm is used, the accuracy of solved SHCF is still bad.

It should be mentioned that all the infiltration fluxes in test C7, M8, and S9 are high, i.e., around $0.1k_{\text{s}}$. If the infiltration flux is lower, e.g., arcund $0.001k_{\text{s}}$, the results from IPM will be better. The effect of infiltration flux will be discussed in the next section.

4.4.4.3 Comparison of IPM and WFAM with Using Numerical Test Results

Effect of Inflow Flux

The effect of infiltration flux on IPM and WFAM is studied by numerical tests No. S9–S12 (Table 4.1). The SHCFs obtained with various infiltration flux values are

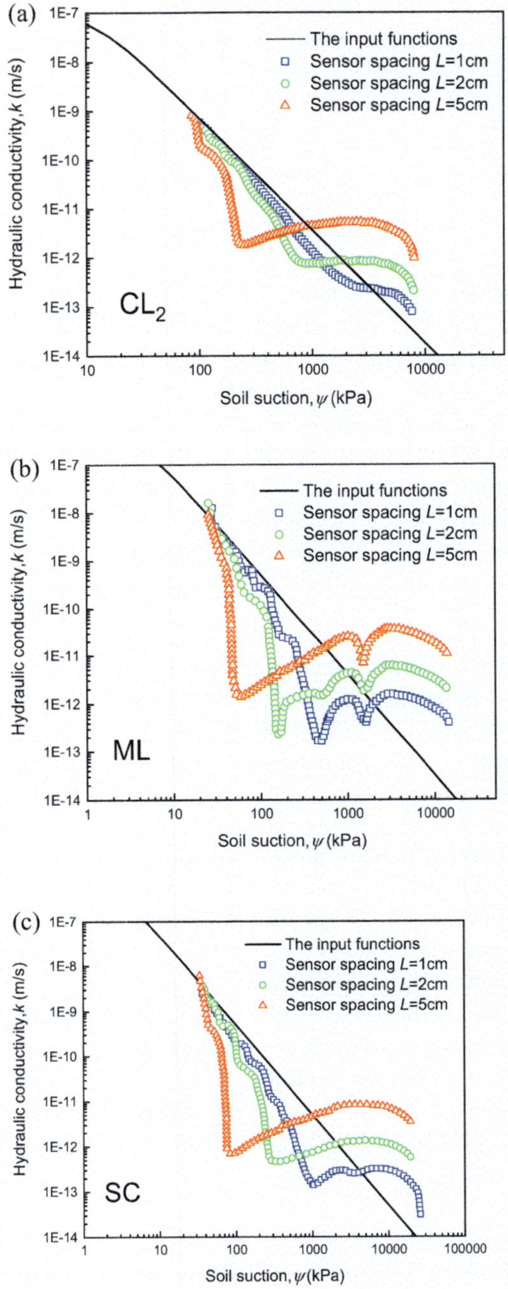

Fig. 4.22 Unsaturated SHCF calculated by IPM: **a** Test No. C7 (clay). **b** Test No. M8 (silt). **c** Test No. S9 (sand)

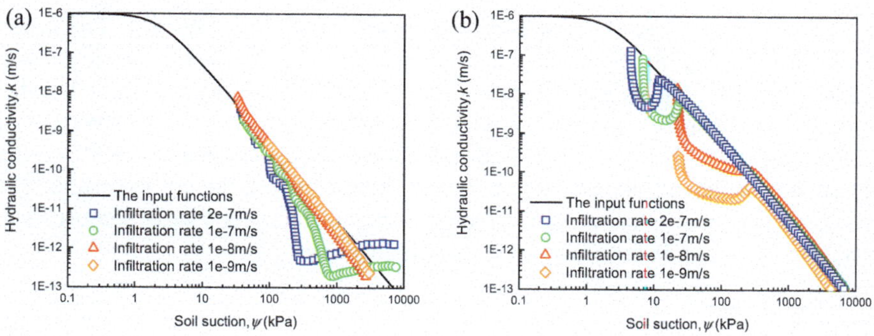

Fig. 4.23 Effect of the inflow flux (Test No. S9, S10, S11, and S12): **a** SHCF solved by IPM. **b** SHCF solved by WFAM

plotted in Fig. 4.23. The result is very interesting that: the smaller the flux is, the higher the accuracy of SHCF solved by IPM is (Fig. 4.23a). This trend is just contrary to the SHCF solved by WFAM (Fig. 4.23b). As shown in Fig. 4.23b, the larger the flux is, the higher the accuracy is for the solved SHCF by WFAM.

Comparison of Maximum Error

To evaluate the precision of the method, the calculated SHCF is compared with the exact SHCF. In the numerical test, the input SHCF can be regarded as the correct one and used to examine the accuracy of WFAM.

The maximum error is used to compare the accuracy of IPM and WFAM. It is defined as,

$$\varepsilon = \left| \lg \frac{k}{kc} \right| \tag{4.23}$$

where k represents the input SHCF and k_C represents the calculated SHCF, i.e., k' in WFAM or k^* in IPM. The maximum errors from numerical tests with various infiltration flux values are listed in Table 4.2. Clearly, for lower infiltration flux values, IPM has higher accuracy. On the contrary, at the higher infiltration flux values, WFAM has higher accuracy. At the condition of high flux, WFAM has much higher accuracy than IPM. At the condition of extremely low flux, IPM has much higher accuracy than WFAM.

Table 4.2 Maximum error and time cost

Infiltration flux	0.2 k_s	0.1 k_s	0.01 k_s	0.001 k_s
Maximum error of k^* in IPM	2.09	1.61	0.54	0.36
Maximum error of k' in WFAM	0.12	0.52	0.98	1.97
Total test time cost (day)	5.3	10.8	102.3	158.5

k_s is 1e−6 m/s for the clayey sand used

The reason is due to the different assumptions of IPM and WFAM. The faster the infiltration flux, the higher the nonlinearity degree, subsequently the lower the accuracy of IPM. On the contrary, the faster the infiltration flux, the smoother the wetting front advancing process, subsequently the higher the accuracy of WFAM.

Referring to Table 4.2, if a large infiltration flux is used, the calculated SHCF from digital WFAM has a maximum error of 0.12. That is to say, the error is within 1.3 times. Because SHCF has a relatively large inherent uncertainty, such measurement precision is enough in both laboratory and engineering.

4.4.4.4 Comparison of IPM and WFAM with Using Laboratory Test Results

With the digital WFAM, the data in the literature can be reanalyzed. Then, the SHCFs from WFAM and IPM can be compared.

***Reanalysis of the Data of Column Test Reported by Cui et al.* (2008)**
Cui et al. (2008) reported a valuable flux-controlled column test. The test data also can be analyzed by wetting front advancing method to obtain hydraulic conductivity. Cui et al. (2008) did not record the observed wetting front curve, but reported the suction profiles recorded at different times and soil–water characteristic curve. Based on the suction profiles and soil–water characteristic curve, the water content profiles at different times (Fig. 4.24a) can be obtained.

Since this test is proposed for IPM, the infiltration rate is extremely low and the test lasts for over 100 days. In this condition, the water content duration curve in Fig. 4.24a is not as sharp as that in Fig. 4.16 and the test is stopped in the second stage of wetting process (water content rising stage).

Based on Fig. 4.18 in Cui et al. (2008), there exists a range where the suction varies dozens of MPa within a height of 100 mm. This region can be recognized as wetting front. Obviously, in this test, the wetting front is not visible to human eyes and only the digital WFAM is applicable. When WFAM is applied to this test, θ_d is just a little above the initial water content.

Further, the wetting front advancing process obtained $h(t)$ is illustrated in Fig. 4.24b, and the wetting front advancing velocity is calculated by deriving $h(t)$, as shown in Fig. 4.24c. Further, $v(t)$ is used to solve the hydraulic conductivity by Eq. (4.18).

In this case, the local initial water content at a monitoring section is used as θ_0 and varies with the location of the monitoring section. This is an upward capillary test, and a is used as -1 in Eq. (4.18).

For the data at section D ($h = 95$ mm), the hydraulic conductivities calculated by wetting front advancing method (WFAM) is compared with those calculated by instantaneous profile method (IPM), which is conducted and reported by Cui et al. (2008). As shown in Fig. 4.24d, the results from IPM and WFAM are close, i.e., most of the data points are in the same order of magnitude. However, the hydraulic coefficient by IPM has some unexpected trend. In fact, such an unexpected trend is

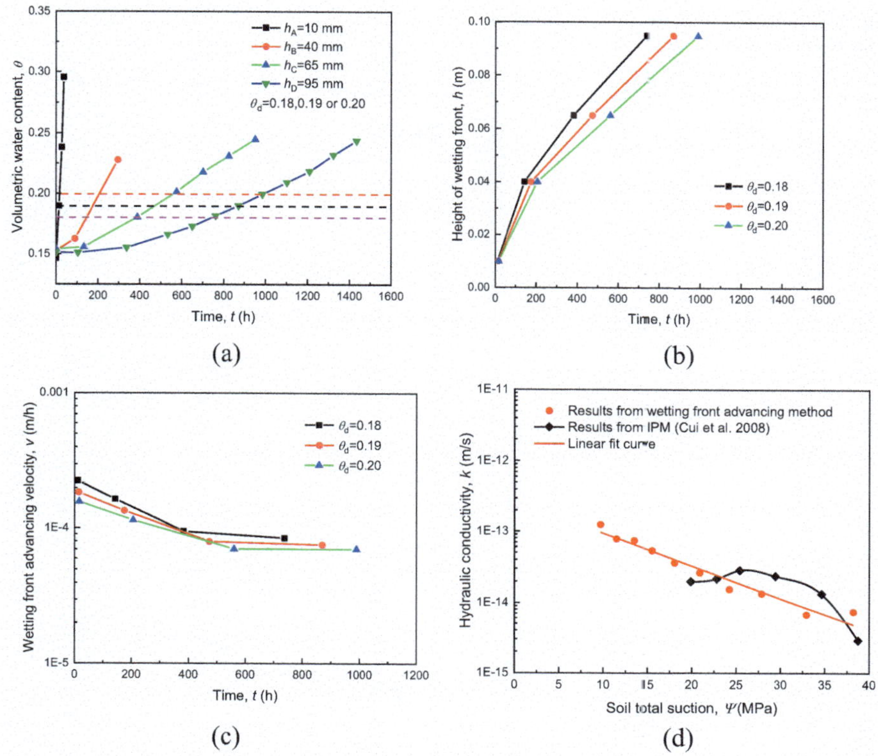

Fig. 4.24 Reanalysis of the test data in Cui et al. (2008) by digital WFAM: **a** Recorded moisture profile in Cui et al. (2008), **b** $h(t)$, **c** $v(t)$, **d** SHCF at section A ($h_D = 95$ mm)

just as same as the trend of IPM results in the numerical tests (as shown in Fig. 4.22). That is to say, the data in this test is reliable and the inflow flux in this test is not low enough for applying IPM to such a dry soil column. On the contrary, the results from WFAM are more reasonable, as shown in Fig. 4.24d.

***Reanalysis of the Data of Column Test Reported by Ng and Leung* (2012)**

Ng and Leung (2012) reported another valuable data of column test where both the suction profiles and water content profiles at various time are reported [as Figs. 5 and 6 in Ng and Leung (2012)]. Based on these data [SC0 test in Ng and Leung (2012), which is an infiltration test], the wetting front advancing process is calculated and plotted in Fig. 4.25a, b. Similar to the process used in Sect. 4.4.4.4, the wetting front advancing velocity is used to solve the hydraulic conductivity. The local initial water content at monitoring section is also used as θ_0 in the WFAM. This is a downward infiltration test; hence, a will be taken as 1.

The hydraulic conductivities calculated by using the wetting front advancing method are compared with those reported by Ng and Leung (2012) in Fig. 4.25c. It

Fig. 4.25 Reanalysis of the test data in Ng and Leung (2012) by digital WFAM: **a** $h(t)$, **b** $v(t)$, **c** SHCFs

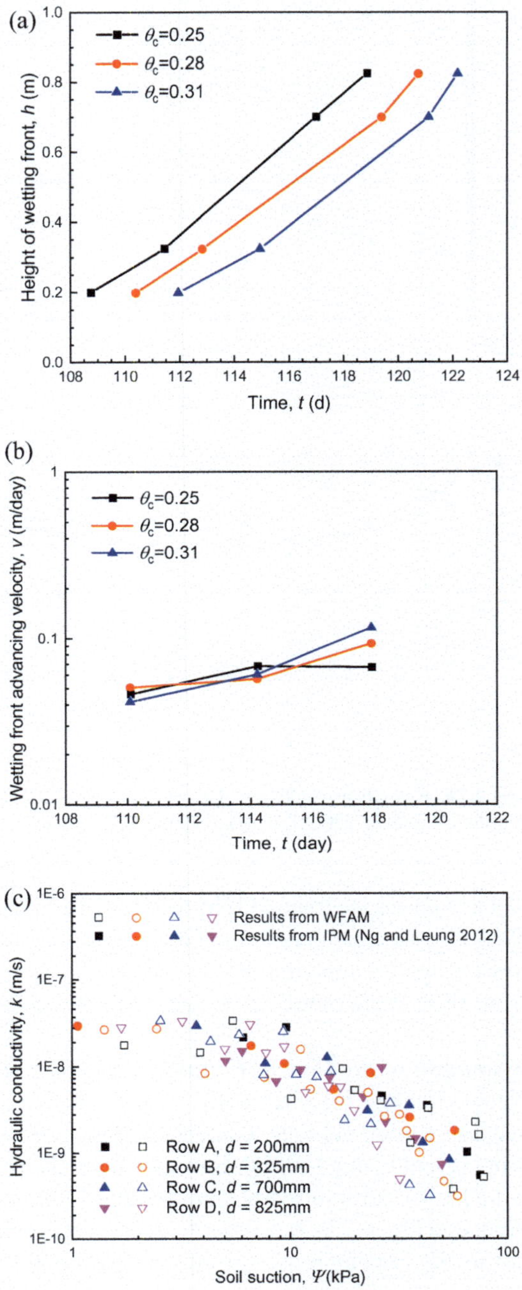

is clear that the results from WFAM and IPM are consistent for this test. Since this test used soil with a high initial water content, both WFAM and IPM are applicable.

4.4.4.5 A Full Comparison of IPM and WFAM

Characteristics of IPM and WFAM

A full comparison of IPM and WFAM is listed in Table 4.3. Clearly, these two methods have different assumptions and applicable conditions. Comparing to IPM, the most important advantage of WFAM is that it can reduce the test time cost from several months to 1–3 days and can improve the precision of SHCF at the same time (as shown in Table 4.2).

Options for Increasing the Accuracy of IPM

Because a linear assumption between two adjacent sections is adopted in the IPM, the only routine for increasing the accuracy of IPM is to reduce the nonlinearity of the infiltration process. Hence, the following options can be used:

(1) Increase the initial water content. This option will be effective as IPM has better performance for measuring SHCF of unsaturated soil with a high degree of saturation.
(2) Decrease the inflow flux to avoid the formation of wetting front. Because IPM assumes that the hydraulic gradient between adjacent sections is linear, inflow flux must be controlled to reduce the nonlinearity of the problem.
(3) Reduce the sensor spacing. In laboratory tests, the closest spacing of measuring sections is limited by the equipment. For example, if TDR sensors are used to measure soil water content, the spacing should be larger than 50 mm. The γ ray method (Wang and Zhang 1990; Chen et al. 2019) is recommended to

Table 4.3 Comparison of IPM and WFAM

	IPM	Digital WFAM
Assumption	Linear approximation is applicable to two adjacent monitored sections	A smooth advancing process is assumed for wetting front (Eq. 4.1)
Device	Soil column test device	
Number of section	>10	3–5
Sensor spacing	As small as possible	No limit
Time–cost	Over several months	1–3 days
Faster infiltration flux	Lower accuracy	Higher accuracy
Lower initial θ_0	Lower accuracy	Wider range of SHCF
Drying test	Available	Not available
Applicable conditions	(i) Wetting process in soil with high initial θ_0 (ii) Extremely low infiltration process in soil with low θ_0	Fast inflow in soil with relatively low θ_0

monitor the water content because it has no disturbance to soil and can measure the moisture content of any section.

Comments on WFAM

In original WFAM, if the wetting front is identified by naked eyes, only one sensor is required to calculate SHCF. This is the advantage of the original WFAM. When the wetting front cannot be identified by naked eyes, the digital WFAM is proposed in this study and requires 3–5 sensors to identify the wetting front advancing process. Based on the digital wetting front advancing process, the wetting front advancing velocity can be determined and used to calculate SHCF using Eq. (4.18).

4.4.5 Error Analysis for WFAM and IPM

4.4.5.1 Error Analysis of the Assumption of WFAM

Referring to Eqs. (4.7) and (4.8), WFAM depends on the assumption of a smooth advancing process. The error of this assumption is analyzed based on the numerical test results. Taking Test No. C1 as an example, the water content and suction profile in the capillary process at various time are plotted in Fig. 4.26.

Based on Eqs. (4.7) and (4.8), the relative error can be defined as,

$$\varepsilon x = \frac{x(h + v\mathrm{d}t, t + \mathrm{d}t) - x(h, t)}{x(h, t)} \tag{4.24}$$

where x represents the volumetric water content θ or soil matric suction ψ; and h is the position of monitor section. Referring to Eq. (4.24), the error can be calculated by the water content (or suction) records with a time difference of Δt. Such time

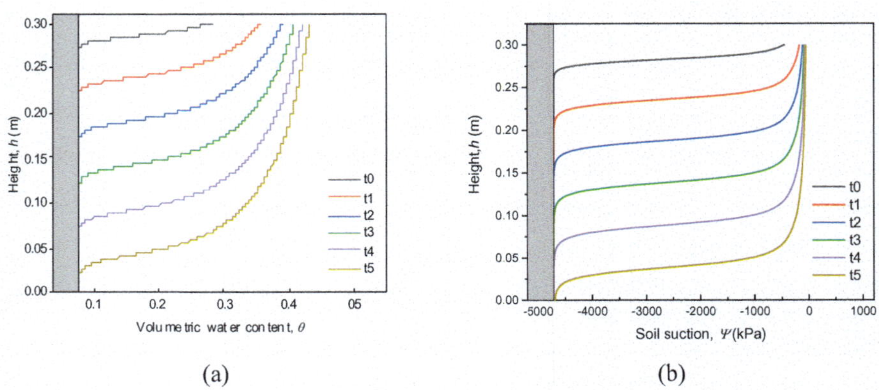

Fig. 4.26 Evolution of the wetting front (Test No. C1): **a** Water content profiles. **b** Suction profiles

difference Δt is the time step for data saving in numerical tests and is the time interval between readings in laboratory tests. Obviously, error ε_x is related to the time interval Δt.

The error at a monitor section with a height of 0.25 m is plotted in Fig. 4.27 and found to increase with time step Δt. Even the time interval Δt is as large as 4×10^5 s, the relative error is still within 8%. When the time interval $\Delta t = 1000$ s, the relative error of water content (or suction) is close to 0 and the solved SHCF is overlapped with the input value (Fig. 4.27c). In the laboratory test, a time interval Δt of the 60 s

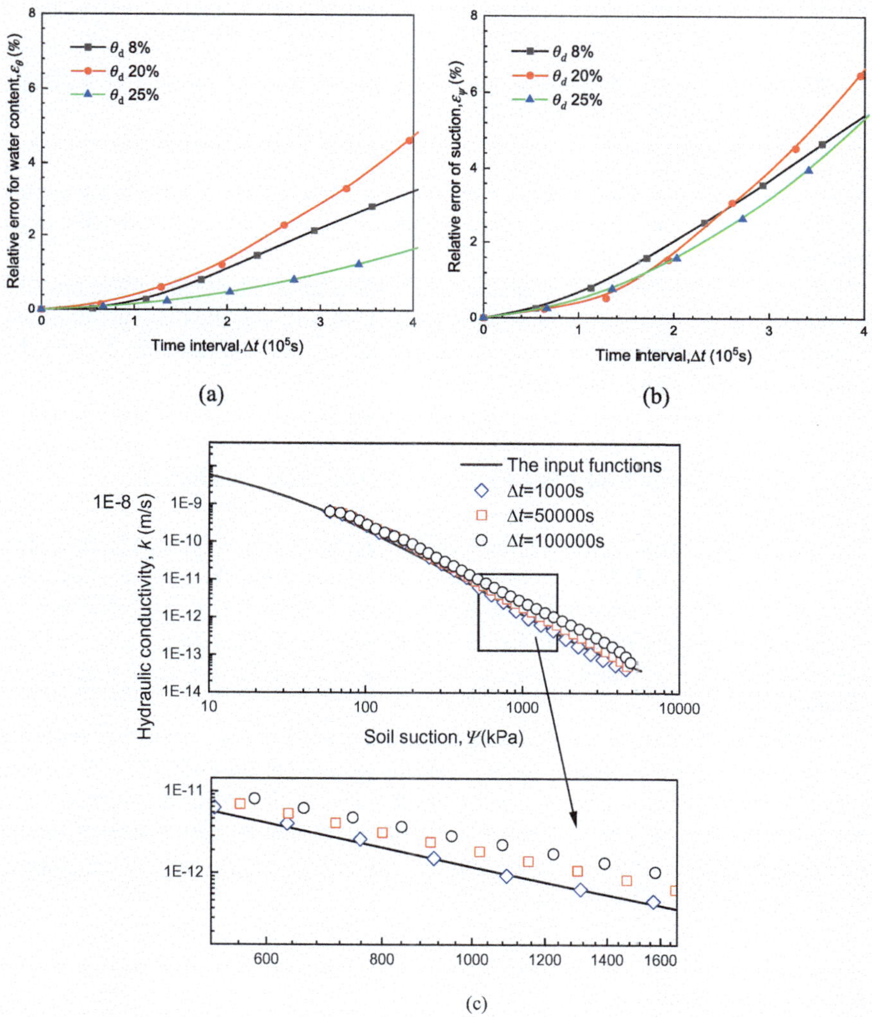

Fig. 4.27 Relative error for the assumption of a smooth advancing process (Test No. C1): **a** Relative error of water content. **b** Relative error of matric suction. **c** Effect of time interval on SHCF calculated by WFAM

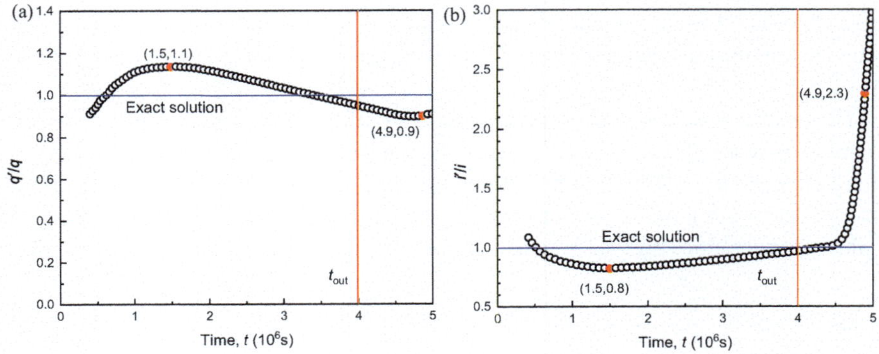

Fig. 4.28 Relative error for the flux and hydraulic gradient in WFAM (Test No. C1): **a** Relative error of flow flux q'. **b** Relative error of i'

usually can be promised for water content measurement sensor. A smooth advancing process is valid in the infiltration process and the accuracy of solved SHCF can be guaranteed in WFAM.

It should be mentioned that measurement error is unavoidable and there must be some inhomogeneity in soil in laboratory tests. That is to say, the data of laboratory test will be much scattered than that of numerical test and the calculated hydraulic conductivity will be also much scattered.

4.4.5.2 Error Analysis of Flow Flux q' and Hydraulic Gradient i'

To determine the error source of the SHCF obtained from WFAM, the flow flux q' calculated by Eq. (4.22) is compared with the true flow flux q that is directly extracted from the numerical program. For Section A of Test No. C1, the flux error is determined as q'/q and plotted in Fig. 4.28a. Obviously, if the ratio of q'/q is closer to 1, the accuracy of q' is higher. According to Fig. 4.28a, q' is very close to the exact solution of q.

The hydraulic gradient i' calculated using Eq. (4.15) is also compared with the true hydraulic gradient i that directly from the numerical program. The hydraulic gradient error is determined as i'/i and plotted in Fig. 4.28b. After the wetting front has passed the soil column, i' becomes incorrect and significantly differs from the exact solution i. However, even the wetting front has passed the whole soil column ($t > t_{out}$), q' is still close to q in a long time.

4.4.5.3 Error Analysis of i^*

For Section B of Test No. C1, the hydraulic gradient i^* calculated using Eq. (4.19) is plotted in Fig. 4.29. In this figure, i^*_- is i^* calculated by Eq. (4.19) before the

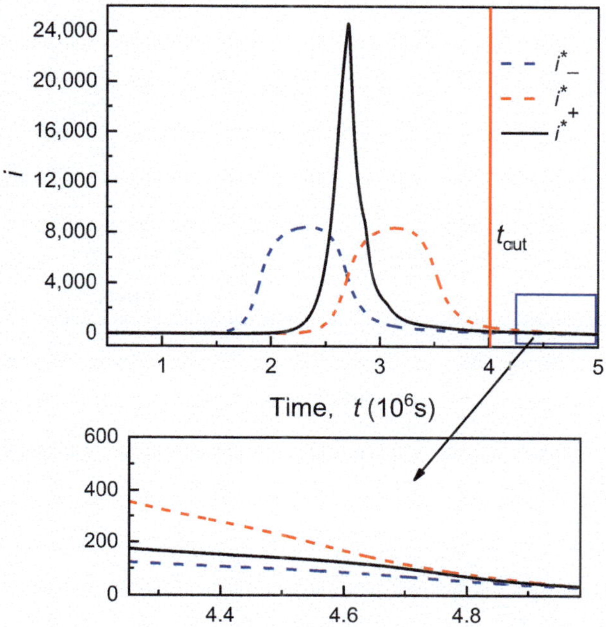

Fig. 4.29 Hydraulic gradient i^*_+, i^*_-, and i^* in IPM (Section *B* of Test No. C1)

monitored section (e.g., between section *A* and section *B*); i^*_+ is i^* calculated by Eq. (4.19) after the monitored section (i.e., between section *B* and section *C*).

According to Fig. 4.29, before wetting front passes the whole soil column, a huge difference exists between i^*_+, i^*_- and the true i^*. This is because the linear assumption is not applicable in such wetting process. After the wetting front has passed the soil column, the difference between i^*_+ and i^*_- becomes smaller. In this stage, the true i^* will be larger than i^*_- and smaller than i^*_+. Finally, i^*_+ and i^*_- are equal to i^* in the quasi-steady state (Fig. 4.29).

4.4.5.4 Correction of i' in WFAM

When $t > t_{out}$, the infiltration process is approaching to a quasi-steady state. There is no wetting front existing in the soil column anymore. According to Fig. 4.28, the unacceptable error is existing for i' while the relative error of q' is still low when $t > t_{out}$. If a reasonable estimation of i' is used, the hydraulic conductivity is still can be properly estimated.

Considering that i' becomes very large after the wetting front has passed the soil column and i^* becomes be close to the true i, i^* can be used instead of i' to calculate the hydraulic conductivity k' in this stage. As shown in Fig. 4.29, i^* is within the range of (i^*_-, i^*_+), that is to say, i^* can be estimated as $\sqrt{i* - i^*_+}$. Hence, a correction formula of the hydraulic gradient can be used as,

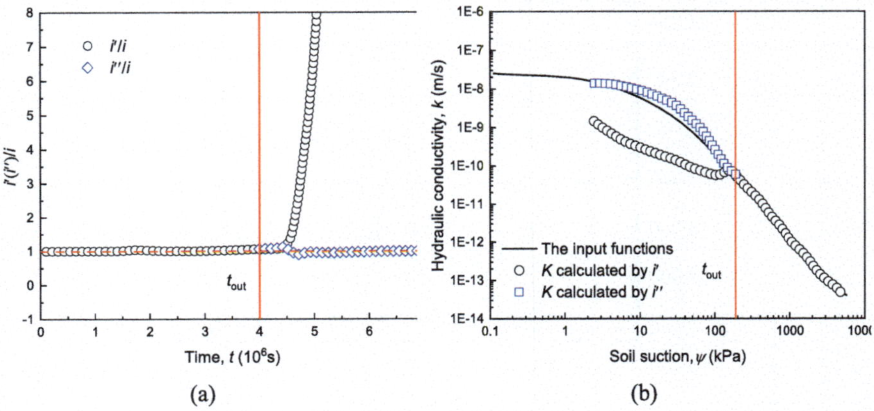

Fig. 4.30 Error of SHCF from WFAM (Test No. C1): **a** Relative error of i''. **b** The solved SHCF with using the corrected i''

$$i'' = \min(i', \sqrt{i^* - i_+^*}) \quad \text{when}\, t > t_{\text{out}} \tag{4.25}$$

When the monitored section is far behind the wetting front, such correction is meaningful and can help to improve the precision of WFAM.

The physical basis of Eq. (4.25) is that when $t > t_{\text{out}}$, the governing process of infiltration is transformed from wetting front advancing between the dry–wet interface (transient state) into the seepage in a wet soil column with nonuniform water distribution (quasi-steady state). That is to say, after t_{out}, the assumption describing infiltration process is changed from WFAMs into IPMs, which inducing an out-of-control of i' in WFAM. Therefore, Eq. (4.25) is proposed to calculate hydraulic gradient when the i' is out of control (as after 4.5e6 s in Fig. 4.28b.

Using the corrected i'', the i''/i is calculated at the middle monitoring section (Section B) and plotted in Fig. 4.30a. Clearly, the correction is effective. That is to say, if a proper estimation of i can be obtained, Eq. (4.18) can still work.

With using q' and i' (or i'') shown in Fig. 4.30a, k' can be calculated by Eq. (4.18) and plotted in Fig. 4.30b. When i' is used, the solved k' is very close to the input k when $t < t_{\text{out}}$ and is wrong when $t > t_{\text{out}}$. With using i'' instead of i' in Eq. (4.18), the solved k is close to the input k in a full range. That is to say, the correction of Eq. (4.25) can enhance the robustness of WFAM and enlarge the valid suction range of WFAM.

4.4.6 A Combined IPM and WFAM for Measuring SHCF in Full Suction Range

4.4.6.1 Basic Idea for Combining IPM and WFAM

As IPM is of higher precision for wetter soil and WFAM is of higher precision for drier soil, a combination of IPM and WFAM can lead to an accurate SHCF in full range. The idea is as follows:

(1) A soil column with dry soil is initially prepared.
(2) A wetting test is conducted on the dry soil column (the 1st wetting stage) and the data is analyzed by WFAM. In this process, the wetting process fully fulfills the assumption of WFAM and the solved k' is accurate. If $t > t_{out}$, the corrected hydraulic gradient i'' (Eq. 4.25) is used instead of i' (Eq. 4.15).
(3) After the wetting front has passed the whole soil column, the accuracy of IPM will gradually increase and the accuracy of WFAM will gradually decrease.
(4) The infiltration process is continued until the whole soil column is close to a saturated condition (the 2nd wetting stage).
(5) If i^*_- is close to i^*_+, the infiltration process fully fulfills the assumption of IPM and the solved k^* is accurate.
(6) Combined the k' solved by WFAM and k^* solved by IPM, an accurate SHCF in a full suction range can be obtained.

4.4.6.2 Design of Test Program

In the design of a test combined WFAM and IPM, the following limitations should be considered:

(1) For WFAM, the faster the infiltration process, the higher the accuracy of SHCF solved by WFAM. However, the infiltration process in the 1st wetting stage should not exceed the response time of the sensor. Otherwise, the monitored volumetric water content or soil suction will be inaccurate. A wetting boundary of 0–10 kPa will be a good choice. For example, for an overconsolidated clayey soil, a relatively higher water head such as 10 kPa can be applied to accelerate the infiltration process. For sandy soil, a relatively lower water head such as 0 kPa should be used to control the infiltration process. Such a boundary condition can be easily implemented in the laboratory and will lead to a suitable infiltration speed, i.e., not too intensive and not time-consuming.
(2) For IPM, the lower the infiltration process, the higher the accuracy of SHCF solved by IPM. So in the 2nd wetting stage, a wetting boundary between the air entry value (AEV) and 0 kPa will be proper.
(3) For IPM, the closer the sensor, the higher the accuracy of SHCF. If the TDR sensor is used in the test, a minimum distance of 5 cm should be used.

With the upper consideration, the test program for combined WFAM and IPM is designed as follows:

(1) A natural dried soil can be compacted into a soil column of 40 cm (or 60 cm).
(2) A wetting boundary of 0 kPa water head is applied to the top of soil column. Five TDR sensors below the top surface with a space of 8 cm (or 12 cm) are used to monitor soil volumetric water content.
(3) The data at a middle sensor will be used to calculate k' with using WFAM in the 1st wetting stage. If wetting front has passed the lowest sensor, i'' is used instead of i' to calculated k' (Eq. 4.18).
(4) if i^*_+ is close to i^*_-, the data at a middle sensor will be used to calculate k^* with using IPM in the 2nd wetting stage. Considering that the hydraulic conductivity is one of the most discrete parameters in geotechnical tests, $i^*_+ < 2\, i^*_-$ can be used as a reasonable criterion for the calculation of k^*.

4.4.6.3 Test Results and the Applicability of the Combined Method to Various Soils

Following the test procedure in Sect. 4.4.6.2, the infiltration processes in 3 types of soil are analyzed, as Test No. C13, M14, and S15 in Table 4.1. The wetting front advancing processes in three tests are illustrated in Fig. 4.31a and used to calculate the wetting front advancing velocity (Fig. 4.31b). The water content and suction profile at the middle monitoring sections are illustrated in Fig. 4.31c.

The k' function solved by WFAM in the 1st wetting stage and k^* solved by IPM in the 2nd wetting stage are plotted in Fig. 4.32. The time cost and maximum error are listed in Table 4.4. For the tested three soils, the maximum ratio between the measured hydraulic conductivity and the exact hydraulic conductivity is smaller than 2. More important, a SHCF in a wider suction range is obtained within 5 days. Clearly, the combined WFAM and IPM is of great success and can measure the whole SHCF with very high precision.

4.4.7 Limitations of This Study

It should be pointed out that this study is based on numerical tests. In the numerical test, the uncertainties of operation, soil inherent inhomogeneity, and sensors can be ignored. Whereas in the laboratory test, these uncertainties may take essential roles and lead to disastrous results.

In a laboratory test, at least the following influence should be considered:

(1) The sensor may affect the water migration in soil column. Electricity-based sensors such as TDR conductance sensor, magnetic-based sensors, or air humidity sensors should be selected to avoid exchange water with soil and reduce the influence of sensors on water movement.
(2) Attention should be paid to the corresponding time of sensors. The response time (time for suction equilibrium) of matric suction sensors based on axis-translation technology is relatively long, generally between minutes and hours.

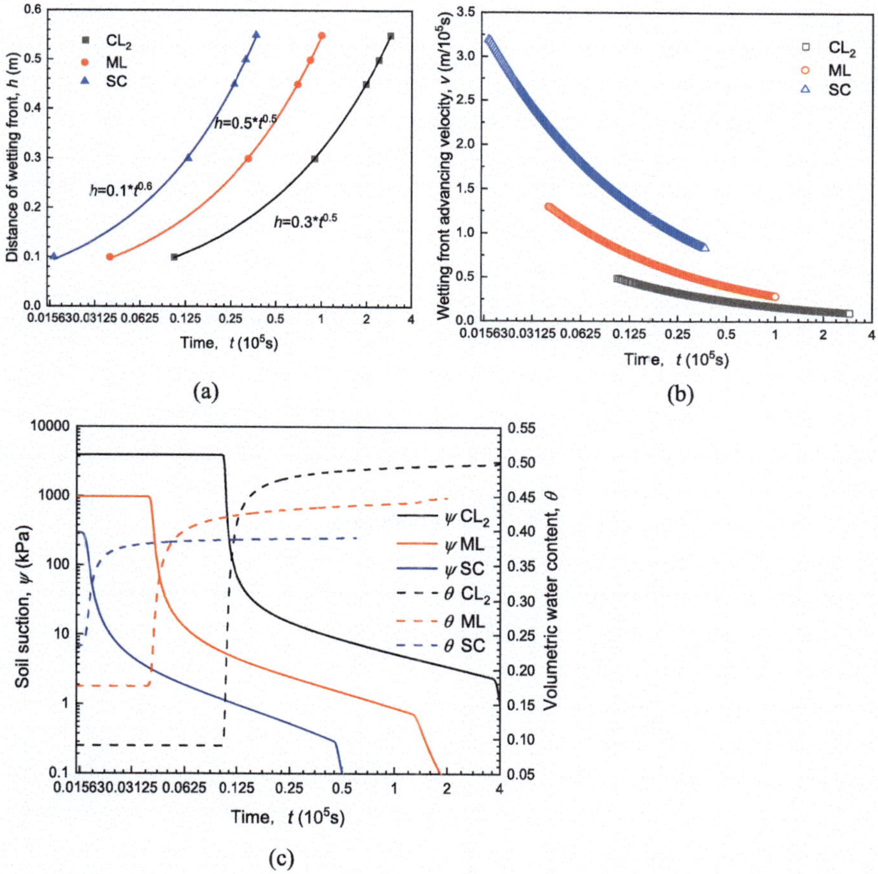

Fig. 4.31 Results of numerical tests (Test No. C13, M14, and S15): **a** Wetting front advancing process $h(t)$. **b** Wetting front advancing velocity $v(t)$. **c** Time sequence records of water content and suction at middle monitoring section

When suction sensors (which is good for the suction measurement in wet soil, especially in coarse-grained soils) are used, the flow rate should be controlled at a reasonable level to leave sufficient response time for suction sensors. When soil is dry and suction sensors are not applicable in this condition, the humidity sensor is a candidate choice in replace.

(3) Due to the uncertainties of dry density, the composition of soil, and the errors of sensor measurement, the monitoring data in laboratory test will be more discrete than the results of numerical simulation. It is necessary to analyze the mean value, variance, and confidence interval of SHCF obtained in the laboratory test.

(4) In a laboratory test, the infiltration flux in soil column is usually less than the hydraulic conductivity of saturated soil. Besides, the infiltration rate in

Fig. 4.32 The SHCF solved by combining WFAM and IPM in the data analysis of one test: **a** SHCF of CL2. **b** SHCF of ML. **c** SHCF of SC

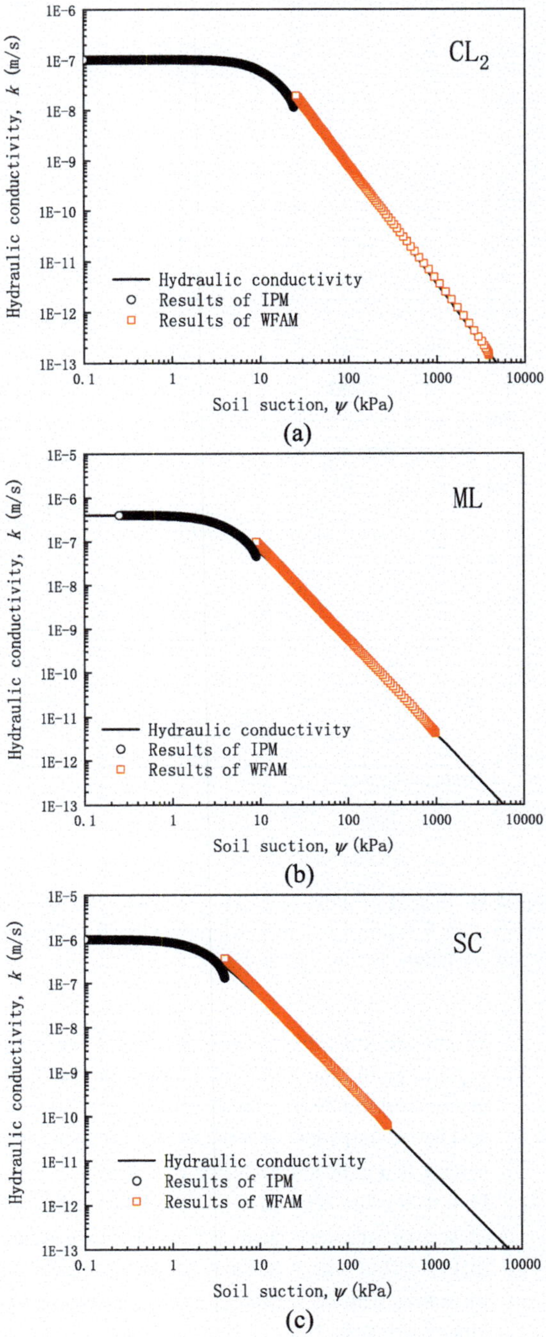

Table 4.4 Maximum error and time cost

Soil type	Clay	Silt	Sand
Test No.	C13	M14	S15
Maximum ratio of k^*/k or k/k^* in IPM	1.46	1.56	1.89
Maximum ratio of k'/k or k/k' in WFAM	1.23	1.41	1.50
Total test time cost (day)	4.7	2.3	0.6

laboratory test is limited by techniques. For example, the penetration rate of needle infiltration devices is around 10–1000 ml/day.

(5) Sometimes soil suction is hard to be monitored in soil column test. In this case, SWCC should be measured separately. Then in soil column test, soil matric suction can be inferred from the known SWCC. (Of course, this will introduce some extra errors and ignore the hysteresis of SWCC.)

(6) This study provides a method for accurately measuring SHCF within several days. Another key problem for the development and application of unsaturated soil mechanics is to quickly measure SWCC in the full suction range (0–100 MPa).

4.5 Summary

A new SWCC device was developed for measuring the SWCC at very low suctions (i.e., smaller than 1 kPa). The device has a large specimen chamber as 300 mm in diameter and 150 mm in height, which is suitable for coarse soils. The device can control the soil suction accurately (i.e., a precision of 0.5 mm suction head) by a water head control method and an axis-translation technique. A new interpretation procedure was proposed to reduce the error induced by sample height and to extend the minimum measurable suction head from centimeters (i.e., 0.1 kPa) to millimeters (i.e., 0.01 kPa).

A wetting front advancing method is developed for the measurement of unsaturated SHCF. The proposed method is able to measure the SHCF in a few days resulting in considerable time-saving as compared to traditional steady-state methods and the instantaneous profile method. The proposed method is applicable to a wide range of soil suctions and soil types. In the wetting front advancing method, an initially air-dried soil column is subjected to a wetting event. During the wetting process, a wetting front advances with time and gives a wetting front advancing velocity function and the soil suction and water content are monitored. The SHCF interpreted based on measurements at different sections agree well. Therefore, the interpreted SHCF can be considered independent of the location of the measurement section. Only one or two measurement sections are needed when using the method proposed. The methodology is less tedious than having to make measurements of water content and soil suction as in the instantaneous profile method. The SHCF obtained from a capillary-rise test and an infiltration test also agree with one another.

A digital WFAM is proposed to overcome the two limitations of the original WFAM, such as the requirement of a visually observed wetting front, and it is only valid in high suction range. The assumption, applicability, and performance of IPM and WFAM are fully compared based on both numerical and experimental results. Further, a combination method that utilizes the digital WFAM to measure SHCF in high suction range and the IPM to measure SHCF in the low suction range is proposed.

References

Askarinejad A, Beck A, Casini F, Springman SM (2012) Unsaturated hydraulic conductivity of a silty sand with the instantaneous profile method. Unsaturated soils: research and applications. Springer, Berlin, Heidelberg, pp 215–220

Bruce RR, Klute A (1956) The measurement of soil moisture diffusivity. Soil Sci Soc Am J 20(4):458–462

Chen B, Huang Y, Zhang K, Wu F (2019) Investigation on unsaturated hydraulic conductivity with nuclear magnetic resonance technology and instantaneous profile method. Appl Magn Reson 50(1–3):441–453

Cui YJ, Tang AM, Mantho AT, Laure ED (2008) Monitoring field soil suction using a miniature tensiometer. Geotech Test J 31(1):95–100

Daniel D (1983) Permeability test for unsaturated soil. Geotech Test J 6(2):81–86

Delage P, Cui YJ (2008) A novel filtration system for polyethylene glycol solutions used in the osmotic method of controlling suction. Can Geotech J 45(3):421–424

Feng M, Fredlund DG (2003) Calibration of thermal conductivity sensors with consideration of hysteresis. Can Geotech J 40(5):1048–1055

Fredlund DG, Rahardjo H (1993) Soil mechanics for unsaturated soils. Wiley, New York

Hamilton DC, Gloeckler G, Krimigis SM, Lanzerotti LJ (1981) Composition of nonthermal ions in the Jovian magnetosphere. J Geophys Res 86:8301

Klute A (1972) The determination of the hydraulic conductivity and diffusivity of unsaturated soils. Soil Sci 113(4):264–276

Krisdani H, Rahardjo H, Leong EC (2009) Use of instantaneous profile and statistical methods to determine permeability functions of unsaturated soils. Can Geotech J 46(7):869–874

Leong EC, Tripathy S, Rahardjo H (2003) Total suction measurement of unsaturated soils with a device using the chilled-mirror dew-point technique. Geothchnique 53(2):173–182

Leung AK, Coo JL, Ng CWW, Chen R (2016) New transient method for determining soil hydraulic conductivity function. Can Geotech J 53(8):1332–1345

Li X, Zhang LM, Fredlund DG (2009a) Wetting front advancing column test for measuring unsaturated hydraulic conductivity. Can Geotech J 46(12):1431–1445

Li X, Zhang LM, Li JH (2009b) Development of a modified axis translation technique for measuring SWCCs for gravel soils at very low suctions. Geotech Test J 32(6):1–11

Likos WJ, Lu N (2003) Automated humidity system for measuring total suction characteristics of clay. Geotech Test J 26(2):179–190

Marinho FAM, Oliveira OM (2006) The filter paper method revisited. Geotech Test J 29(3):250–258

McCarthy EL (1934) Mariotte's bottle. Science 80:100

Meerdink JS, Benson CH, Khire MV (1996) Unsaturated hydraulic conductivity of two compacted barrier soils. J Geotech Eng 122(7):565–576

Miller JD, Gaskin GJ (1997) Theta probe ML2x principles of operation and applications. In: MLURI technical note, 2nd edn. Delta-T Devices Ltd., UK

Monroy R, Ridley A, Dineen K, Zdravkovic L (2007) The suitability of the osmotic technique for the long-term testing of partly saturated soils. Geotech Test J 30(3):220–226
Mualem Y (1976) A new model for predicting the hydraulic conductivity of unsaturated porous media. Water Resour Res 12(3):513–522
Ng C, Leung AK (2012) Measurements of drying and wetting permeability functions using a new stress-controllable soil column. J Geotech Geoenviron Eng 138:58–68
Ni J, Leung AK, Ng CW (2019) Unsaturated hydraulic properties of vegetated soil under single and mixed planting conditions. Geotechnique 69(6):554–559
Nichol C, Smith L, Beckie R (2003) Long-term measurement of matric suction using thermal conductivity sensors. Can Geotech J 40(3):587–597
Plagge R, Scheffler G, Grunewald J, Funk M (2006) On the hysteresis in moisture storage and conductivity measured by the instantaneous profile method. J Building Phys 29(3):247–259
Richards SJ, Weeks LV (1953) Capillary conductivity values from moisture yield and tension measurements on soil columns. Soil Sci Soc Am J 17:206–209
Ridley AM, Burland JB (1996) A pore pressure probe for the in situ measurement of a wide range of soil suctions. In: Advances in site investigation practice. Thomas Telford, London, pp 510–520
Roth CH, Malicki MA, Plagge R (1992) Empirical evaluation of the relationship between soil dielectric constant and volumetric water content as the basis for calibrating soil moisture measurements by TDR. J Soil Sci 43(1):1–13
Scheffler GA, Plagge R (2011) Application of instantaneous profile measurement of moisture content and moisture potential in porous materials. Mater Struct 44(8):1517–1536
Soil Moisture Equipment Corp (2005) Operating instructions: volumetric pressure plate extractor and hysteresis attachments. Soil Moisture Equipment Inc., CA
Su W, Cui YJ, Qin PJ, Zhang F, Ye WM, Conil N (2018) Application of instantaneous profile method to determine the hydraulic conductivity of unsaturated natural stiff clay. Eng Geol 243:111–117
Take WA, Bolton MD (2003) Tensiometer saturation and the reliable measurement of soil suction. Geotechnique 53(2):159–172
Van Genuchten MT (1980) A closed-form equation for predicting the hydraulic conductivity of unsaturated soils. Soil Sci Soc Am J 44(5):892–898
Wang WY, Zhang JF (1990) Test study of measurement of motion parameters of soil water with a horizontal soil column. J Hydraul Eng 7:26–30 (in Chinese)
Watson KK (1966) An instantaneous profile method for determining the hydraulic conductivity of unsaturated porous materials. Water Resour Res 2(4):709–715

Chapter 5
Unsaturated Hydraulic Properties for Widely-Graded Soils

5.1 Introduction

In this chapter, the hydraulic properties of the five design soils listed in Table 2.2 are investigated. The SWCCs were measured in the laboratory using the newly developed low suction SWCC device (Fig. 4.2) and three types' commercial available devices (i.e., Fredlund SWCC device, Model 1600 pressure plate extractor, and WP4 dewpoint potentiometer device). The soil hydraulic conductivity functions (SHCFs) were measured using the wetting front advancing column tests as described in Chap. 4.

5.2 Properties of SWCCs for Widely-Graded Soils

In order to measure the SWCCs over a wide range suction, the laboratory tests were conducted using the newly developed low suction SWCC device (Fig. 4.2) and three types' commercial available devices (i.e., Fredlund SWCC device, Model 1600 pressure plate extractor, and WP4 dewpoint potentiometer device). The newly developed low suction SWCC device has been introduced in Chap. 4; the test procedures for Fredlund SWCC device, Model 1600 pressure plate extractor, and WP4 dewpoint potentiometer device are simply introduced as follows.

5.2.1 Other Experimental Techniques for Measuring SWCCs

5.2.1.1 Fredlund SWCC Device

The Fredlund SWCC device [as shown in Fig. 5.1a, Geotechnical Consulting and Testing Systems Corporation] is a simple unsaturated soil testing apparatus with

X. Li et al., *Fundamental Behavior of Unsaturated Widely-Graded Soil*,
https://doi.org/10.1007/978-981-19-3402-5_5

Fig. 5.1 Existing SWCC devices: **a** Fredlund SWCC device, **b** Model 1600 pressure plate extractor, **c** WP4 dewpoint potentiometer device

great flexibility for applying matric suction. The pore-air pressure, u_a, at the top of the specimen is controlled through the use of dual pressure regulators and precision gauges. The stainless steel cell is constructed with simple knobs and screws for fast soil specimen setups. The device includes a pressure panel with dual gauges and regulators for increasing precision in the low-pressure range. The apparatus also includes necessary plumbing and valves for periodic flushing of diffused air. High-air-entry-value (HAEV) ceramic stone is used in the bottom, which can be passed through by water but not by air below a specific pressure after it is saturated. Hence, the air pressure (u_a) (as also the suction ($u_{a\text{-}}$ u_w), where the pore water pressure is controlled as 0 in experiment) can be controlled.

The Fredlund SWCC device was used to measure the wetting and drying SWCCs in the suction range of 0–500 kPa. The following steps were used to prepare a sample:

(1) A mount of water is added into the oven-dried bulky soil to the optimum water content. Churning up the soil and water until no obvious dried particles can be seen.

(2) The mixed soil is sieved by No. 16 sieve (1.18 mm). The aggregated particles above the sieve are crushed by finger.
(3) Then, the soil is put into a plastic bag for maturing for 7 days.
(4) The water content of the mature soil is measured. If its water content is $\pm$ 1% around the optimum water content, then the soil is accepted for soil specimen compaction.
(5) The soil sample is compacted to the specific density according to the BS.
(6) For sample saturation, the compacted soil specimen is immersed in a water tank for 7 days, where vacuum is maintained by a laboratory vacuum source.
(7) After the soil specimen is saturated, it is put into the apparatus for soil water characteristic curve testing.

During the test, the soil specimen is applied with a pore-air pressure, u_a, in the soil sample chamber. The pore-water in the soil specimen is at equilibrium with the water in the volume tube through the ceramic disk. Hence, the pore-water pressure u_w is equal to the water pressure in the volume tube. Under a specific air pressure, the corresponding suction is $u_a - u_w$. The detailed steps are as follows:

(1) Set up the device.
(2) To saturate the ceramic disk, a certain amount of water is added into the soil specimen chamber and a specific air pressure (for example 500 kPa) is applied to the chamber to let water seeping through the ceramic disk.
(3) Maintaining the chamber pressure close to the air entry value of the ceramic disk for 2 h, if the diffused air bubbles are few (i.e., the change of the readings in the water volume tubes is smaller than 0.1 cc), the ceramic disk is considered to be saturated.
(4) After the ceramic disk is saturated, put the saturated soil sample into the soil specimen chamber, and record its initial weight W_{ini}.
(5) The air pressure is increased step by step. For each air pressure step, the air pressure is remained until equilibrium (i.e., the difference of the water volume readings in 8 h is smaller than 0.1 cc).
(6) After the air pressure reaches the maximum pressure of the range of ceramic disk, it is decreased step by step. At each step, the air pressure is also remained until no reading change in the water volume tube and the water volume readings are taken.
(7) After the air pressure decreases to the water pressure in the water volume tube (i.e., 1–2 kPa), the experiment is terminated and the weight of the soil specimen is measured as W_{final}.
(8) Using W_{ini} and the cumulative water volume change to calculate the final weight of the soil specimen, which should be about $\pm$ 1 g around W_{final}. If the difference is larger than 1 g, the measurement is of unacceptable accuracy.
(9) Oven-drying the soil specimen and calculate the water content corresponding to each air pressure step.
(10) Plot the suction ($u_a - u_w$) versus water content relationship on log-suction scale.

5.2.1.2 Model 1600 Pressure Plate Extractor

The Model 1600 pressure plate extractor [refer to Fig. 5.1b, Soil moisture Equipment Corporation] consists of a pressure-controlled vessel, a vessel lid, clamping bolts, O-ring seals, and an outflow tube. With the convenience of a large sample container, it is able to measure the SWCCs of large size soil samples and suitable for gravel soils. The Model 1600 pressure plate extractor can only be used to measure the drying SWCC because of the lack of a diffused air flushing system. The diffused air bubbles cumulate under the ceramic disk and induce limit contact between the water and the ceramic disk. It is hard to maintain continuous inflow channels in the wetting SWCC measurement for The Model 1600 pressure plate extractor.

The experimental procedure of Model 1600 pressure plate extractor is the same as that of Fredlund cell device except for water content calculation method. For the Fredlund cell device, the water volume change in the soil specimen can be reflected by the water volume tube readings. For Model 1600 pressure plate extractor, the soil sample is taken out for weight measurement; the change in soil weight is equal to the water weight change in the soil specimen.

5.2.1.3 WP4 Dewpoint Potentiometer

WP4 dewpoint potentiometer [refer to Fig. 5.1c, Decagon Devices Inc.] is only suitable for measuring the SWCC of a small size sample or bulky soil, because of the small sample container (30 mm in diameter × 15 mm in height). WP4 dewpoint potentiometer is limited to the measurement in high suction range (above 500 kPa), with an accuracy of ± 100 kPa.

In this study, WP4 dewpoint potentiometer was used to measure the suction of CL with sand in the high suction range. In the experiment, independent soil samples with different water contents were prepared following the same procedure and put into the soil sample container for suction measurement.

The following steps were used to prepare a sample:

(1) First, the dry soil is mixed with an amount of water to specific water content.
(2) Take some soil for water content measurement (oven-dry method).
(3) Then, the soil is compacted to specific relative compaction (0.85).
(4) The soil sample is cut into a flat disk, such as 5 mm in height and 12 mm in radius, for suction measurement.

The suction of each sample is measured as the following steps:

(1) Before the measurement, KCL solution is used to calibrate the WP4 dewpoint potentiometer.
(2) The soil specimen is put into the measurement cup and is used to measure the suction continuously for 5–10 times.
(3) If the environment temperature changes more than 0.5° in the measurement process, the WP4 dewpoint potentiometer should be recalibrated.
(4) Plot the average suction vs. the water content relation on log-suction scale.

5.2.2 *Effect of Coarse Fractions on SWCC*

The SWCCs of the five colluvial soils at different dry densities (as listed in Table 5.1) were measured and are shown in Figs. 5.2, 5.3, 5.4, 5.5, 5.6, and 5.7. As shown in Fig. 5.2a, the combined SWCC for CL with sand based on the SWCC measured using the 5-Bar SWCC device and the SWCC measured using the WP4 dewpoint potentiometer device covers a wide suction range, 0–100,000 kPa. It is effective to combine the chilled-mirror dew-point technique and the axis-translation technique for measuring a full-range SWCC.

Table 5.1 SWCC experiment list and estimated AEV for each experiment

Group symbols (ASTM D2487)	Dry density (kg/m^3)	AEV (kPa)	Sample size ($D \times H$, mm × mm)	Test device
CL with sand	1320	10	50 × 8	Model 1600 extractor
	1320		24 × 8	WP4 dewpoint potentiometer device
	1380	30	70 × 32	Fredlund cell
	1470	70	70 × 32	Fredlund cell
Sandy ML	1390	4	120 × 100	Model 1600 extractor
	1450	2	70 × 32	Fredlund cell
	1660	40	70 × 32	Fredlund cell
SC with gravel	1550	0.5	70 × 60	Fredlund cell
	1620	6	120 × 100	Model 1600 extractor
	1740	10	120 × 100	Model 1600 extractor
SM with gravel	1630	0.3	230 × 50	Low suction SWCC device
	1810	0.1	230 × 50	Low suction SWCC device
	1630		120 × 100	Model 1600 extractor
	1810		120 × 100	Model 1600 extractor
GW-GM with sand	1660	0.07	230 × 50	Low suction SWCC device
	1780	0.2	230 × 50	Low suction SWCC device
	1660		120 × 100	Model 1600 extractor
	1850		120 × 100	Model 1600 extractor

D is the diameter of a soil sample in cylinder shape; H is the higher of a soil sample in cylinder shape

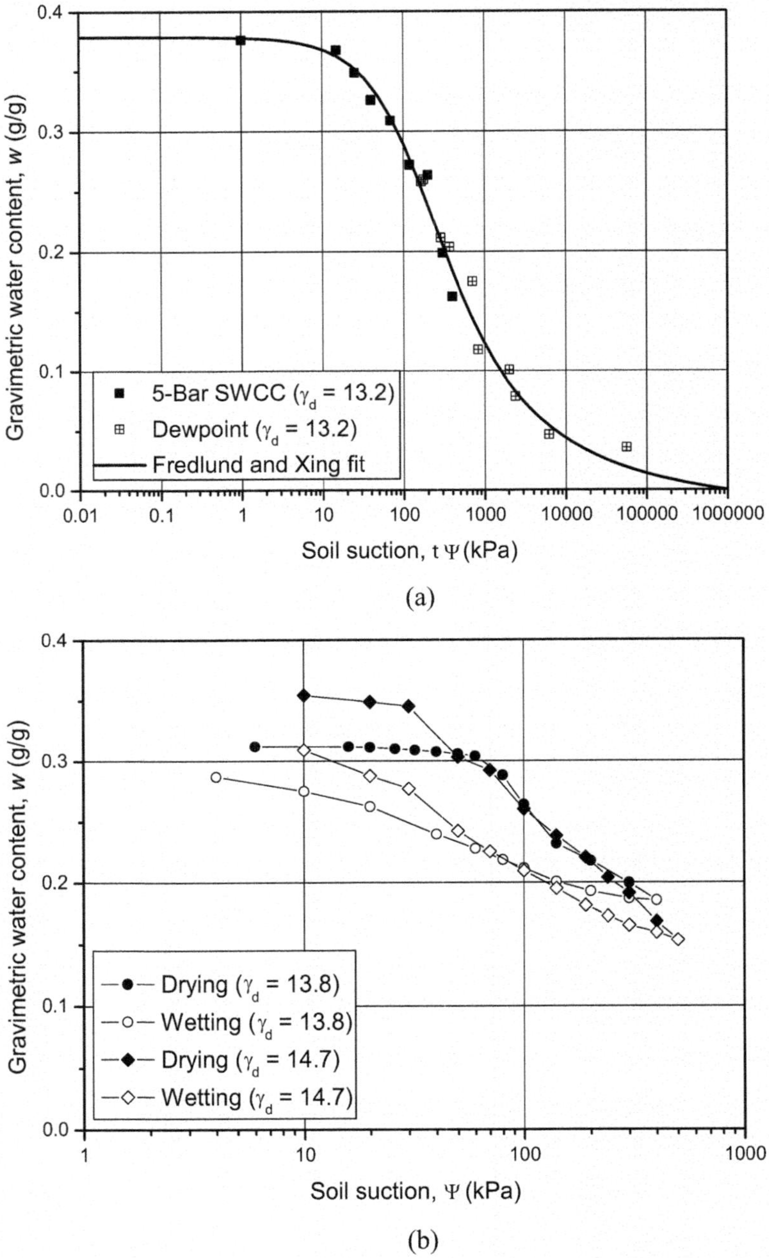

Fig. 5.2 SWCCs for CL with sand (w_s is saturated water content, γ_d is the soil unit weight, kN/m^3): **a** SWCC covering a wide suction range, **b** Hysteretic SWCCs

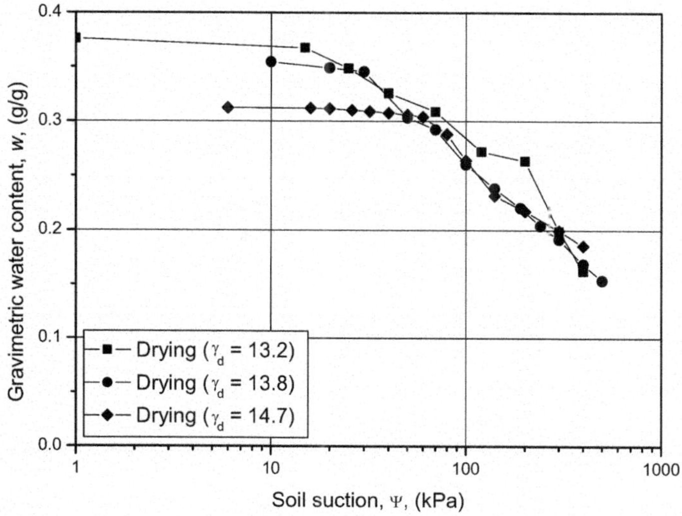

(a)

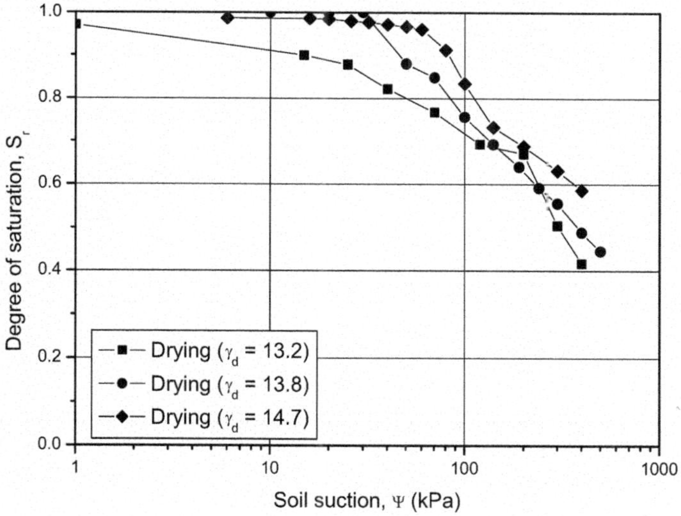

(b)

Fig. 5.3 Drying SWCCs for CL with sand (w_s is saturated water content, γ_d is the soil unit weight, kN/m^3): **a** Gravimetric water content, **b** Degree of saturation

5.2.2.1 SWCC for Fines-Controlled, Widely-Graded Soils

The measured SWCCs for CL with sand are show in Fig. 5.2. The measured drying SWCCs of CL with sand are shown in Fig. 5.3. The measured drying SWCCs of

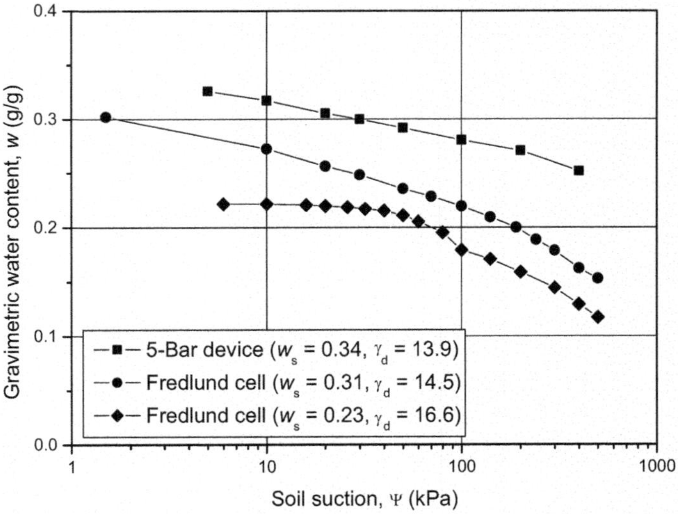

(a)

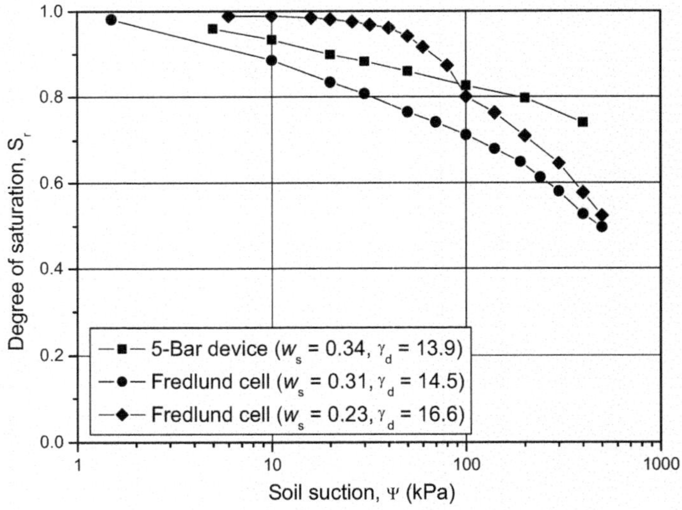

(b)

Fig. 5.4 Drying SWCCs for sandy ML (w_s is saturated water content, γ_d is the soil unit weight, kN/m^3): **a** Gravimetric water content, **b** Degree of saturation

sandy ML are shown in Fig. 5.4. The SWCCs of CL with sand and the sandy ML are smooth and seem to be unimodal curves.

Little water drains at suctions smaller than 10 kPa for CL with sand and sandy ML [i.e., the degree of saturation, S_r, is above 0.9 at suction = 10 kPa, as shown in Figs. 5.3b and 5.4b]. This behavior is consistent with the porosity structure measured

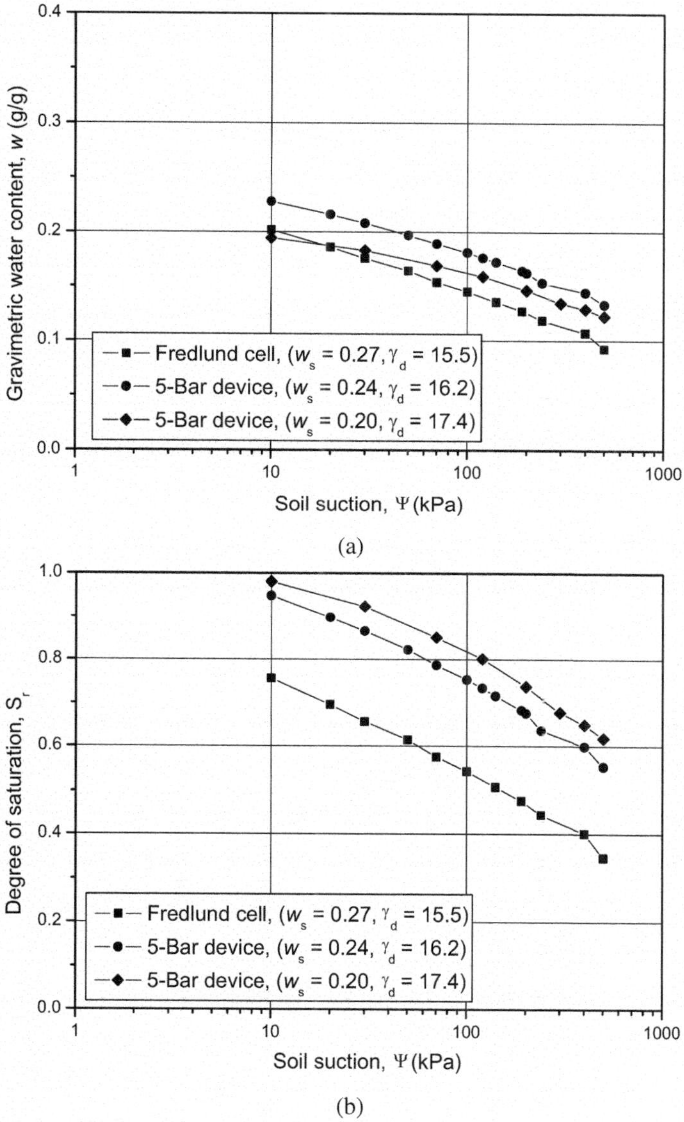

Fig. 5.5 Drying SWCCs for SC with gravel (w_s is saturated water content, γ_d is the soil unit weight, kN/m^3): **a** Gravimetric water content, **b** Degree of saturation

using MIP. The compacted and saturated samples of CL with sand and sandy ML show an intra-aggregate pores structure (as shown in Chap. 3, Fig. 3.20), where the volume fractions of pore with radii larger than 2 μm (i.e., ccrresponding to suctions smaller than 72 kPa) are smaller than 20%.

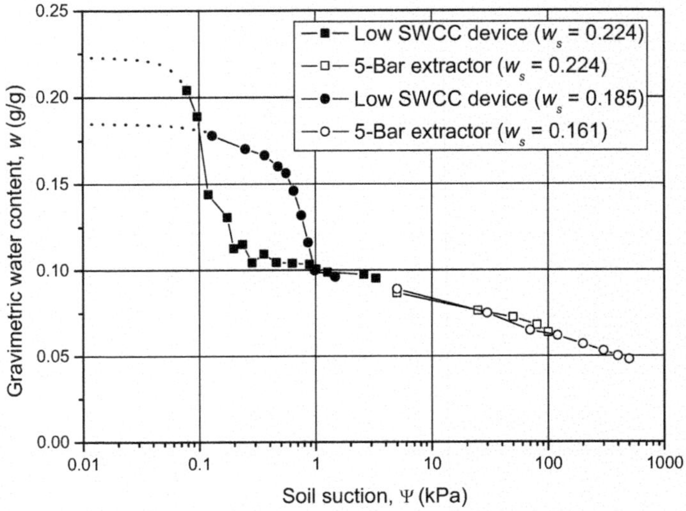

Fig. 5.6 Drying SWCCs for GW-GM with sand (w_S is saturated water content)

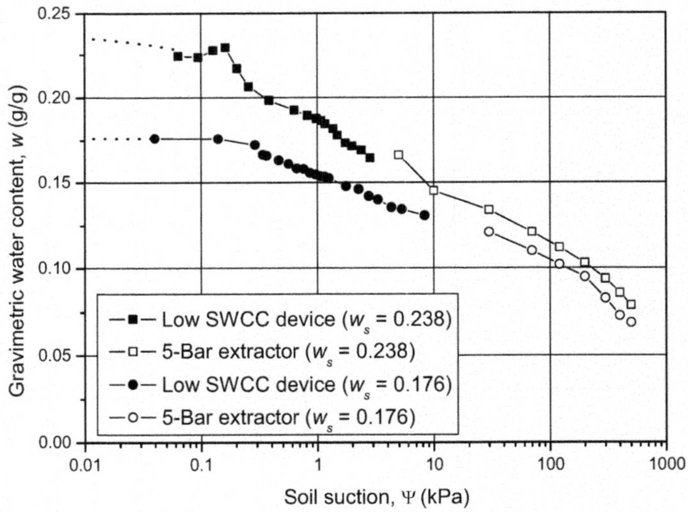

Fig. 5.7 Drying SWCCs for SM with gravel (w_S is saturated water content)

The CL with sand and the sandy ML are considered as fines-controlled, Widely-graded soils. It can be concluded that the fines-controlled, Widely-graded soils have unimodal SWCCs and show limit water content changes at low suctions (smaller than 10 kPa).

5.2.2.2 SWCC for Coarse-Controlled, Widely-Graded Soils

The drying SWCCs, two tested using the new SWCC device and another two tested using the 5-Bar pressure plate extractor, for GW-GM with sand are shown in Fig. 5.6. Four drying SWCCs for SM with gravel were also measured and are shown in Fig. 5.7. These drying SWCCs demonstrate that the drainage is significant in the low suction range for gravel soils. For GW-GM with sand, over 50% of water drains when the soil suction increases to 1 kPa. For SM with gravel, over 20% of water drains when the soil suction increases to 1 kPa. The measured pore volume with radius above 72 μm (corresponding to a suction of 2 kPa) in GW-GM with sand are about 47% (referring to Chap. 3, Fig. 3.19). The measured pore volume with radii above 72 μm (corresponding to a suction of 2 kPa) in GW-GM with sand is about 15% (referring to Chap. 3, Fig. 3.18). Although the values are not exactly the same, the existence of abundant macro-pores in gravel soils can be verified.

As shown in Fig. 5.6, the SWCCs for GW-GM with sand show strong bimodal features, where the water content decreases sharply in the first drainage period (i.e., smaller than 1 kPa) and decreases gradually in the second drainage period (i.e., from 1 to 500 kPa or values above 500 kPa). In Fig. 5.7, the SWCC for SM with gravel at a dry density of 1.63 g/cm^3 shows a weak bimodal feature, which is similar to the case of GW-GM with sand. The SWCC for SM with gravel at a dry density of 1.81 g/cm^3 does not show any bimodal feature.

The compacted and saturated samples of SM with gravel and GW-GM with sand show a dual-porosity structure (referring to Chap. 3, Figs. 3.18, 3.19, and 3.20). In the gravelly colluvial soil, sand and fines adhere to large gravel granules and form aggregates sometimes (as shown in Fig. 5.8). The macro-pores are formed between the large gravel granules or aggregates [Fig. 5.8b]. The micropores are formed between the large gravel granules and the sand or the fine contents, or within the sand and fine contents [Fig. 5.8b]. It is considered that the pore water drains from the macro-pores in the first drainage period and from the micropores in the second drainage period.

For gravel soils, considering the possible bimodal feature of SWCCs and the large amount of water drained or absorbed at low suctions (i.e., smaller than 1 kPa), the SWCC in the low suction range cannot be ignored or extended from the SWCC measured in the high suction range. The measurement of SWCCs for gravel soils at low suctions is therefore necessary and important.

The SM with gravel and the GW-GM with sand are considered as coarse-controlled structure Widely-graded soils. It can be concluded that the coarse-controlled structure Widely-graded soils have typical bimodal SWCC and show significant water content change at low suctions (smaller than 10 kPa). The water content change at very low suctions (smaller than 1 kPa) is important and necessary for coarse-controlled structure Widely-graded soils.

Fig. 5.8 Dual-porosity structure of a GW-GM with sand sample: **a** Soil surface photograph, **b** Macro-pores and micropores

5.2.2.3 SWCC for Widely-Graded Soils with Variable Structure

The measured drying SWCCs for SC with gravel are shown in Fig. 5.5. For the loose SC with gravel sample (i.e., $\gamma_d = 1550\ \text{kg/m}^3$ and relative compaction = 0.82), about 25% water drains as suction increases to 10 kPa. But for the denser SC with gravel sample, little water drains as suction increases to 10 kPa [referring to Fig. 5.5b]. The dense SC with gravel sample shows also an intra-aggregate pore structure (as shown in Chap. 3, Fig. 3.17).

The SWCC of loose SC with gravel sample shows features similar to that of SM with gravel. The SWCC of dense SC with gravel sample shows features similar to that of sandy ML. When the SC with gravel sample is compacted from a loose state to a dense state, the soil structure may change from a fine-controlled structure to a coarse-controlled structure.

5.2.3 *Influence of Soil Density on SWCC*

As shown in Figs. 5.3, 5.4, 5.5, 5.6, and 5.7, the SWCCs at different dry densities or void ratios for the same type of soil are quite different. The air entry values of different samples were estimated by the tangent method proposed by Brooks and Corey (1964). As shown in Fig. 5.6, the air entry value of the soil sample at a dry density of 1.66 g/cm^3 (i.e., $w_s = 0.224$), which is about 0.07 kPa as listed in Table 5.1, is smaller than the air entry value of the soil sample at a dry density of 1.78 g/cm^3 (i.e., $w_s = 0.185$), which is about 0.2 kPa. Also as shown in Fig. 5.7, the air entry value (about 0.1 kPa) of the soil sample at a dry density of 1.63 g/cm^3 is smaller than the air entry value (about 0.3 kPa) of the same soil at a dry density of 1.81 g/cm^3. It seems that the SWCCs for higher densities also show higher AEVs. Similar results are found in Ng and Pang (2000) and Yang et al (2004), which are summarized in Table 5.2.

For the GW-GM with sand in Fig. 5.6, the SWCCs for the soil samples at different dry densities are almost identical as the soil suction is above 1 kPa. For SM with gravel, the gravimetric water content curves for the soil samples at different dry densities become closer as the soil suction is larger than 10 kPa. Similar phenomenon is found in Yang et al (2004). This phenomenon is clearly related to the pore structures of the soil samples. In the laboratory tests, the soil samples used for SWCC testing were first compacted under different compaction stresses and then saturated. During the soil compaction process, soil samples formed different soil pore structures. The compaction efforts have a substantial influence on the macro-pores (referring to Sect. 3.4.3), but a limited influence on the micropores. The soil volume reduction caused by soil compaction is mainly induced by the reduction of macro-pores. Therefore, the compacted soil samples exhibit different macro-pore distributions, but

Table 5.2 Air entry values of different soils (from the literature)

Soil	Dry density (g/cm^3)	Estimated AEV (kPa)	References
Clayey sand II	1.47	~ 4	Yang et al (2004)
	1.70	~ 20	
Completely decomposed volcanic	1.23	0.8–1.5	Ng and Pang (2000)
	1.24	2–5	
	1.27	6–20	

similar micropore distributions. As a result, the SWCCs for soil samples at different dry densities are quite different at low suctions but become closer at high suctions.

Huang et al. (1998) analyzed some data from Labiberte and Corey (1996) and gave a log-linear relation between the air entry value and void ratio, which is expressed as,

$$\log \psi_{aev} = \log \psi_{aev0} + a(e - e_0) \tag{5.1}$$

where ψ_{aev0} is the air entry value at a reference void ratio, e_0; e is the void ratio; and a is a parameter indicating the slope related to the soil type. The relation between the AEV and void ratio is plotted in Fig. 5.9a. The data seem to follow the trend described by Eq. 5.1 for each soil. However, the data for each soil are limited. Further, the relation between the AEV and the relative compaction, which is defined as the ratio between the sample dry density and the soil maximum dry density, is plotted in Fig. 5.9b. The fines-controlled soils, i.e., SC with gravel, sandy ML, and CL with sand, show a similar relation between the relative compaction value and the air entry value. The coarse-controlled soils, i.e., SM with gravel and GW-GM with sand, also show a similar relation between the relative compaction value and the air entry value. These relations can be expressed as [as shown in Fig. 5.9b],

$$\begin{cases} \log \psi_{aev} = 17.4(RC - 0.84) & \text{for fines - controlled colluvial soils} \\ \log \psi_{aev} = 7.1(RC - 1) & \text{for coarse - controlled colluvial soils} \end{cases} \tag{5.2}$$

where RC is the relative compaction.

5.2.4 Influence of Sample Size on SWCC

In Fig. 5.4, the AEV of the soil sample at dry density of (1.39 cm^3/g) is larger than the AEV of the soil sample at dry density of (1.45 cm^3/g). This is contrary to the common phenomenon that a looser sample has a lower AEV on the same conditions. One possible reason relates to that the two samples were measured at different sample size. The soil sample at dry density of (1.39 cm^3/g) was measured by Model 1600 extractor and had a larger dimension of 120 mm in diameter × 100 mm in height. The soil sample at dry density of (1.45 cm^3/g) was measured by Fredlund SWCC device and had a small dimension of 72 mm in diameter × 45 mm in height. In soil sample, a larger pore is normally surrounded by smaller passages and drains through the smaller passages. The drainage suction for water stored in the large pore is determined by the largest radii of entrance connected to the large pore, but not the radii of the large pore. Further, the drainage suction for water stored in a large pore inside soil sample is determined by the largest radii of throttles (i.e., the radii of a dominant throttle) on these passages. In a larger size sample, the drain passages are longer. A longer passage will result in more throttles and a smaller, dominant

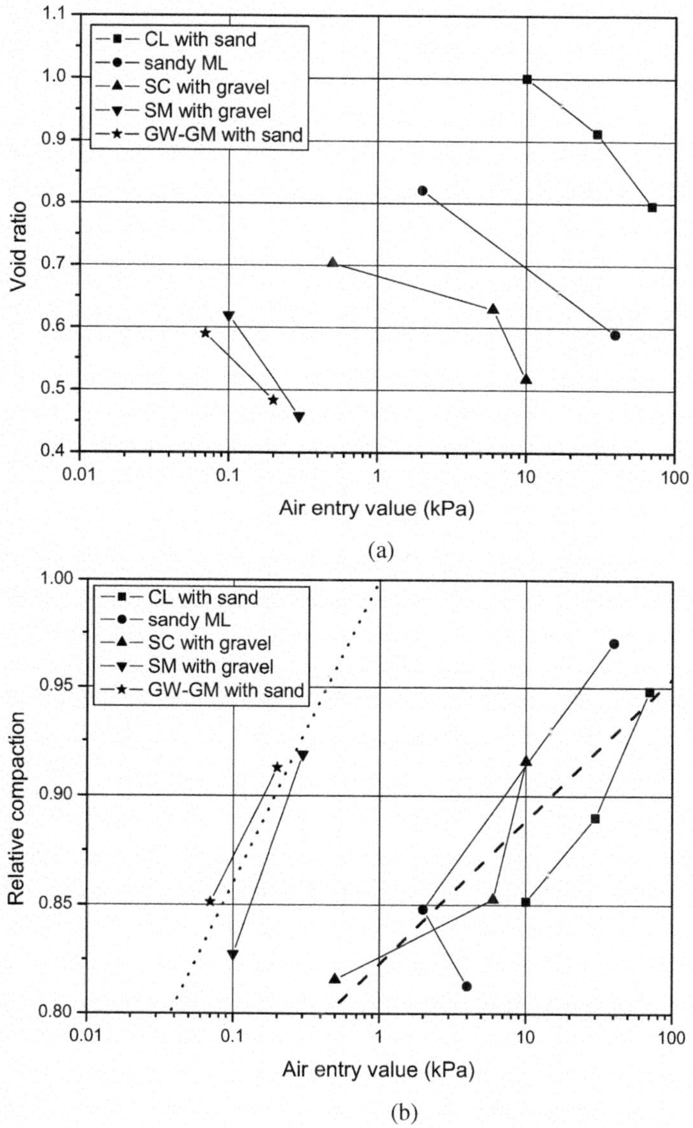

Fig. 5.9 Influence of soil dry density on air entry value: **a** AEV versus void ratio, **b** AEV versus relative compaction

throttle, which controls the drainage suction for water stored in the large pore inside soil sample. Therefore, a larger size sample will drain smaller proportion of water at a specific suction and shows a higher AEV.

However, the AEV of the larger soil sample at dry density of (1.39 cm^3/g) is still smaller than the AEV of another smaller soil sample at dry density of (1.66 cm^3/g).

The soil density is still considered as the major influence parameter on AEV for same type soil samples.

5.2.5 *Hysteresis in SWCCs*

The drying and wetting SWCCs for colluvial soils are shown in Figs. 5.10, 5.11, 5.12, and 5.13. Hydraulic hysteresis is common in these SWCCs whenever for different soils or for different suction ranges. In these SWCCs, a drying curve and a wetting curve has similar process.

Four possible reasons of hydraulic hysteresis were discussed by Hillel (1998): (1) the geometric non-uniformity of the individual pores, which results in the "ink bottle" effect; (2) hysteresis of contact angle; (3) encapsulation of air in "blind" or "dead-end" pores; (4) swelling, shrinkage, or aging phenomena.

For GW-GM with sand and SM with gravel, considering the weak shrinkage or swell ability of gravel soils and continuous water phase at high degrees of saturation, the authors believe the hysteresis presented at low suctions is mainly caused by the "ink bottle" effect. Consider a large pore (e.g., 288 μm, which corresponds to a desorption/absorption suction of 0.5 kPa), which is interconnected by a smaller passage (e.g., 144 μm, which corresponds to a desorption / absorption suction of 1 kPa). During the drying process, the water stored in the large pore will drain as the suction increases to the desorption suction of the smaller passage (e.g., 1 kPa). During the wetting process, water will fill the large pore as the suction decreases to the absorption suction of the large pore (e.g., 0.5 kPa), but not the absorption suction of the smaller passage (e.g., 1 kPa).

For SC with gravel, sandy ML, and CL with sand, the other three reasons may also take important roles in the hysteresis in addition to the "ink bottle" effect.

5.3 SHCFs for Widely-Graded Soils

The unsaturated hydraulic properties of five colluvial soils are measured by the newly developed wetting front advancing column tests (referring to Chap. 4, Fig. 4.8). A series of soil hydraulic conductivity tests have been conducted to measure the saturated coefficient of hydraulic conductivity of these five colluvial soils. Falling head tests on the Flexpanel [as shown in Fig. 5.14a, GENEQ Inc.], which offers fast and flexibility pressure control, were conducted to measure the saturated hydraulic conductivity of CL with sand, sandy ML, and SC with gravel. Constant head tests were conducted on a sand/gravel permeameter [as shown in Fig. 5.14b, QCQA Media, LLC], which is designed to determine the coefficient of hydraulic conductivity for granular soils, to measure the saturated hydraulic conductivity of SM with gravel and GW-GM with sand.

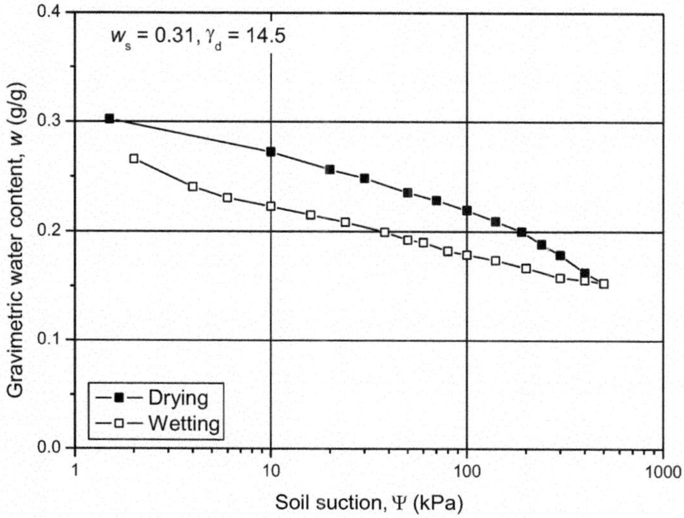

(a)

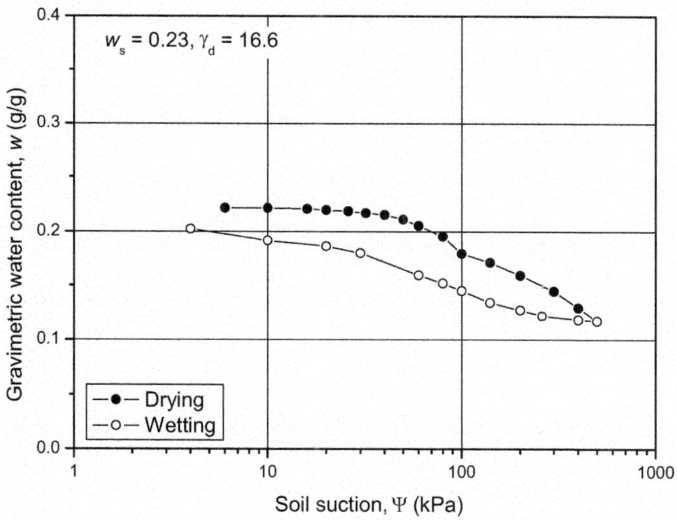

(b)

Fig. 5.10 Hydraulic hysteresis in the SWCCs for sandy ML (ws is saturated water content, γ_d is the soil unit weight, kN/m^3): **a** Loose soil sample, **b** Dense soil sample

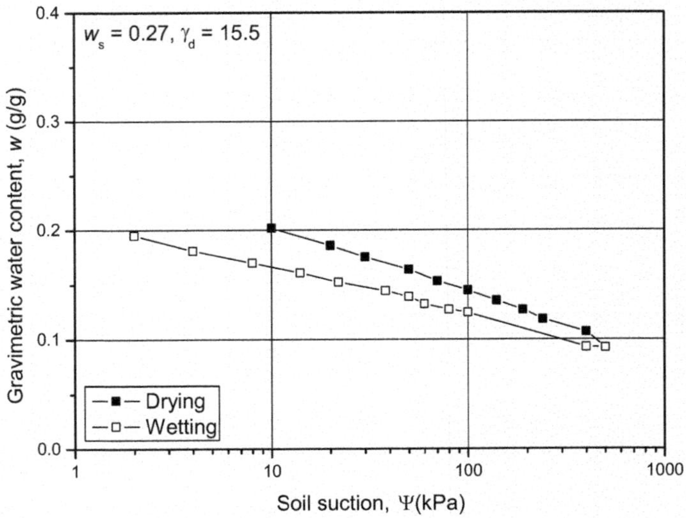

Fig. 5.11 Hydraulic hysteresis in the SWCC for SC with gravel (w_s is saturated water content, γ_d is the soil unit weight, kN/m^3)

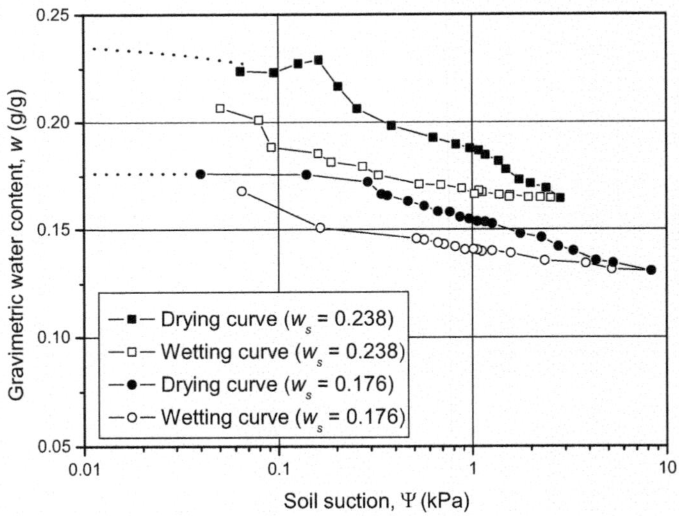

Fig. 5.12 Hysteretic SWCCs for SM with gravel (w_s is saturated water content)

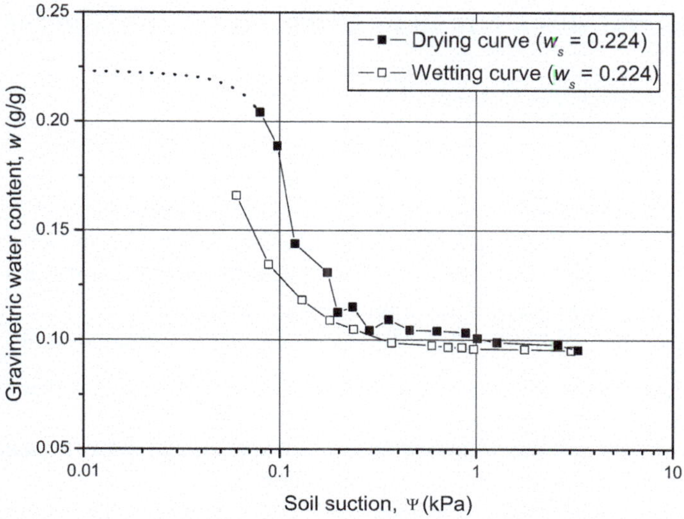

Fig. 5.13 Hysteretic SWCC for GW-GM with sand (w_s is saturated water content)

(a) Flexpanel (b) Sand/gravel permeameter

Fig. 5.14 Devices used for saturated hydraulic conductivity test: **a** falling head test device; **b** constant head test device

5.3.1 Saturated Soil Hydraulic Conductivity Measurement

5.3.1.1 Experimental Technique

Falling head test

The falling head tests were performed on specimens with a diameter of 70 mm and a height of 60 mm. To prepare uniform specimens, the specimens were statically compacted to the objective density in three layers at the optimum water content. After the compaction, the specimens mounted in triaxial cells.

To saturate the soil specimen, the following procedure was conducted:

(1) Carbon dioxide was introduced through the soil specimen for one hour so as to replace the air inside.
(2) The triaxial cell was connected to the Flexpanel, and the cell chamber was filled with de-aired water at a pressure of 5 kPa.
(3) Two water supply lines were connected to the top and bottom of soil specimen.
(4) The water pressure in the top water supply line was adjusted as 5 kPa higher than that connected to the cell for pressing the membrane against the soil specimen closely. The water pressure in the bottom water supply line was 5 kPa higher than that in the top water supply line so as to ensure water flew through from the base to the top. This process was remained for 24 h.
(5) Back pressure saturation technique, in which the cell pressure and pore water pressures were increased synchronously in steps, was taken until the pore water pressure reached 300 kPa. The pressure increment was of 50 kPa and continued for half an hour. After the back pressure reached 300 kPa, it was maintained for 24 h.
(6) After the soil specimen was saturated, the B values were measured, which were above 0.99.

After the soil specimens were saturated, the saturated coefficients of hydraulic conductivity were measured for 3 times for each specimen. Then, the soil specimens were released and used to measure the final water contents by the oven-dry method. During the saturation process, the soil dry density will change on time and not equal to the dry density at compaction. Therefore, the soil dry density is calculated from the final water content with the assumption that the sample is fully saturated.

Constant head test

The constant head tests were performed on specimens with 63 mm in diameter and 200 mm in height. The soil specimens were first dynamic compacted into the device by layers at the optimum water content. Then, water was filled into the device and remained a head of 2 kPa above soil top for 24 h. The saturated coefficients of hydraulic conductivity were measured for 6 times for each sample.

5.3.1.2 Saturated Hydraulic Conductivity Measurement Results

The saturated hydraulic conductivity of CL with sand was measured using the Flex-panel, and the results are shown in Fig. 5.15. Generally speaking, the saturated coefficients of hydraulic conductivity of CL with sand increase with the void ratios, but show hydraulic conductivity variations of about 0.3 log scales at a specific void ratio. As the void ratio increases from 0.98 to 1.10, the coefficient of hydraulic conductivity increases from 3.73×10^{-8} to 2.97×10^{-7} m/s. The hydraulic conductivity depends on the soil flow network, i.e., the soil porosity structure. As presented in Chap. 3, the porosity of CL with sand at a larger void ratio shows a larger average pore radius and a larger volume of pores (referring to Chap. 3, Figs. 3.8 and 3.9). Therefore, the saturated hydraulic conductivities of CL with sand at larger void ratios will also be larger. But even for soil samples of the same void ratio, they may have different porosity structures, especially different flow channels even on the condition of similar pore-size distribution. Different flow networks will induce variations of the hydraulic conductivity.

The saturated hydraulic conductivity of sandy ML was measured by the Flexpanel and is shown in Fig. 5.16. The coefficient of hydraulic conductivity for sandy ML also increases with void ratio. The variations of the coefficient of hydraulic conductivity for sandy ML at same void ratio are about 0.8 log scales. As the void ratio increases from 0.76 to 0.86, the saturated hydraulic conductivity increases from 2.23×10^{-8} to 3.65×10^{-6} m/s.

The saturated soil hydraulic conductivity of SC with gravel is shown in Fig. 5.17. As the void ratio increases from 0.54 to 0.71, the saturated hydraulic conductivity

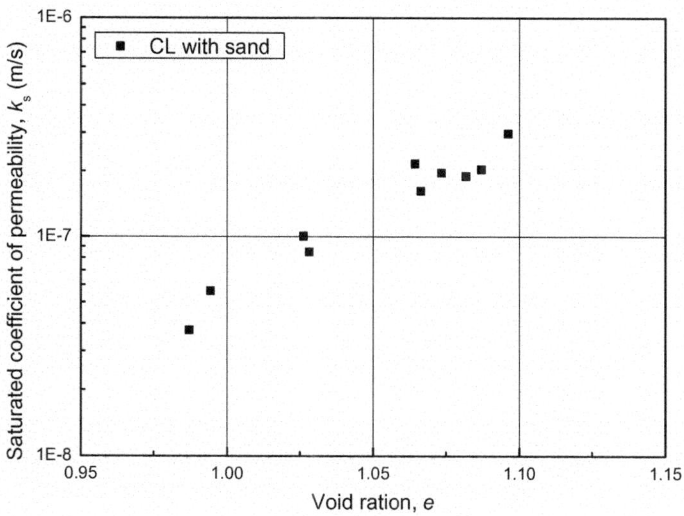

Fig. 5.15 Measured saturated hydraulic conductivity for CL with sand using falling head test

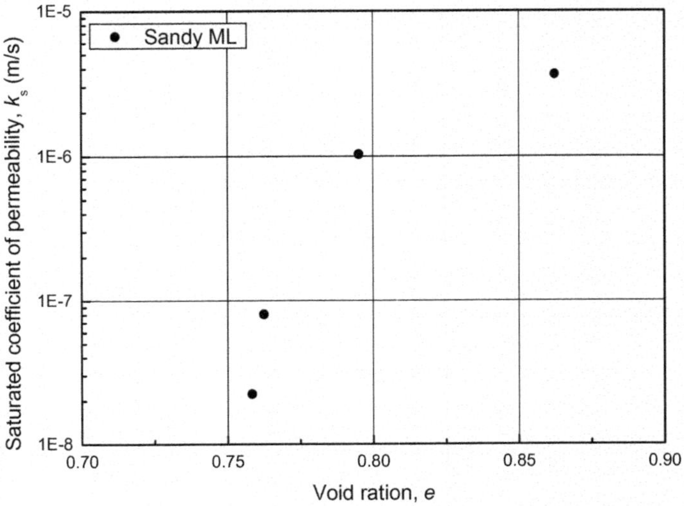

Fig. 5.16 Measured saturated hydraulic conductivity for sandy ML using falling head test

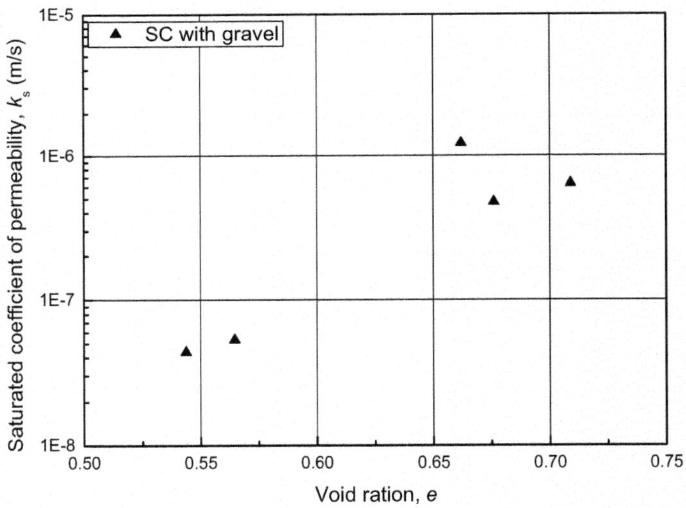

Fig. 5.17 Measured saturated soil hydraulic conductivity for SC with gravel using falling head test

increases from 4.23×10^{-8} to 1.26×10^{-6} m/s. The variations of the coefficient of soil hydraulic conductivity at the same void ratio are about 0.5 log scales.

The saturated soil hydraulic conductivity of SM with gravel and GW-GM with sand was measured using the sand/gravel permeameter, and the results are shown in Figs. 5.18 and 5.19. The results show spatial and temporal variations, which are about 0.8 log scales for a same soil sample. Possible reasons are changes in the porosity

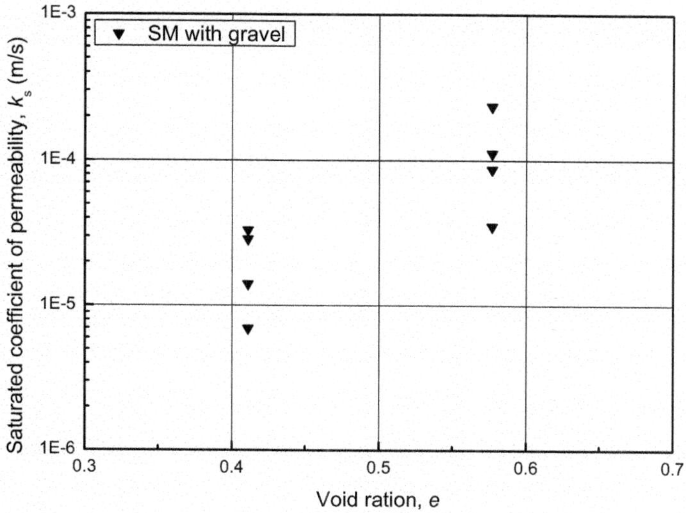

Fig. 5.18 Measured saturated soil hydraulic conductivity hydraulic conductivity for SM with gravel using constant head test

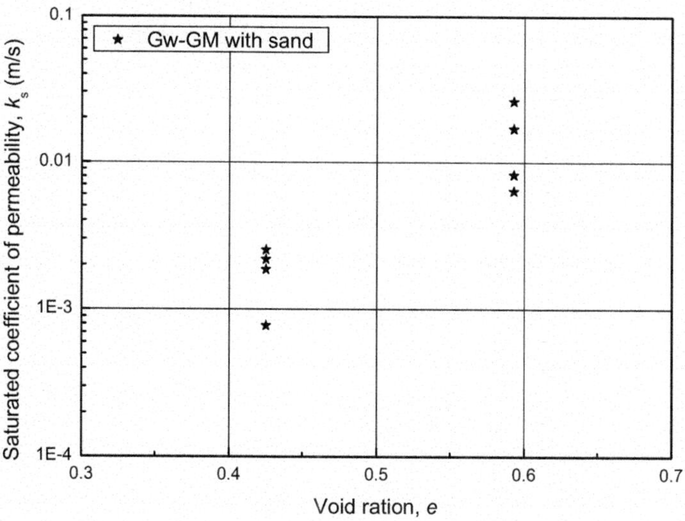

Fig. 5.19 Measured saturated soil hydraulic conductivity for GW-GM with sand using constant head test

structures during the flow process and the spatial heterogeneity of soil samples. Also, the soil samples at higher void ratios have higher hydraulic conductivity.

5.3.2 *Infiltration Process in Column Tests of Colluvial Soils*

5.3.2.1 Wetting Process

Capillary-rise tests were conducted using the GW-GM with sand, SC with gravel, SM with gravel, sandy ML, and CL with sand. The capillary-rise processes in the column cell for the five soils are presented in Fig. 5.20a. The elapsed time was taken as zero when the water supply was first connected to the soil column. The wetting front in the GW-GM with sand rose to 113 mm height above the water table and remained constant. This behavior indicated an extremely slow wetting front advancing velocity at the 113 mm-high section. In other words, the unsaturated soil hydraulic conductivity above that section was extremely low. The GW-GM with sand contained 35% sand, 58% gravel, and 7% fines. The GW-GM with sand contained large particles that formed a structure with many macro-pores. The macro-pores in the soil were filled with air and few capillary flow channels were available above the 113 mm-high section, where the corresponding soil suction was 1.1 kPa at steady-state conditions.

The wetting front advancing process in the other four soils (i.e., SC with gravel, SM with gravel, Sandy ML, and CL with sand) showed similar behavior when capillary water rose to the top of the soil column (i.e., 700–1000 mm) in 2–4 weeks. The fines content in these four soils was larger than 24%; numerous micropores were formed and filled with water under low soil suctions (i.e., smaller than 100 kPa in the capillary-rise process). There was an abundance of continuous flow channels for the capillary-rise process.

The wetting front advancing velocity [as shown in Fig. 5.20b] generally increased as the fines content increased. In particular, the advancing velocity in the GW-GM with sand was much lower than (i.e., 1/5–1/10 at same time series) in the four finer soils. The wetting front advancing velocity shows a log-linear relationship with time,

$$\lg v = a + \lambda \lg t \tag{5.3}$$

where λ is $-$ 0.63; and a is 0.18 for the four finer soils [when using units of mm/s, with reference to Fig. 5.20b]. Based on Eq. 5.4, the rising height of wetting front, h, is,

$$h = \frac{10^{a} t^{\lambda+1}}{\lambda + 1} \tag{5.4}$$

According to Eq. 5.4, if the column length is doubled, the time needed for the wetting front to reach the column top will be $2^{1/(\lambda+1)}$ times longer (i.e., 6.5 times as long when $\lambda = -0.63$). Consequently, the column length will significantly influence the time required for the experiment. Figure 5.20a shows that the time for an experiment in a soil column 400 mm high is about 3–6 days. The experiment time

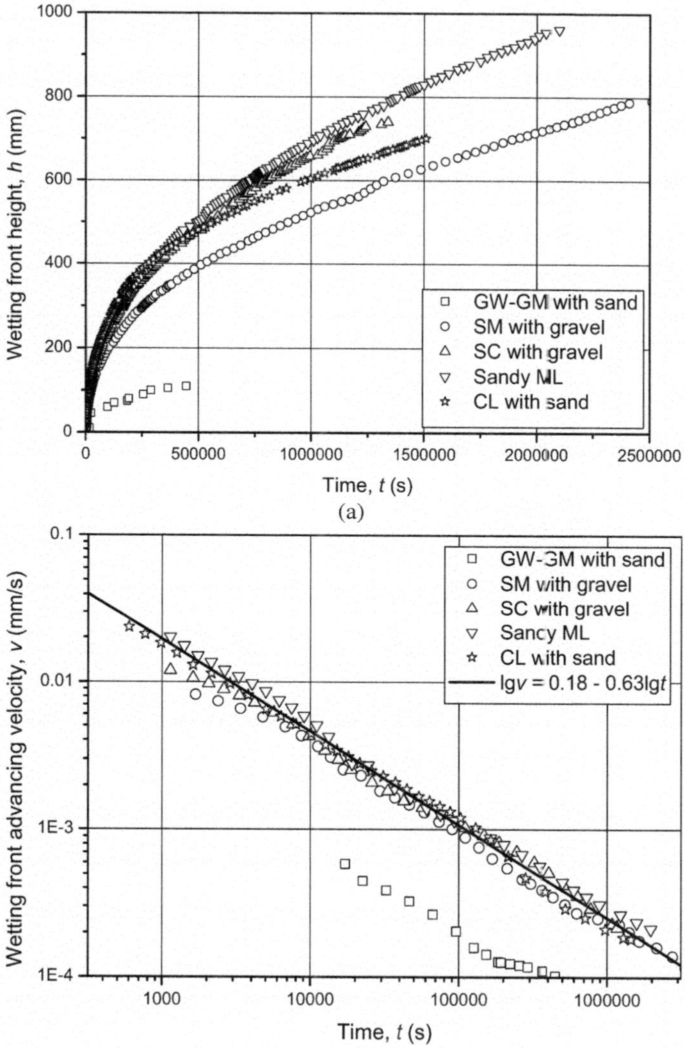

(a)

(b)

Fig. 5.20 Capillary-rise processes in five soils: **a** Wetting front height recorded over time, **b** Wetting front advancing velocity

for a soil column 600 mm high is about 7–14 days, while the experiment time for a soil column of 800 mm high will increase to 15–30 days.

5.3.2.2 Water Content Changes During Wetting Process

The water content changes at heights of 100, 300, 600, and 800 mm were monitored during the experiments. The water content changes are shown in Figs. 5.21, 5.22,

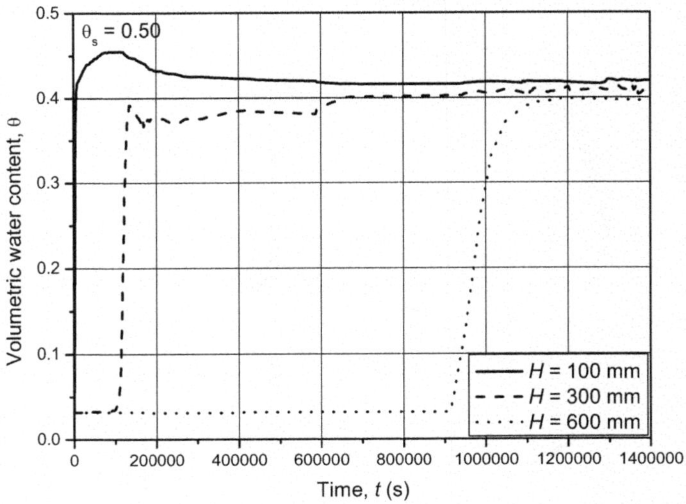

Fig. 5.21 Monitored water content profiles at different sections during the capillary-rise process in CL with sand

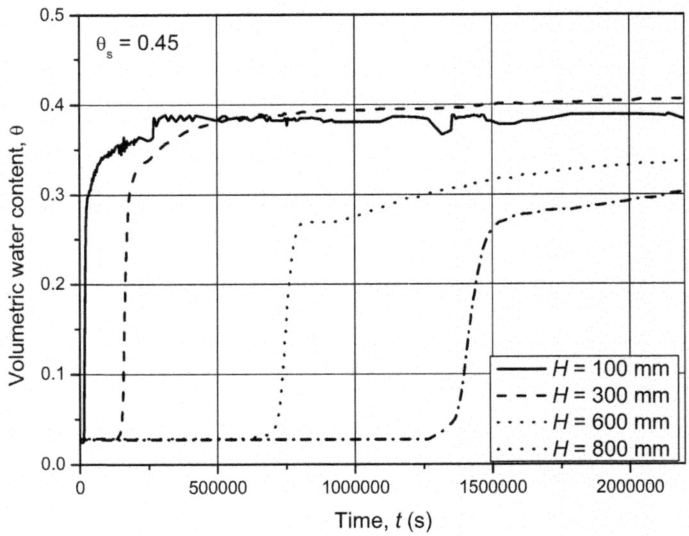

Fig. 5.22 Monitored water content profiles at different sections during the capillary-rise process in sandy ML

5.23, 5.24, and 5.25. The water contents in the soil columns show three distinct stages: namely, the initial stage, the wetting stage, and the wetted stage.

In the initial stage, before the wetting front reached a particular section, the water content in the section remained at the initial water content corresponding to the placement compacted conditions. Since the soil was initially oven-dried and had a

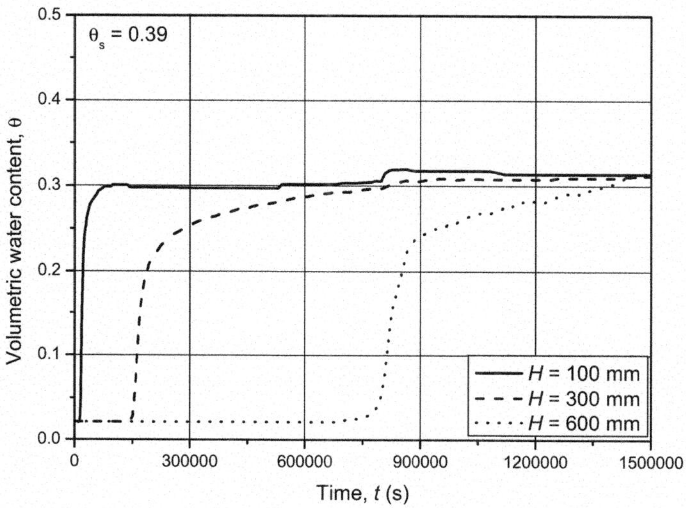

Fig. 5.23 Monitored water content profiles during the capillary-rise process in SC with gravel

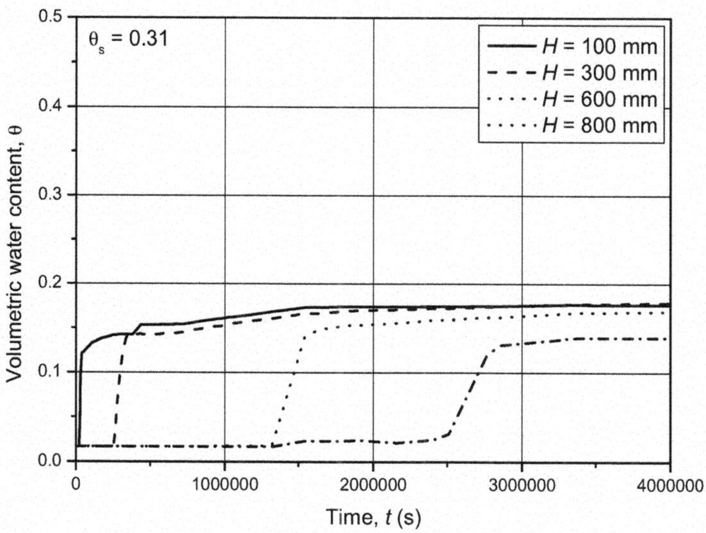

Fig. 5.24 Monitored water content profiles during the capillary-rise process in SM with gravel

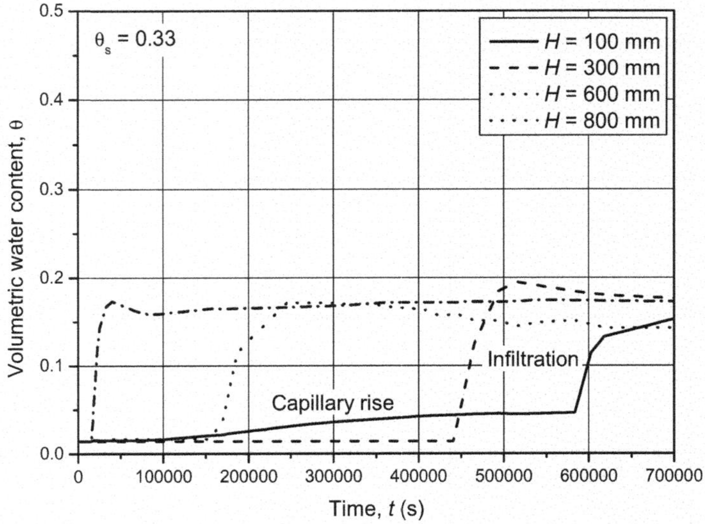

Fig. 5.25 Monitored water content profiles during the capillary-rise and infiltration processes in GW-GM with sand

very low water content (and an extremely low coefficient of soil hydraulic conductivity), the hydraulic gradient in the soil was approximately 1.0, (i.e., the gradient was controlled by elevation change).

During the wetting stage, after the wetting front reached a particular section, the suctions in the wetted soil decreased rapidly and large hydraulic gradients were present. Therefore, water was able to enter into the initially dried soil quickly and the water content increased rapidly.

During the wetted stage, after the soil at a particular section was wetted, the suction at that section would change to a similar value at a lower adjacent section. The hydraulic gradient would decrease to a low value. Therefore, the water content in the wetted soil could only increase slowly and approach the steady-state water content. The water contents in the wetted zone corresponded to the degrees of saturation of about 80% for CL with sand and Sandy ML, about 75% for SM with gravel, and about 40% for SC with gravel.

Because of the extremely slow advance of the wetting front in the GW-GM with sand, an infiltration test was also conducted in the soil in addition to a capillary-rise test. Figure 5.20 shows also three stages that are similar to those observed during the capillary-rise tests of four finer soils.

Tensiometers were inserted into the soil column after the wetting front passed through the respective sections. The monitored suctions were relatively stable and decreased slowly with time. One of the monitored suction profiles is shown in Fig. 5.26, where suctions decreased with time and approached a steady state. At the 100 mm-high section, suction decreased to a constant value of 1.2 kPa, which was close to the steady-state value of 1.0 kPa at that section.

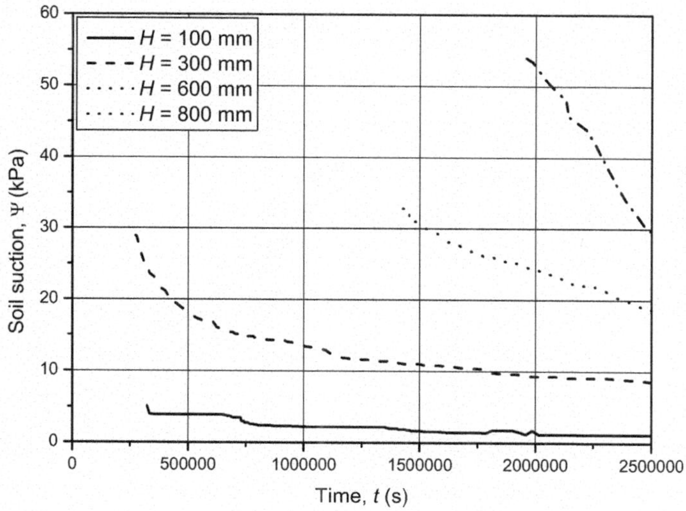

Fig. 5.26 Measured suction profiles during the capillary-rise process in sandy ML

5.3.3 *Calculated Unsaturated SHCFs*

5.3.3.1 SHCFs for Five Types of Soil

The water contents and soil suctions during the infiltration versus capillary-rise process are used to calculate the SHCFs using Eq. 4.18. If measured suction values are inserted directly into Eq. 4.18, the SHCFs are calculated as shown in Figs. 5.27, 5.28, 5.29, 5.30, and 5.31 (i.e., using open symbols). Soil suctions larger than 80 kPa could not be monitored due to the limited range of tensiometers. Therefore, the directly calculated SHCF values are limited in a narrow suction range. The soil–water characteristic curves can also be used in Eq. 4.18 to extend the suction range. If the suctions are estimated from the SWCCs (as shown in Fig. 5.32) based on the monitored water contents, the SHCF can be calculated over a much wider suction range as shown in Figs. 5.27, 5.28, 5.29, 5.30, and 5.31 (i.e., using solid symbols).

The computed SHCFs for CL with sand based on the monitored and inferred suctions are shown in Fig. 5.27. There is an overlap between the SHCFs computed from the suction measurements and that from the inferred suctions on the SWCC. The SHCFs from the measurements at the 100, 300, and 600 mm sections can be seen to be consistent and found to form a log-linear band. The upper value is about 5–10 times that of the lower value at the same suction. The Gardner model (Gardner 1958) was used to fit the calculated data. A saturated coefficient of hydraulic conductivity of 3×10^{-7} m/s was used to fit Gardner's model. The range of the laboratory measured saturated coefficients of hydraulic conductivity for the CL with sand ranged from 3×10^{-8} to 3×10^{-7} m/s. The slope in the log-linear part is 2.18. As the soil suction

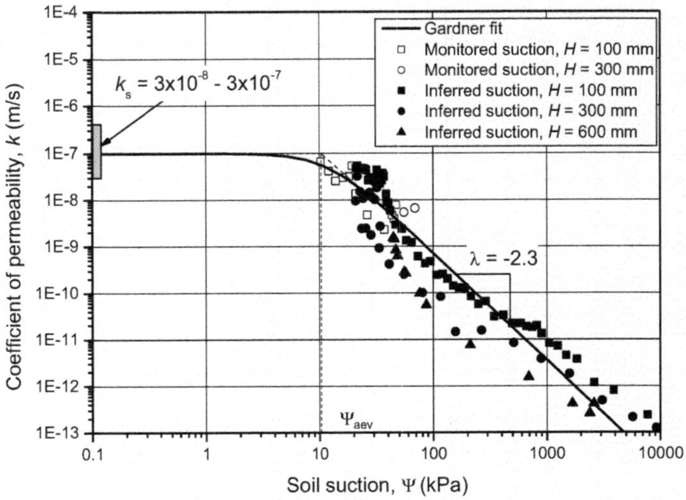

Fig. 5.27 Unsaturated SHCFs for CL with sand

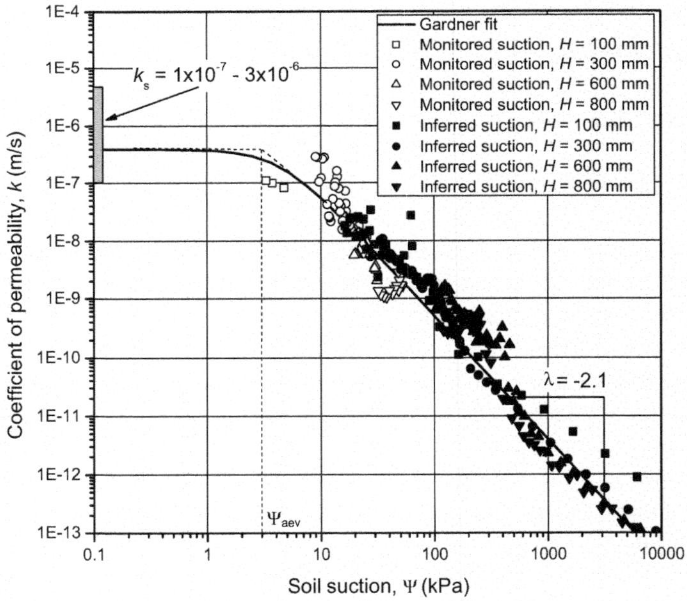

Fig. 5.28 Unsaturated SHCFs for sandy ML

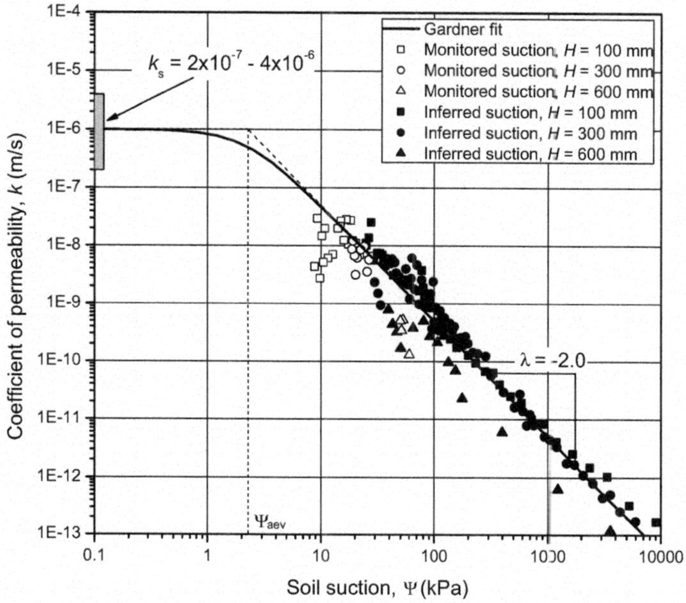

Fig. 5.29 Unsaturated SHCFs for SC with gravel

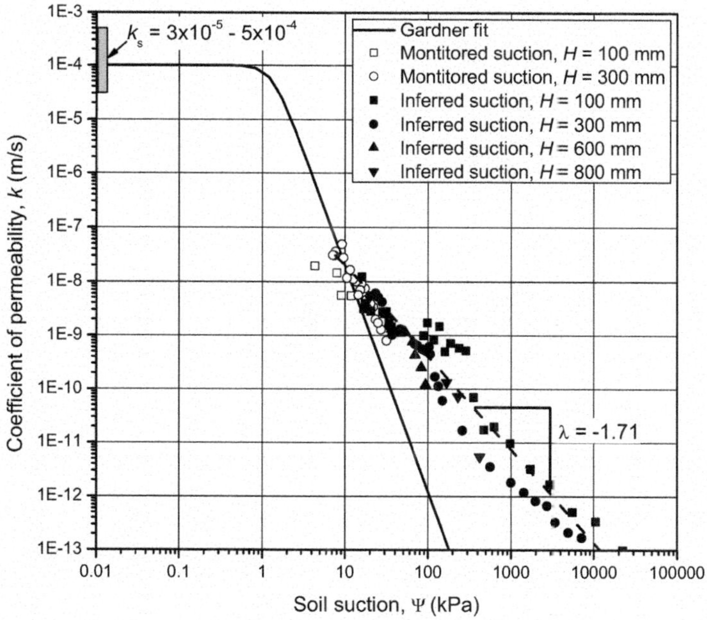

Fig. 5.30 Unsaturated SHCFs for SM with gravel

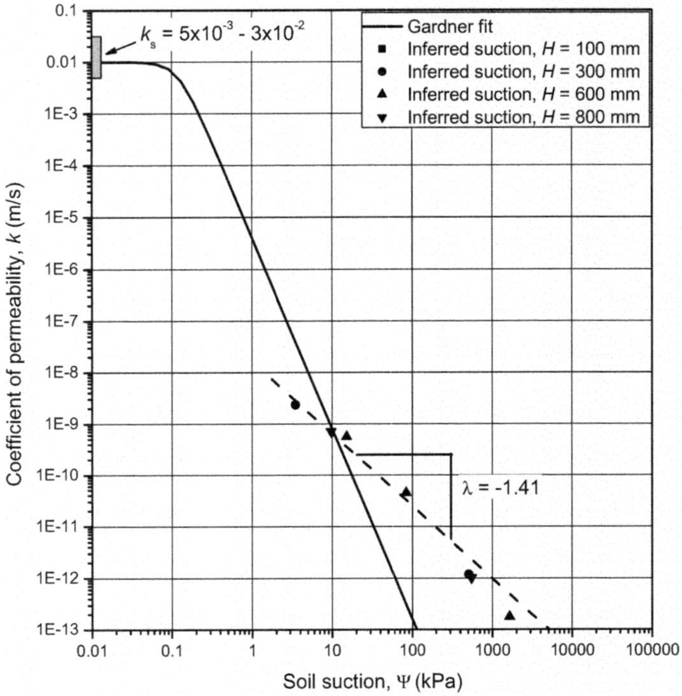

Fig. 5.31 Unsaturated SHCFs for GW-GM with sand

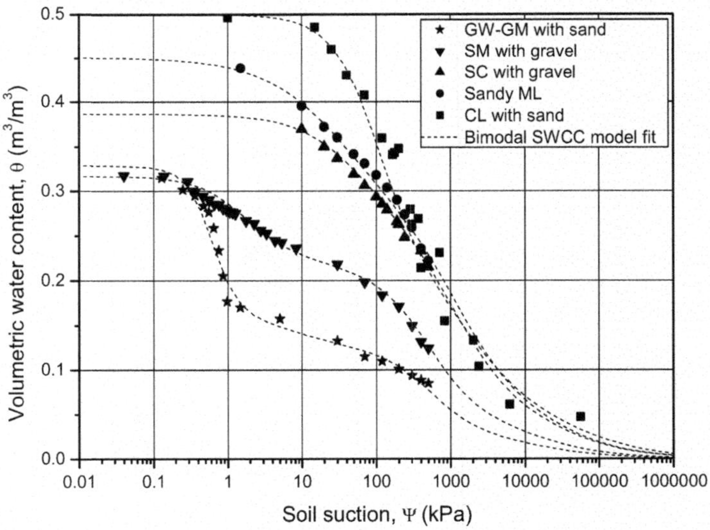

Fig. 5.32 SWCCs for five soils used in hydraulic conductivity calculation

increases from 10 to 10,000 kPa, the hydraulic conductivity decreases from 9×10^{-8} to 1×10^{-13} m/s.

The SHCFs for sandy ML and SC with gravel are shown in Figs. 5.28 and 5.35, respectively. The SHCFs calculated from the monitored suctions and the suctions inferred from the SWCC agree well. In addition, the SHCFs interpreted based on the measurements at the four monitoring sections at different heights are similar. The calculated SHCFs were also found to locate along a log-linear hydraulic conductivity band, where the upper value was about 3–8 times of the lower value at the same suction. The Gardner model was also used to best-fit the calculated data with saturated coefficients of hydraulic conductivity of 2×10^{-6} m/s for sandy ML and 6×10^{-7} m/s for SC with gravel.

The SHCFs for the two coarse soils (i.e., SM with gravel and GW-GM with sand) are shown in Figs. 5.30 and 5.31, respectively. Note that only the suctions inferred from the SWCCs are used to calculate the SHCF for GW-GM with sand because suctions were not measured during the capillary-rise test. Similar to the three finer soils, the SHCFs calculated from the suction measurements at different monitoring sections agree quite well and the curves are located along a log-linear hydraulic conductivity band. The Gardner model, however, did not fit the SHCF well when using the measured saturated coefficients of hydraulic conductivity (i.e., 3×10^{-5}–5×10^{-4} m/s for SM with gravel and 5×10^{-3}–3×10^{-2} m/s for GW-GM with sand).

In summary, the measured unsaturated SHCFs for the five soils show the following features:

(1) The wetting front advancing method can be used to measure the SHCF for different kinds of soils over a wide suction range (i.e., from several kPa to several 1000 kPa) and a wide hydraulic conductivity range.
(2) The hydraulic conductivity values interpreted based on measurements at different sections follow a consistent trend. Therefore, the interpreted SHCF can be considered to be independent of the location of section used to take measurements. It appears that only one or two measuring sections are required when using this method. In other words, it is not necessary to have a series of measurement sections to provide both a water content profile and a suction profile as is the case for the instantaneous profile method.
(3) The calculated hydraulic conductivity function based on measured suctions agrees well with the calculated SHCF based on suctions inferred from the SWCC.
(4) The SHCFs show a log-linear relationship in the suction range between 10 and 10,000 kPa. The slope of the log-linear portion of the SHCF appears to increase with an increase in the gravimetric proportion of fines content in the soil. The fines contents of the five soils, classified as CL with sand, sandy ML, SC with gravel, SM with gravel, and GW-GM with sand, are 7, 24, 42, 60, and 80%, respectively, and the slopes of the log-linear portion of the SHCFs are 2.18, 2.06, 1.84, 1.71, and 1.41, respectively.

(5) The SHCFs obtained are located in a relatively narrow log-linear hydraulic conductivity band. The upper hydraulic conductivity values are about 5–10 times greater than the lower hydraulic conductivity values at the same suction.
(6) A column length of 400–600 mm is recommended taking into consideration factors such as sample disturbance, the wetting front advancing process, testing time, and data sufficiency. Several series of SHCFs tests can be performed in one or two weeks with such a soil column.

5.3.3.2 Comparison with Instantaneous Method

The instantaneous profile method can also be used to calculate the SHCFs. The calculated SHCFs for CL with sand and sandy ML are compared in Fig. 5.33. The hydraulic conductivity data calculated using the instantaneous profile method is similar to those obtained when using the wetting front advancing method. However, the results from the instantaneous profile method are more scattered than those obtained using the wetting front advancing method. The interval of the water content measurement sensors in the test device was about 200 mm. This separation is too high to obtain an accurate water content profile when calculating the total flow volume and hydraulic gradient by the instantaneous profile method. Therefore, the hydraulic conductivity data calculated using the instantaneous profile method shows more scatter. However, the trend lines of SHCF calculated using the instantaneous profile method and the wetting front advancing method are similar.

5.3.3.3 Comparison of Infiltration Test and Capillary-Rise Test

The water content change during infiltration into the SC with gravel is shown in Fig. 5.34. The same three stages (i.e., stable stage, wetting stage, and wetted stage) can also be observed during the infiltration process. The volumetric water content in the wetted soil reaches a higher level (i.e., about 0.36) in the infiltration process comparing to that in the capillary-rise process (i.e., about 0.30). The SHCFs for SC with gravel obtained in the capillary-rise process and the infiltration process are compared in Fig. 5.20. In general, the measured SHCF in the capillary test and the infiltration test are similar; therefore, both the capillary test and the infiltration test are suitable for measuring the SHCF. However, for gravel soils (e.g., GW-GM with sand), the infiltration test is preferable because of the extremely low wetting front advancing velocity in a capillary-rise process in gravel soils.

5.4 Summary

In this chapter, a new low suction SWCC device and three commercial available SWCC devices were used to measure the SWCCs for five colluvial soils. The SWCCs

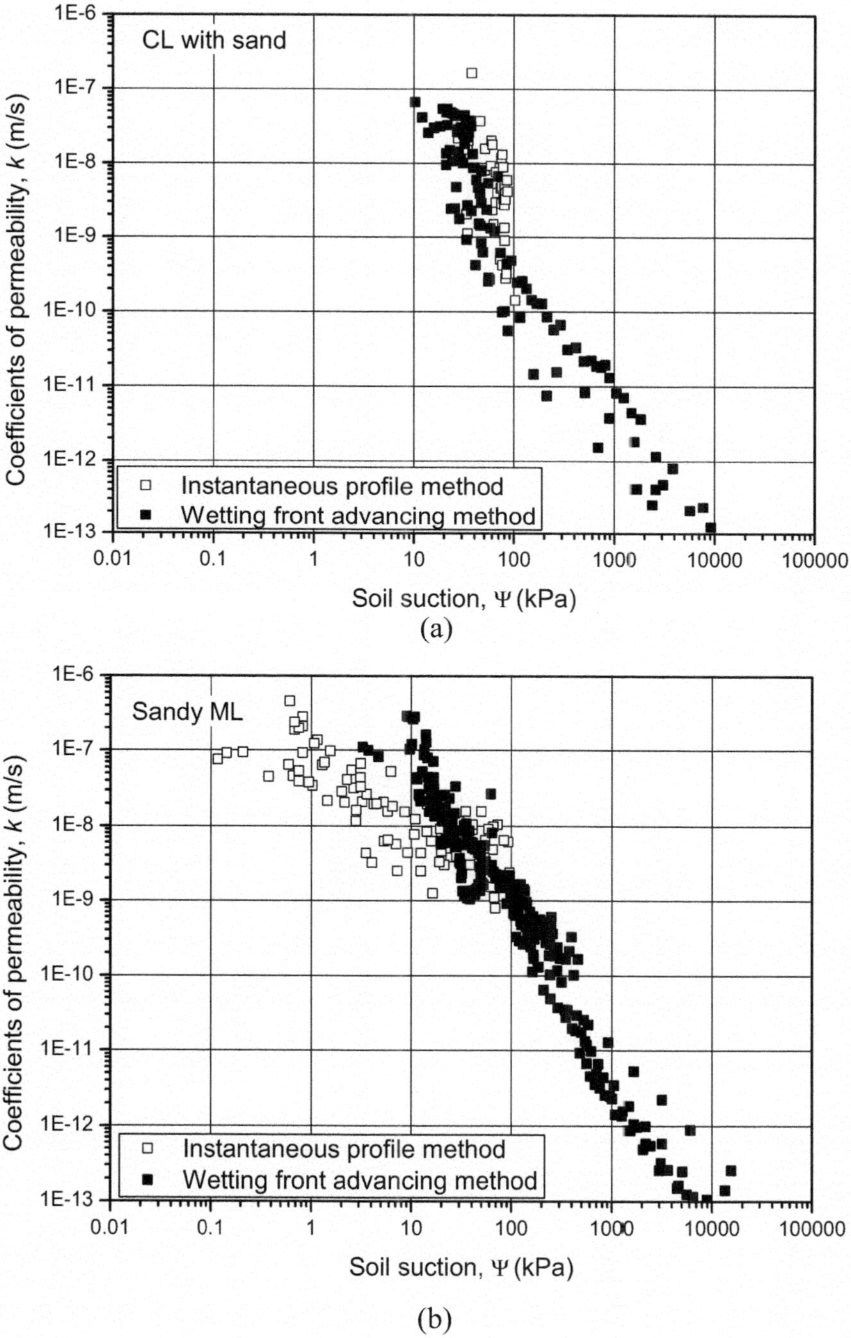

Fig. 5.33 Comparison of SHCFs calculated using the instantaneous profile method and the wetting front advancing method: **a** CL with sand, **b** Sandy ML

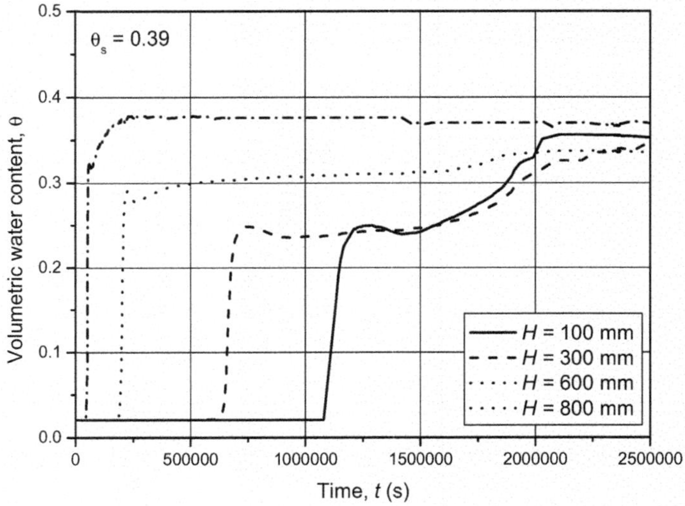

Fig. 5.34 Monitored water content profiles in the infiltration process in SC with gravel

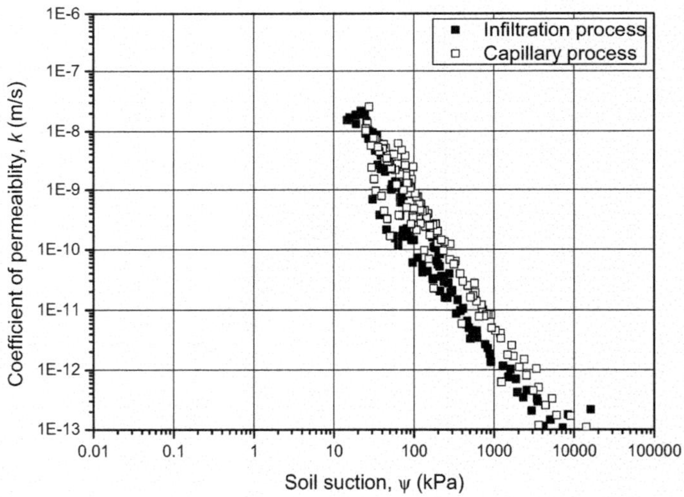

Fig. 5.35 Comparison of SHCFs for SC with gravel from a capillary test and from an infiltration test

for GW-GM with sand show a strong bimodal feature, and about 50% of water drains sharply at suctions smaller than 1 kPa. The SWCCs for SM with gravel show a weak bimodal feature, and 20% of water drains at suctions smaller than 1 kPa. These curves show that the measurement of SWCCs at very low suctions for gravel soils is necessary and important. While the water content of the gravel soil decreases sharply in the first drainage period (i.e., smaller than 1 kPa), it decreases slowly in the second drainage period followed (i.e., from 1 to 500 kPa or values above 500 kPa). No obvious bimodal feature is found in the SWCCs for SC with gravel, sandy ML, and CL with sand.

The fines-controlled Widely-graded soils, such as CL with sand and sandy ML, have unimodal SWCCs and show limit water content changes at low suctions (smaller than 10 kPa). The coarse-controlled Widely-graded soils, such as SM with gravel and GW-GM with sand, have typical bimodal SWCCs and show significant water content changes at low suctions (smaller than 10 kPa). The SWCC for the loose SC with gravel sample shows features similar to that of the coarse-controlled Widely-graded soils. The SWCC for the dense SC with gravel sample shows features similar to that of the fines-controlled Widely-graded soils. When the SC with gravel sample is compacted from a loose state to a dense state, the soil structure may change from a fine-controlled structure to a coarse-controlled structure.

The SWCCs for the same type of soil at different dry densities show that the SWCC for a denser soil sample has a larger air entry value. The gravimetric water contents of soil samples at different dry densities are quite different at low suctions, but become closer as the soil suction increases to above 10 kPa. Hydraulic hysteresis is common in measured SWCCs whenever for different soils or for different measured suction ranges. In a hysteretic SWCC, a drying curve and a wetting curve have similar processes.

The SHCFs for the five soils tested show a log-linear relationship in the suction range of 10 kPa–10,000 kPa. The slopes of the log-linear portion of SHCF are from 1.41 to 2.3 and appear to increase with the fines content in soil.

Tremendous differences are found between the measured saturated hydraulic conductivity and the measured unsaturated hydraulic conductivity at small suctions for GW-GM with sand and SM with gravel. These differences may be induced by the dual-porosity structure of GW-GM with sand and SM with gravel and related to the bimodal SWCCs. The new bimodal SHCF model proposed in Chap. 4 is used to fit experimental data.

References

Brooks RH, Corey AT (1964) Hydraulic properties of porous media. Hydrology Paper, No. 3, Colorado State University, Fort Collins, Colorado

GENEQ inc. Flexpanel I 3-burette testing panel with master controller, 220V, 50/60Hz. Jarry East, Montreal, QC, Canada

Gardner WR (1958) Some steady-state solutions of the unsaturated moisture flow equation with application to evaporation from a water table. Soil science 85(4):228-232

Hillel D (1998) Environmental soil physics. Academic Press, San Diego, CA, USA

Huang S, Barbour SL, Fredlund DG (1998) Development and verification of a coefficient of permeability function for a deformable unsaturated soil. Can Geotech J 35(3):411–425

Laliberte GE, Corey AT (1996) Brooks RH. Properties of unsaturated porous media. Colorado State University. Libraries

Li X, Zhang LM, Fredlund DG (2009a) Wetting front advancing column test for measuring unsaturated hydraulic conductivity. Can Geotech J 46(12):1431–1445

Li X, Zhang LM, Li JH (2009b) Development of a modified axis translation technique for measuring SWCCs for gravel soils at very low suctions. Geotech Test J 32(6):1–11

Ng CWW, Pang YW (2000) Experimental investigations of the soil-water characteristics of a volcanic soil. Can Geotech J 37(6):1252–1264

Yang H, Rahardjo H, Leong EC et al (2004) Factors affecting drying and wetting soil-water characteristic curves of sandy soils. Can Geotech J 41(5):908–920

Chapter 6
A Unified Framework for Unsaturated Hydraulic Behavior of Widely-Graded Soils

6.1 Introduction

The microporosity structure of a soil provides important information of hydraulic conductivity and soil–water characteristics of the soil, as described in Chaps. 3 and 5. The microporosity structures of fine soils may change significantly during wetting and drying (Simms & Yanful 2001; Cuisinier & Laloui 2004; Boivin et al. 2004; Braudeau et al. 2004). Such variations of soil microporosity structures are inherently related to the drying and swell/shrinkage behavior of the soil (Ferber et al. 2009; Aung et al. 2001) and are the main difficulty in establishing quantitative relations between soil fabric and soil shrinkage/swell behavior.

With technological developments, the measurement of soil fabric becomes more and more mature. Scanning electron microscopy (Markgraf and Horn 2006) and optical microscope, e.g., X-ray scan (Masad et al. 2002), are used to study the soil fabric qualitatively. The nitrogen adsorption (Prost et al. 1998) and mercury intrusion porosimetry (MIP) (Simms and Yanful 2001; Cuisinier and Laloui 2004) are widely used to characterize the soil fabric quantitatively. In recent years, numerical algorithms were also developed to analyze the images from scanning electron microscopy and optical microscope tests and identify the soil fabric quantitatively (Masad et al. 2002; Willson et al. 2012). While the measurement of soil fabric is no longer an obstacle, the interpretation of test data remains to be a problem. It is necessary to quantitatively unify soil fabric and soil macroscopic properties, e.g., water content, void ratio, and hydraulic conductivity.

During the drying process of a soil under constant total stresses, the drainage of water and the shrinkage of soil is largely attributed to the increase of soil suction and changes in the soil structure (Boivin et al. 2006; Thom et al. 2007). In this process, soil suction not only controls the change of contractile skin but also affects the change of soil fabric. Soil suction is a key to understanding the soil shrinkage behavior as well as the water retention ability. The soil shrinkage behavior is traditionally characterized as a void ratio-water content curve (e-w curve) (Tripathy et al. 2002; Peng and Horn 2005, 2013; Cornelis et al. 2006; Chertkov 2007; 2008; 2012a; Stewart

X. Li et al., *Fundamental Behavior of Unsaturated Widely-Graded Soil*,
https://doi.org/10.1007/978-981-19-3402-5_6

et al. 2012). As suction is the essential driving force, soil shrinkage behavior can also be interpreted and modeled in the void ratio-suction domain (Baumgartl and Koeck 2004; Stange and Horn 2005; Salager et al. 2010, 2013). Baumgartl and Koeck (2004) distinguished the preshrinkage line and the virgin-shrinkage line on an e-lnψ curve, which have a trend similar to the soil consolidation process as described by an e-lnp curve. The variation of soil fabric can affect the soil shrink/swell curve (Chertkov 2012b) and the soil water characteristic curve (Chertkov 2004). The shrinkage curve of a soil with a dual-porosity structure, in which both intra-aggregate and inter-aggregate pores are dominant, shows different features from those of a soil with an intra-aggregate structure (Chertkov 2008).

In this chapter, the soil fabric is regarded as a "variable" and is characterized by pore-size distribution (PSD). In a monotonic dehydration process, soil fabric is considered as a variable depending on soil suction (Simms and Yanful 2001; Li and Zhang 2009; Stange and Horn 2005), just like the water content and soil void ratio. Then the drying process of soil can be analyzed under an integrated framework where soil fabric, water content, and void ratio are mechanically linked by soil suction. Under this framework, soil dehydration process can be re-recognized or interpreted in the following domain: (1) fabric-suction domain, (2) void ratio-suction domain, i.e., another description of soil shrinkage curve; (3) water content-suction domain, i.e., traditional soil water characteristic curve, (4) void ratio-water content domain, i.e., the traditional soil shrinkage curve, (5) void ratio—fabric domain, and (6) water content—fabric domain. Based on these analysis, the drying process of slurry samples and overconsolidated samples can be distinguished and discussed separately, including their drying mechanisms and behaviors, i.e., soil fabric variation, soil shrinkage curve (SC), and soil water characteristic curve (SWCC). Theoretical relations can be constructed among soil fabric, water content, void ratio, and soil suction. Using these relations, the PSD can be used to estimate the SWCC and the SC. Further, a case study involving clay with sand will be carried out to verify these relations.

6.2 Experimental Data

In this chapter, experimental data of a silty clay with sand, which is classified as CL containing 30% of sand, 50% of silt, and 20% of clay, will be used to interpret the relations among soil fabric, SC and SWCC. The experimental data of CL include pore-size distributions (PSD), soil water characteristic curves, and soil void ratios. Before measuring the SWCC and PSD, the soil samples were first compacted to the target dry density at the optimum water content and then saturated in a vacuumed water tank for 48 h. The axis-translation technique was used to measure the SWCC in the suction range of 0–500 kPa and the dew-point technique was used to measure the SWCC in a high suction range (> 300 kPa). Soil samples at suction states of 0, 100, and 500 kPa were first dried using freeze drying and later subject to MIP testing,

Table 6.1 Experimental data used in this study

Soil description	Sample description	Test	Fig. No	References
Clay with sand	Compacted sample	MIP	2	Li (2009)
Clay with sand	Saturation samples dehydrated to different suctions	MIP	3	Li (2009)
Sandy slit (ML)	Slightly compacted sample	SC	4, 5	Krisdani et al. (2008)
Clay with illite main contained	Compacted sample	SC	6	Peng et al. (2009)
Clay with sand	Saturation samples	SWCC	8	Li (2009)

and their PSDs were measured. The details of sample preparation and experimental procedure have been described by Li (2009).

Other experimental data used include the shrinkage curve of a sandy slit (ML) in Singapore (Krisdani et al. 2008), which had a liquid limit of 41% and a plastic limit of 27% and contained 45% of sand, 30% of silt and 25% of clay, and the shrinkage curve of a clay with illite (Peng et al. 2009). The experimental data are listed in Table 6.1.

6.3 Fabric—A State-Dependent Variable of Soil

Soil fabric varies with soil state and depends on soil stresses and the suction history. Soil fabric is often characterized by a pore-size distribution (PSD), which is a function of soil state. In this chapter, soil fabric is regarded as a "variable" and expressed using a PSD. In a monotonic dehydration process, soil fabric is considered as a monotonic function of soil suction.

6.3.1 Definition of Pore-Size Distribution

Soil is a porous medium with variable pore geometry. From the macroscopic point of view, a single parameter, such as void ratio, e, or total porosity, n, can be used to characterize the soil pore volume. In the soil pore structure, the cumulative volume of pores with radii smaller than r in 1 g of soil solid is defined as the cumulative pore volume curve, $F(r)$. The pore geometry is often characterized by a pore-size distribution (see Fig. 6.1), in which the differential pore volume function, $f(r)$, is defined as $\Delta F(r)/\Delta lg(r)$. For a soil at a particular state, the microstructure determines its void ratio, e. The void ratio is related to the pore-size distribution as

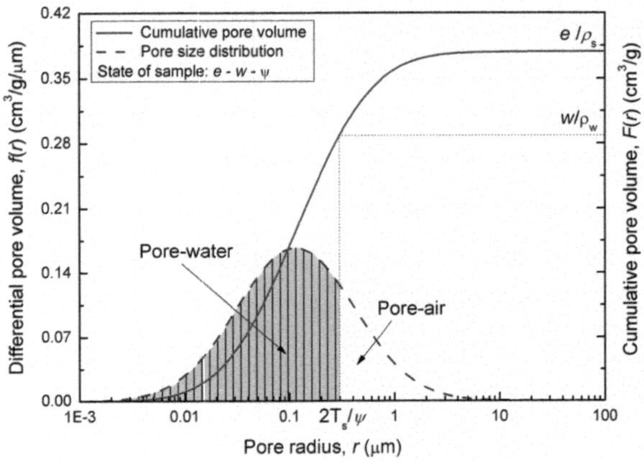

Fig. 6.1 Soil pore-size distribution of a sample at a particular state of $e - w - \psi$

$$\int_{0}^{+\infty} f(r) d\,\lg r = \frac{e}{\rho_s} \tag{6.1}$$

where ρ_s is the density of soil particle. $F(r)$ is preferably in cm^3/g and r in μm.

It well known that the pore geometry can change due to various physical or chemical processes such as soil compaction, consolidation, aging effect, acid and alkali treatment, and soil wetting/drying processes (Li and Zhang 2009). If a soil sample is subject to a monotonic process, its differential pore volume function, $f(r)$, will change continuously. A state parameter can be involved in the description of PSDs, i.e., $f(r, \Omega)$, where Ω is related to the physical or chemical process.

Obviously, if a non-monotonic process is considered, the soil PSD variation history should also be addressed in which hysteresis and other complex mechanisms may be present. In this study, only a monotonic process is considered. In the soil wetting/drying process, Ω can be soil suction ψ or void ratio e. And in the soil compaction/consolidation process, Ω can be pre-consolidation stress p_c or void ratio e. In this study, Ω is chosen as void ratio e and the variable pore geometry during a specific process is expressed as $f(r, e)$.

In this chapter, the PSDs of a clay with sand (classified as CL) are used in the analysis. The PSDs of samples compacted at different compaction efforts were measured by the MIP technique, so were the PSDs of duplicate compacted and saturated samples dehydrated to different soil suctions. The details of sample preparation and experimental procedure have been described by Li (2009).

6.3.2 *Fabric Variation upon Mechanical Actions*

The PSDs of a clay with sand (classified as CL) at three soil compaction efforts are shown in Fig. 6.2. In Fig. 6.2, the volume fractions of the macro-pores with radii

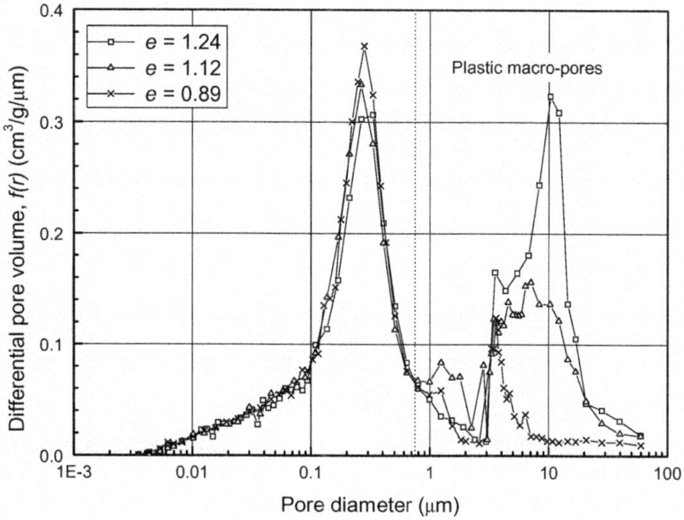

Fig. 6.2 Variations of macro-pore volumes due to soil compaction (compacted samples of CL with sand)

larger than 0.8 μm decrease significantly after compaction. Here, the lower boundary of macro-pores is arbitrarily determined according to the PSD and SEM photos (Li and Zhang 2009) and equal to 1/11 of the mean radius of the soil aggregates in CL ($0.5d_{50}$), which is quite close to one half of effective aggregate diameter d_{10} (~2 μm). Such pores can be termed as plastic macro-pores (Alonso et al. 1999), or non-swelling pores (Braudeau et al. 2004) or rigid pores (Peng et al. 2009), which can be easily destroyed under stresses (Li and Zhang 2009; Peng et al. 2009). Images from scanning electron microscopy and optical microscope tests show that the soil pores can be divided into two groups: the inter-aggregate pores (or inter-particle pores in coarse soils) and the intra-aggregate pores. Normally, inter-aggregate pores are larger than the intra-aggregate pores and match the size of the macro-pores in the PSDs measured by MIP (Fig. 6.2). So most macro-pores are considered as inter-aggregate pores (or inter-particle pores in coarse soils). The geometry of inter-aggregate pores depends on the arrangement of soil aggregates/particles and is easily affected by mechanical actions. The pores with radii smaller than 0.8 μm are stable under soil compaction (Fig. 6.2). Such pores can be termed as elastic micropores (Alonso et al. 1999), or swelling pores (Braudeau et al. 2004) or non-rigid pores (Peng et al. 2009). Most micropores are considered as intra-aggregate pores.

There are also some constant nanopores (Figs. 6.2 and 6.3), e.g., those with radii smaller than 0.05 μm, which are quite stable upon mechanical stress or water actions. The stability of nanopores may be because of the high electricity forces between clayey particles at very close distances. However, it is still hard to distinguish the plastic/non-recoverable macro-pores, elastic/recoverable micropores and constant nanopores in the present. First, the community still has not reached a common radius

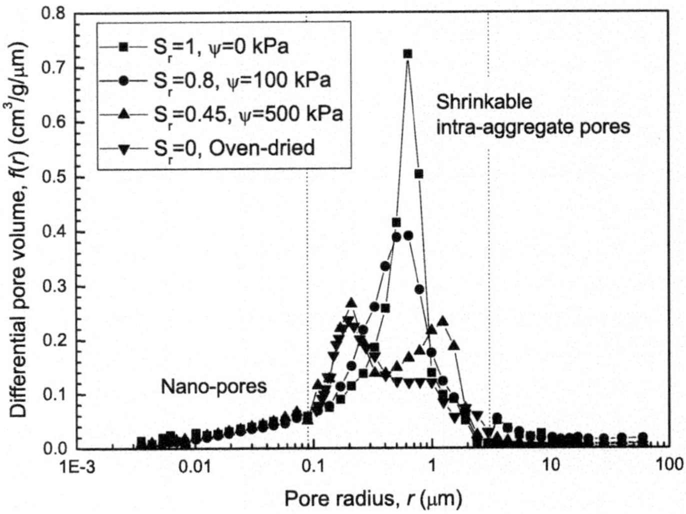

Fig. 6.3 Variable pore-size distribution upon water action (duplicate saturation samples dehydrated to different suctions)

criterion for these pores. Second, the ranges of the radii of the macro-pores and micropores, and those of the micropores and nanopores may overlap. Third, in spite of that the majority of macro-pores are unrecoverable /plastic, some macro-pores are still recoverable/elastic. Theoretically speaking, the proportion of the elastic macro-pores can be estimated by the ratio between the swelling index and compression index of the soil.

If both inter-aggregate pores and intra-aggregate pores are dominant, two series of pores with different radius ranges, i.e., a dual-porosity structure, may be present. A compacted sample of fine soils often shows a dual-porosity structure (Fig. 6.2). Li and Zhang (2009) used two log-normal distributions to characterize the PSDs of inter-aggregate pores and intra-aggregate pores separately,

$$f(r) = V_1 Log\_N(\mu_1, \sigma_1) + V_2 Log\_N(\mu_2, \sigma_2) \tag{6.2}$$

where Log_N indicates the log-normal distribution function; μ_1 and μ_2 are the mean radii of the inter-aggregate pores and intra-aggregate pores, respectively; σ_1 and σ_2 are the standard deviations of the radii of the inter-aggregate pores and intra-aggregate pores, respectively; V_1 and V_2 are the cumulative volumes of the inter-aggregate pores and intra-aggregate pores in the soil with 1 g of solid, having a unit of cm^3/g.

Empirical relations can be established between the parameters in Eq. 6.2, i.e., $\mu_1, \mu_2, \sigma_1, \sigma_2, V_1$ or V_2 and state parameter *e*. As the micropores are stable in the compaction process, the varied PSDs for CL during compaction can be written as,

$$f(r, e) = \left(\frac{e}{\rho_s} - V_2\right) \text{Log\_}N(a10^{be}, \sigma_1) + V_2 \text{Log\_}N(\mu_2, \sigma_2) \tag{6.3}$$

where e is the void ratio of soil; a, b, V_2, μ_2, σ_1, σ_2, are empirical parameters. Obviously, Eq. 6.3 is a rough conceptual model to indicate the major contribution of macro-pores on soil deformation in compression process. Equation 6.3 is still an engineering fitted curve and of limited theoretical meaning. In Eq. 6.3, coefficients a and b are different in compression and swelling processes. In a compression process, a and b empirically reflect both the contribution of plastic macro-pores and elastic macro-pores. In a swelling process, only the contribution of elastic macro-pores exists and smaller values of a and b should be used.

6.3.3 *Fabric Variation upon Water Actions*

Both the macro-pores and micropores are sensitive to water actions. The size of clayey aggregates is non-sensible upon mechanical actions but sensitive to water actions. Accordingly, the shape of intra-aggregate pores and the associated PSD do not change much upon mechanical actions but may change significantly upon water actions, i.e., wetting/drying processes. The change in the size of clayey aggregates will also result in rearrangements of the soil aggregates, thus affect the volume fraction of macro-pores. For example, upon saturation of a compacted sample in a rigid mold, the swelling of the clayey aggregates will cause the intra-aggregate pores to expand and occupy the space of the inter-aggregate pores. Finally, the compacted sample with a dual-porosity structure may evolve into a structure dominated by intra-aggregate pores as observed in Li and Zhang (2009). For the soils with an intra-aggregate pore structure, V_1 in Eq. 6.2 will be small. If V_1 is taken as 0 cm^3/g, Eq. 6.2 reduces to,

$$f(r) = \frac{e}{\rho_s}\text{Log\_}N(\mu, \sigma) \tag{6.4}$$

Large pores are found to shrink in laboratory tests (Simms and Yanful 2001; Li and Zhang 2009). As another example, the variations of the PSDs of CL during dehydration are shown in Fig. 6.3. During soil dehydration, active clayey aggregates shrink into smaller ones. Subsequently, both the total sample volumes and the mean pore radii (the intra-aggregate pores in majority) of the PSDs decrease. The varying PSD during the dehydration process for CL (Fig. 6.3) can be described as,

$$f(r, e) = \frac{e}{\rho_s}\text{Log\_}N(a10^{be}, \sigma) \tag{6.5}$$

where a, b, and σ are empirical parameters.

It should be mentioned that the volume of inter-aggregate pores can be small, but will never be 0. Inter-aggregate pores will not completely disappear. If Eqs. 6.4 or 6.5 is used, the volume of inter-aggregate pores will be underestimated. Equation 6.2 is more general for all soils.

6.4 Drying Process in Void Ratio-Water Content and Void Ratio–Suction Domain

6.4.1 Definition of Soil Shrinkage Curve

During the wetting/drying process of clayey soils, swelling/shrinkage of intra-aggregate pores will occur, resulting in volume increase/decrease of the soils. Consider a fine soil sample under a constant total stress state. The drainage of pore water is driven by the increase of soil suction, ψ, accompanied by some decrease of the soil total volume. A number of studies have been conducted to extend the traditional shrinkage curve (i.e., the e-w curve) into a water content-suction-pore volume three-dimensional space, including Baumgartl and Koeck (2004), Stange and Horn (2005), Salager et al. (2010), Salager et al. (2013) and many others. Following their tracks, both the e–ψ curve and the e–w curve are regarded as soil shrinkage curves in this chapter.

Extensive research on the shrinkage curves for soil samples undergoing drying from the saturated state has been conducted (Tripathy et al. 2002; Chertkov 2003; Groenevelt and Grant 2002; Cornelis et al. 2006, etc.). In the soil dehydration, the volume changes depend on the amount and type of clay minerals (Cornelis et al. 2006). The fines attain a unified, repeatable soil shrinkage characteristic curve (SC) after three to five swell–shrink cycles (Tripathy et al. 2002). That is to say, a small part of the unrecoverable micropores is destroyed in the initial several swell–shrink cycles, leaving the dominating recoverable micropores in the soil.

In additional to the conceptual model, the SC of a slightly compacted sandy silt sample [which is classified as ML following Krisdani et al. (2008)] is used for better interpretation of the soil shrinkage process, as well as the SC of compacted sample [clay with illite main contained, referring to Peng et al. (2009)].

6.4.2 Shrinkage Curve of Sample Drying from a Slurry State or Slightly Compacted State

As an example, the e-w curve of a slightly compacted sample and some common features are shown in Fig. 6.4, where three shrinkage stages can be distinguished (Crescimanno and Provenazano 1999; Braudeau et al. 1999): (1) normal shrinkage where the slope of e–wG_s is nearly 1.0 (Mitchell 1992; Olsen and Haugen 1998; Tripathy et al. 2002; Chertkov 2003; Groenevelt and Grant 2001; 2002; Peng and Horn 2005; Cornelis et al. 2006), (2) residual shrinkage where the e-wG_s curve changes its slope, and (3) zero shrinkage, or termed as non-shrinkage, where the void ratio maintains at a value of e_d as drainage continues. Some experimental data (Birle et al. 2008; Lan et al. 2008; Krisdani et al. 2008) shows that the residual shrinkage starts at S_r of 0.5–0.6.

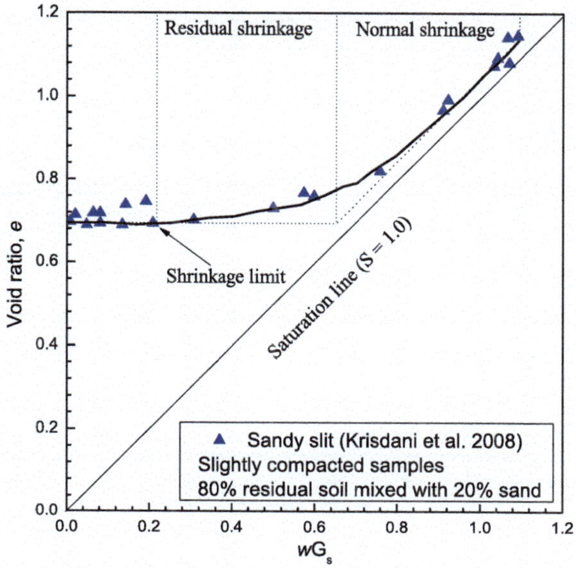

Fig. 6.4 Shrinkage curve for a slightly compacted sample of sandy silt (Krisdani et al. 2008)

The shrinkage limit (SL) is defined as the water content beyond which further loss of moisture will not result in any more volume reduction (Krisdani et al. 2008). That is to say, SL is with respect to the starting point of the zero shrinkage stage on a three-stage shrinkage curve (Krisdani et al. 2008; Chertkov 2007). Some researchers also consider the *e*-*w* curve as a two-stage shrinkage curve, i.e., bilinear curve, and the SL as the water content at turn point (Birle et al. 2008; Cornelis et al. 2006; Cerato and Lutenegger 2006). In ASTM D4943 (ASTM 2008), the shrinkage limit is calculated as,

$$w_{\text{lim}} = e_{\text{d}}/G_{\text{s}} \tag{6.6}$$

where e_{d} is the void ratio of dry soil. Obviously, Eq. 6.6 implies a bilinear shrinkage curve and gives a very rough estimate of the "real" SL (referring to Fig. 6.4), which may be much larger than w_{lim} calculated by Eq. 6.6.

The shrinkage curve in the e–ψ space is shown in Fig. 6.5. The shrinkage curve in the e–ψ space can also be divided into three stages: the normal, residual, and zero shrinkage stage. For comparison, the wG_{s}–ψ curve is also drawn in Fig. 6.5, which indicates the full saturation line. The area between the e–ψ curve and wG_{s}–ψ curve indicates the void occupied by air.

For sample drying from a slurry state or slightly compacted state, an increase of soil matric suction can cause an increase of effective stress *p*, resulting in significant volume change of the soil sample. The deformation behavior in this process can be characterized by the normal consolidation line in *e*–*p* space. In this process, the decease of soil volume is owing to the destruction of plastic macro-pores. Another

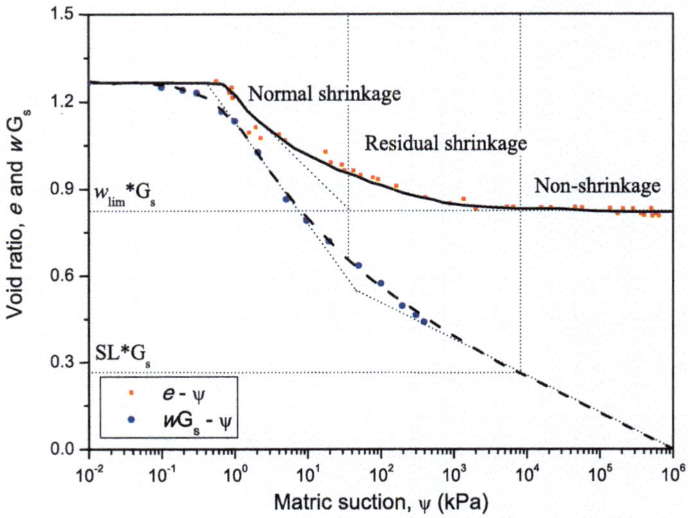

Fig. 6.5 Shrinkage curve in e–ψ scale and soil water characteristic curve in $w\mathrm{G_s}$–ψ scale

mechanism of normal shrinkage is owing to the shrinkage of active clayey aggregates, which results in reduction of the intra-aggregate pores.

6.4.3 *Shrinkage Curve of Overconsolidated or Compacted Sample*

For overconsolidated samples under a constant total stress state, the shrinkage curves may show an "S" shape (Fig. 6.6). In such soils, an additional stage, i.e., the structural shrinkage stage (Cornelis et al. 2006), can be distinguished. The structural shrinkage starts before the normal shrinkage. In this stage, a quantity of water drains as the soil suction ψ increases from 0 to a specific value, ψ_n. Meanwhile, the soil effective stress p increases because of the suction increment, but is still lower than the preconsolidation stress, p_c. Thus, only limited elastic deformation and limited decrease of void ratio will occur. The e–w curve in the structural shrinkage stage has a gentle slope than that in the normal shrinkage stage (Fig. 6.6).

In the structural shrinkage stage, the soil structure is relatively stable while a remarkable mass of pore-water drains associating with a limited decrease of the soil volume. One contribution to the structural shrinkage stage is from the water stored in the macro-pores (inter-aggregate pores). Before the water stored in the micropores drains, the soil aggregate size will not change much. Chertkov (2007, 2008) also addressed the contribution of intra-aggregate pores to the soil structural shrinkage. Chertkov (2007, 2008) proposed "a new concept of intra-aggregate structure", in which silt and sand particles are coated by a layer of clay, termed as an interface layer,

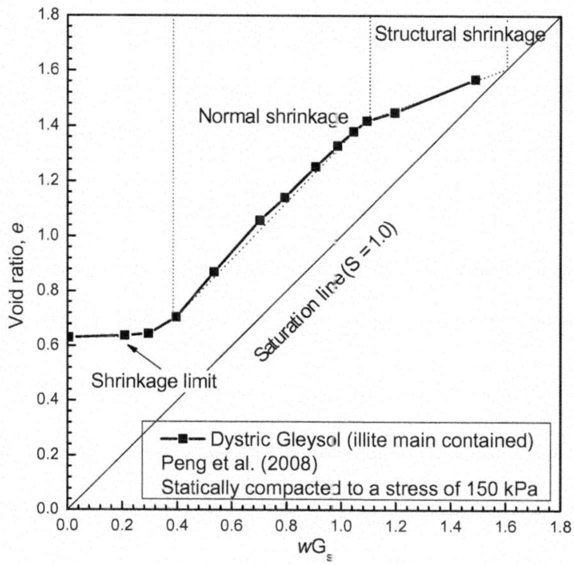

Fig. 6.6 Shrinkage curve of a compacted clay sample

if the clay content in the soil is above a critical percentage. The soil deformation is limited when the pore-water drains from the clay interface layer, which is considered to be more rigid upon drainage comparing to the inner clay particles. Whatever where the pore-water drains from in the structure shrinkage stage, the soil structure will be relatively stable if the effective stress caused by the increase of suction is smaller than a critical suction ψ_n. ψ_n corresponds to the turning point in the e–ψ curve (Baumgartl and Koeck 2004) and may be related to preconsolidation effective stress p_c. Thus, no significant decrease of soil volume occurs when water drains from the macro-pores.

Structured soils such as undisturbed clayey samples in the field (Cornelis et al. 2006) and compacted samples in the laboratory (Peng et al. 2009) are found to have a structural shrinkage stage while dehydration.

6.5 Drying Process in Water Content—Suction Domain

6.5.1 Definition of Soil Water Characteristic Curve

In the drying/wetting process, the water content varies with matric suction and reflects the soil–water retention ability at that state. The whole process is described by a soil–water characteristic curve (SWCC). A SWCC differs with soil history, including wetting/drying and loading/unloading history. A SWCC is often characterized by: the air-entry value (AEV), the residual water content, and the residual soil suction. The air-entry value (AEV), ψ_{aev}, is defined as the matric suction at which air first

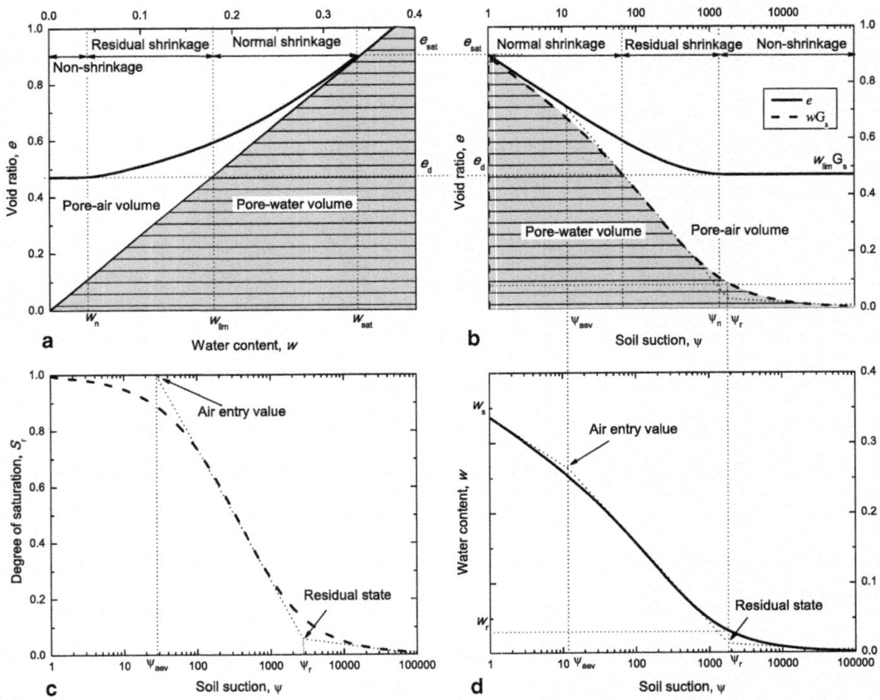

Fig. 6.7 Shrinkage behavior of a fine soil sample draining from a slurry state

enters the largest pores of the soil during a drying process (Brooks and Corey 1964). The residual water content w_r is the water content where a large suction change is required to remove additional water from the soil. The residual soil suction, ψ_r, which indicates the state the water phase becomes discontinuous. For a coarse soil, after the soil suction reaches the residual suction, only a small amount of water drains from the pendulum ring pertaining to the contact zone between two particles. For a fine soil, after the soil suction reaches the residual suction, only a small amount of water is stored in the nanopores, which is regarded as absorbed water. The characteristic values are often determined empirically from the measured SWCC [as shown in Fig. 6.7c], as recommended by Brooks and Corey (1964), Lu and Likos (2004), and many others.

6.5.2 *Soil Water Characteristic Curve Affected by Soil Consolidation or Compaction*

Stress history, wetting and drying cycle, confining pressure, and so on also affect the shape of the soil–water characteristic curves (Fredlund and Xing 1994). For example,

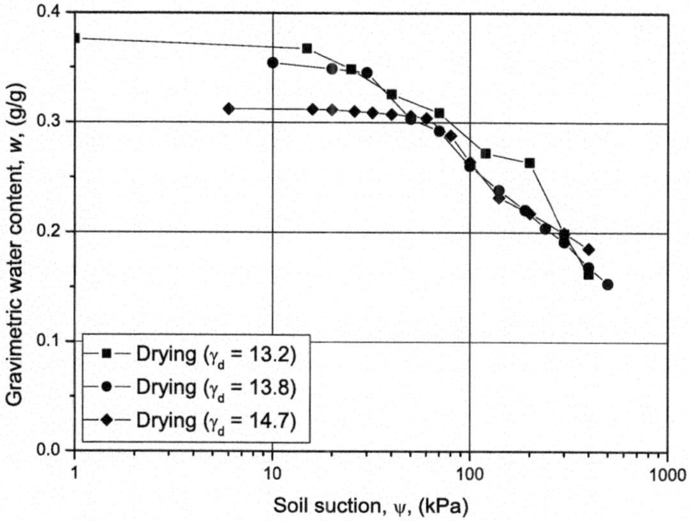

Fig. 6.8 Drying SWCCs for CL with sand of three densities (γ_d is the soil dry unit weight, kN/m^3)

the soil water characteristic curves also show different features for sample drying from a slurry state or slightly compacted state and overconsolidated or compacted sample (Fig. 6.8).

For sample drying from a slurry state or slightly compacted state, most macro-pores may be destroyed due to the increased effective stress. Consequently, the volume fraction of the macro-pores decreases as well as the mass of water stored in macro-pores. Such a phenomenon also affects the SWCC. First, the saturated water content will decrease with the increase of compaction effort. Second, a higher air-entry value (AEV) is expected on the SWCC of a sample with a higher compaction effort.

As experimental evidence, the measured SWCCs for CL are shown in Fig. 6.8 where the AEVs of the compacted samples increase from 15 to 60 kPa when the dry density increases from 1320 to 1470 kg/m^3. Similar findings were also reported by Ng and Pang (2000), Yang et al. (2004), Huang et al. (1998), Baumgartl and Koeck (2004), Li et al. (2009) and Salager et al. (2013), where the SWCCs for denser samples show higher AEV values. As mechanical stresses cause relatively little changes in the micropores and nanopores, the SWCCs for soils of different densities merge at high suctions, as shown in Fig. 6.8. Because the micropores are stable upon mechanical actions, the soil samples compacted at different compaction efforts, i.e., different initial dry densities at compaction, have similar PSDs in the micropore part and hence similar SWCCs in the high suction range.

6.6 Drying Process in Void Ratio—Water Content—Suction Domain

SC and SWCC can be used together to describe the soil dehydration process. A conceptual shrinkage diagram of sample drying from a slurry state or a slightly compacted state is shown in Fig. 6.7. The authors recommend drawing the SC in the e–ψ scale and the SWCC in the wG_s–ψ scale [Fig. 6.7b]. Then the SC and SWCC can be compared directly. The area between the e–ψ curve and the wG_s–ψ curve refers the volume of void occupied by air. The three shrinkage stages are illustrated in Fig. 6.8. In the normal shrinkage stage, the soil is in a quasi-saturated state; remarkable decrease of void ratio takes place as water drains. Thus, the SWCC in the S_r–ψ space is nearly horizontal, but the SWCC in the w–ψ space has a slope. The SWCC in the wG_s–ψ space is quite close to the SC in the e–ψ space in this stage [Fig. 6.7b]. In the residual shrinkage stage, the e–ψ curve starts to separate from the wG_s–ψ curve because some void is occupied by air. The zero shrinkage stage begins after most of the water stored in the inter-aggregate pores and the intra-aggregate pores are drained, leaving some water in the stable nanopores. The residual water content, w_r, which refers the total volume of nanopores, is considered to be the real shrinkage limit. After the soil suction reaches ψ_r, the soil turns into a zero shrinkage stage.

Similarly, the conceptual shrinkage diagrams of an overconsolidated sample are shown in Fig. 6.9. For the overconsolidated fine soil sample, a structural shrinkage stage (Cornelis et al. 2006) is present. In the structural shrinkage stage, the water stored in the large inter-aggregate pores and some intra-aggregate pores (Chertkov 2007, 2008) drains with limited soil volume change, as discussed in the previous section. Before the effective stress which involve some contribution from soil suction reaches the pre-consolidated pressure, the volume change is relatively low and follows the assumption of elastic recompression. After the soil suction increases to a certain value where the effective stress reaches the pre-consolidated pressure, the shrinkage curve for overconsolidated sample merges with that of the slurry sample.

The structural shrinkage is also present in mixed soils with a dual-porosity structure (Zhang and Li 2010), where both inter-aggregate pores and intra-aggregate pores are dominant. When the bulky water drains from the large inter-aggregate pores with pore diameters larger than 29.1 mμ, the suction increment is rather low (e.g., lower than 10 kPa). Hence, no distinct volume change can be induced by such a low suction increment in an initially stable sample. The conceptual shrinkage diagram shown in Fig. 6.9 is also applicable to coarse–fine mixed soils with a dual-porosity structure.

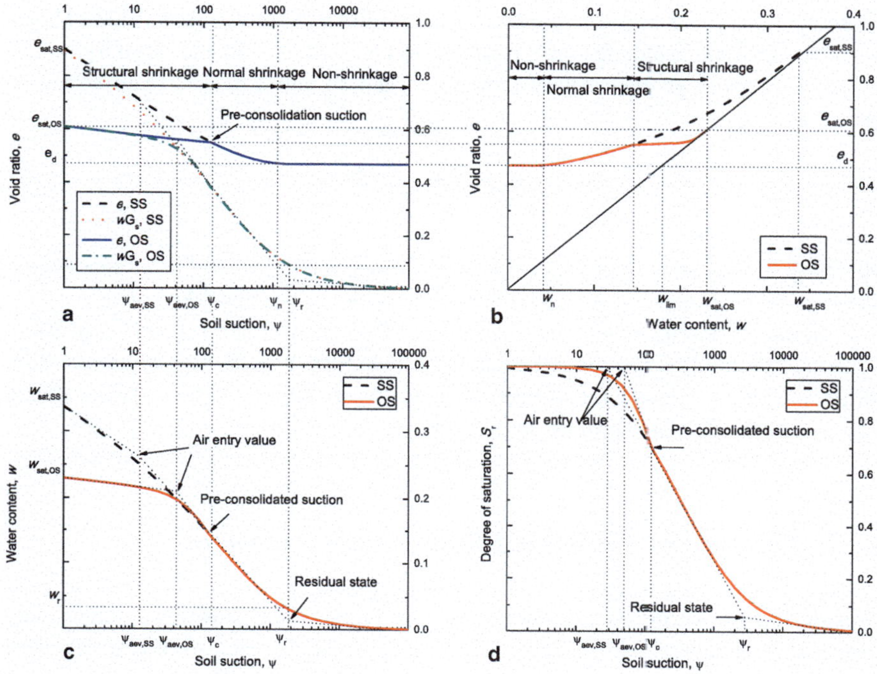

Fig. 6.9 Shrinkage behavior of a fine overconsolidated sample (SS refers slurry sample; OS refers overconsolidated sample)

6.7 Drying Process in Water Content—Fabric Domain and Void Ratio—Fabric Domain

6.7.1 *Water Retention Ability Controlled by PSD*

As water is stored in pores, the water content can be calculated by

$$w(\psi) = \rho_w \int_0^{r_c(\psi)} f(r) d \lg r \tag{6.7}$$

where $r_c(\psi)$ is the maximum pore radius filled by water in a soil sample with a suction of ψ (as shown in Fig. 6.1).

At high suctions, the soil macro-pores (inter-aggregate pores) are filled by air and water is stored in the micropores (intra-aggregate pores) or absorbed in the minerals. Hence at high suctions, the water retention ability is primarily controlled by surface area and surface charge of the clay minerals (Reddi and Poduri 1997; Prost et al. 1998). When the matric suction is low (e.g., less than 3000 kPa), the variations of the intra-aggregate pores and adsorbed water are minimal (Saiyouri et al. 2000). Hence at low suctions, the water desorption/adsorption is controlled by the capillary forces. If cylindrical pore geometry is assumed, the relation between $r_c(\psi)$ and ψ can be written as,

$$r_c(\psi) = \frac{2T_s \cos\beta}{\psi} \tag{6.8}$$

where the surface tension of water is $T_s = 72.8$ mN/m at 20 °C and the contact angle between the air–water interface and the soil particle, β, is often taken as 0°.

It should be mentioned that the shape effect (Tuller et al. 1999; Lu and Likos 2004; Willson et al. 2012) is not considered in Eq. 6.8 and the hysteresis of water retention ability, e.g., the presence of air-filled pores, is also not addressed here. These effects may affect the shrinkage process and water retention ability.

6.7.2 Drying Process in Void Ratio—Fabric Domain

Considering the shrinkage of intra-aggregate pores, three shrinkage stages, i.e., normal shrinkage, residual shrinkage, and zero shrinkage, can also be distinguished in the PSDs at different drying states, as shown in Fig. 6.10a. Two assumptions are made to consider the shrinkage of soil pores: (1) The soil pores will not shrink until the pore water starts to drain; and (2) when suction increases from ψ to $+\,\psi \mathrm{d}\psi$, part of the pores with radius r will shrink, resulting in a decrease of void ratio from e to e'. Here pore radius r is a function of ψ according to Kelvin equation (Eq. 6.8). Hence, the following relations can be obtained:

$$\begin{gathered} f\left(r = \frac{2T_s}{\psi + d\psi}, e'\right) > f\left(r = \frac{2T_s}{\psi + d\psi}, e\right) \\ f\left(r = \frac{2T_s}{\psi}, e'\right) < f\left(r = \frac{2T_s}{\psi}, e\right) \end{gathered} \tag{6.9}$$

Thus, the cumulative pore volume will follow

$$F\left(r = \frac{2T_s}{\psi + d\psi}, e'\right) > F\left(r = \frac{2T_s}{\psi + d\psi}, e\right) \tag{6.10}$$

Due to the pore shrinkage, the total volume of pores will decrease after the soil suction increases. Hence, the following equation can be obtained,

$$F\left(r = \frac{2T_s}{\psi}, e'\right) < F\left(r = \frac{2T_s}{\psi}, e\right) \tag{6.11}$$

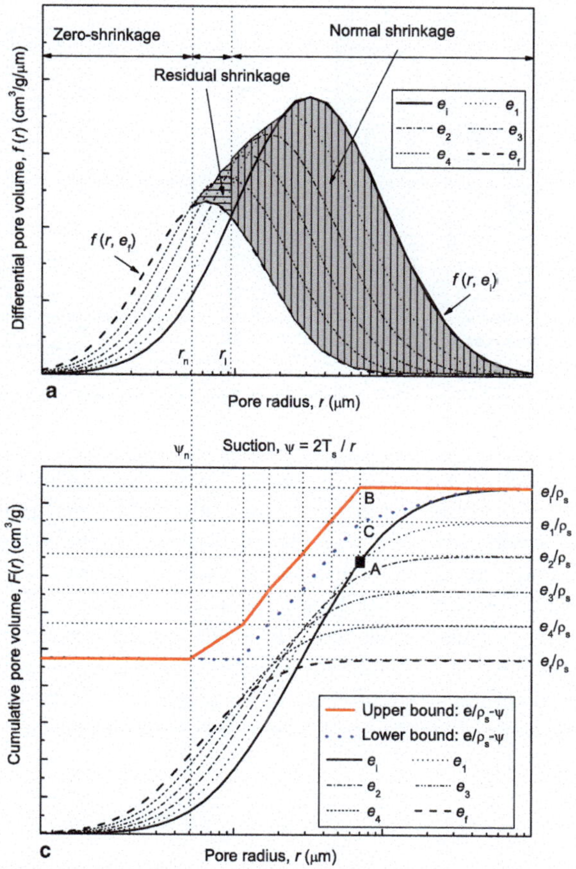

Fig. 6.10 Prediction of soil shrinkage characteristic curve based on pore-size distribution variations

Based on Eqs. 6.6 and 6.7, there exists an intersection point between the two PSDs, i.e., $f(r, e')$ and $f(r, e)$, where

$$F\left(r = \frac{2T_s}{\psi_j}, e'\right) = F\left(r = \frac{2T_s}{\psi_j}, e\right) \quad \text{where, } \psi_j \in (\psi, \psi + d\psi) \tag{6.12}$$

For the soil at a suction of ψ_j, its void ratio should be

$$e_j \in (e, e') \tag{6.13}$$

An example of Eq. 6.13 is shown in Fig. 6.10b. Referring to the intersection, A, between $F(r, e_i)$ and $F(r, e_1)$, the cumulative pore volume at radius of r is F_A, and the radius is r_A that corresponds to a suction of $\psi_A = 2T_s/r_A$. The void ratio, e_A, which represents the total void volume, is larger than the cumulative pore volume,

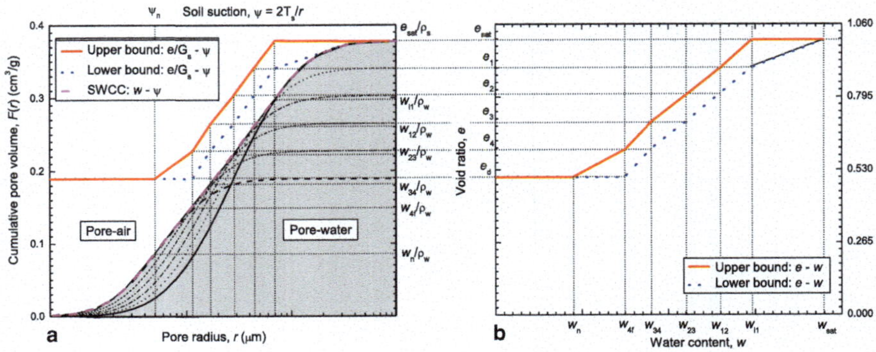

Fig. 6.11 Prediction of soil water characteristic curve based on pore-size distribution variations

F_A, and should be between (e_1, e_i). This means that e_A/ρ_s is located between points B and C in Fig. 6.10b. Following this procedure, an upper bound point and a lower bound point can be determined for each intersection of two cumulative pore volume curves. Consequently, an upper bound and a lower bound of e/ρ_s–ψ curves can be determined based on a series of cumulative pore volume curves. Because there is no further volume change in the zero shrinkage stage, the variations of PSD are considered to be insignificant after the soil suction reaches the zero shrinkage suction, ψ_n. Thus, the zero shrinkage suction, ψ_n corresponds to the critical pore radius at which all the PSDs turn to be same. A reasonable band of e–ψ curve can then be determined (Fig. 6.11).

6.7.3 Drying Process in Water Content—Fabric Domain

Another concern is the relation between the soil water content, w, and the pore-size distribution, $f(r)$. At the state of e–w–ψ, w can be determined using the $f(r, e)$ function as

$$w = \int_0^{\frac{2T_s}{\psi}} f(r, e)\mathrm{d}r = F\left(\frac{2T_s}{\psi}, e\right) > F\left(\frac{2T_s}{\psi}, e'\right) \tag{6.14}$$

At the state of $e' - w' - (\psi + \mathrm{d}\psi)$, w can be determined using the $f(r, e')$ function as

$$w' = \int_0^{\frac{2T_s}{\psi + d\psi}} f(r, e')\mathrm{d}r = F\left(\frac{2T_s}{\psi + d\psi}, e'\right) > F\left(\frac{2T_s}{\psi + d\psi}, e\right) \tag{6.15}$$

Based on Eqs. 6.10 and 6.11, w at a specific suction of ψ is,

$$w = \max\left[F\left(\frac{2T_s}{\psi}, e \in (e_{sat}, e_d)\right)\right] \tag{6.16}$$

Referring to Eq. 6.16, the w-ψ curve, i.e., the SWCC, should be the envelope of the cumulative pore volume curves as shown in Fig. 6.12a. Based on the unique SWCC, w can be inferred from ψ and substituted into the e–ψ bands, resulting in an e-w band as shown in Fig. 6.12b.

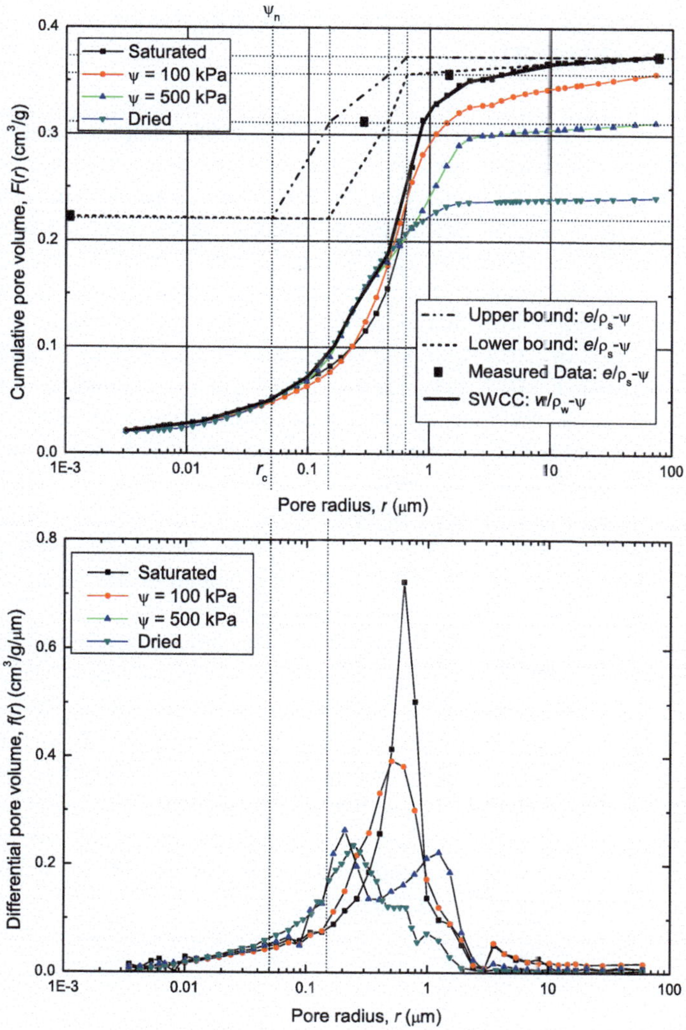

Fig. 6.12 SC and SWCC predictions based on cumulative pore volume curves for CL with sand

It should be mentioned that the quantification of PSD in this section is based on a simple assumption of a cylindrical pore cluster, i.e., the shape effect (Tuller et al. 1999; Lu and Likos 2004; Willson et al. 2012) is not considered. The hysteresis of water retention ability is also not considered. They may affect the shrinkage process and water retention ability. The aforementioned deviation is associated with the assumption that only a part of pores shrinks upon drying. In fact, the drainage of the pores of a given size should lead to simultaneous "self-consistent" shrinkage of all other pores around (Chertkov 2000, 2003).

6.8 Case Study: SWCC and SC Prediction Based on Variable PSDs

Li and Zhang (2009) measured the PSDs of several CL samples during a drying process. These samples were initially compacted at the optimum water content of 20.5% and to the same void ratio of $e_s = 0.99$. Four samples were tested using the mercury intrusion porosimetry technique, resulting in the pore-size distributions shown in Fig. 6.12a. The upper band and lower band of the e/ρ_s–ψ curve are predicted based on Eq. 6.13. For example, $F(r, 0.36)$ intersects with $F(r, 0.31)$ at a radius of 0.45 μm, corresponding to a suction of 310 kPa. Thus the void ratio at a drying state with a suction of 310 kPa should be between 0.31 and 0.36. Similarly, $F(r, 0.31)$ intersects with $F(r, 0.22)$ at a radius of 0.14 μm, corresponding to a suction of 1000 kPa. Thus, the void ratio at a drying state with a suction of 1000 kPa should be between 0.22 and 0.31. The zero shrinkage suction ψ_n should be larger than 1000 kPa. The variation of cumulative volumes of pores with radii smaller than 0.05 μm, corresponding to a suction of 3000 kPa is insignificant. Therefore, the zero shrinkage suction ψ_n should be smaller than 3000 kPa. The real suction state of $F(r, 0.36)$ is 100 kPa and that of $F(r, 0.31)$ is 500 kPa, plotted as the measured data in Fig. 6.12.

As discussed in Eq. 6.16, the envelope of the cumulative pore volume curves can be considered as the SWCC curve (Fig. 6.12), which is compared with the measured SWCC data (Li 2009) in Fig. 6.13. At high suction range (above 200 kPa), the difference of prediction and measured w is small, no more than 3%. The difference between the predicted SWCC and the measured data in low suction range may be due to uncertainty of measured PSD. A small sample (3 cm^3 in volume) used to measure PSD may differ with that (123 cm^3 in volume) used in SWCC measurement. The connectivity of soil pores differs significantly in the two conditions and may have some effect on the measured SWCC and PSD.

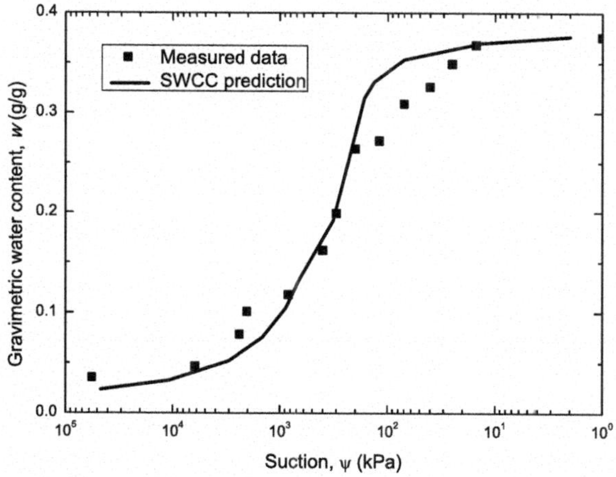

Fig. 6.13 Comparison of predicted SWCC and measured data for CL with sand

6.9 Summary

In this chapter, the drying process of soil is studied under an integrated framework where soil fabric, water content, void ratio are mechanically linked by soil suction. Considering a monotonic dehydration process, soil fabric varies with soil suction and the PSD is considered as a function of soil suction. Soil dehydration process is re-recognized or interpreted in the following domain: (1) fabric-suction domain; (2) void ratio-suction domain, i.e., another description of soil shrinkage curve; (3) water content-suction domain, i.e., traditional soil water characteristic curve; (4) void ratio-water content domain, i.e., the traditional soil shrinkage curve; (5) void ratio—fabric domain; (6) water content—fabric domain. Based on these analyses, the following conclusions can be drawn:

(1) The shrinkage behavior of a slurry sample can be distinguished into normal shrinkage, residual shrinkage and zero shrinkage stages both on the e–w curve and the e–ψ curve. The shrinkage limit is theoretically defined as the water content at the start of the zero shrinkage stage. ASTM D4943 provides a simple method for measuring the shrinkage limit, which is a rough estimation and do not refer to the real SL, i.e., the start of the zero shrinkage stage. The zero shrinkage suction on an e–ψ curve is close to the residual suction on a SWCC.

(2) Soil fabric varies during the soil drying process. If a monotonic drying process is considered, the soil fabric as characterized by a pore-size distribution has a unique relation with soil void ratio, water content and matric suction. Assuming that the soil pores do not shrink until the pore water starts to drain and that only part of the pores shrinks upon drying, the variable cumulative pore volume curves can be used to predict the SWCC and SC. The envelope of the cumulative pore volume curves is exactly the SWCC and a narrow band of SC can be determined using Eq. 6.13.

(3) As an example, the PSDs at four drying states of a CL with sand are used to predict the SC and SWCC using Eqs. 6.13 and 6.16, respectively, resulting in agreeable predictions with the measured data.

References

ASTM D4943 (2008) Standard test method for shrinkage factors of soils by the wax method. Annual Book of ASTM Standards, American Society for Testing and Materials, Philadelphia, Pa

Alonso EE, Vaunat J, Gens A (1999) Modeling the mechanical behavior of expansive clays. Eng Geol 54:173–183

Aung KK, Rahardjo E, Leong EC, Toll DG (2001) Relationship between porosimetry measurement and soil-water characteristic curve for an unsaturated residual soil. Geotech Geol Eng 19:401–416

Birle E, Heyer D, Vogt N (2008) Influence of the initial water content and dry density on the soil–water retention curve and the shrinkage behavior of a compacted clay. Acta Geotech 3:192–200

Boivin P, Garnier P, Vauclin M (2006) Modeling the soil shrinkage and water retention curves with the same equations. Soil Sci Soc Am J 70:1082–1093

Boivin P, Garnier P, Tessier D (2004) Relationship between clay content, clay type, and shrinkage properties of soil samples. Soil Sci Soc Am J 68:1145–1153

Braudeau E, Costantini JM, Bellier G, Colleuille H (1999) New device and method for soil shrinkage curve measurement and characterization. Soil Sci Soc Am J 63:525–535

Baumgartl T, Koeck B (2004) Modeling volume change and mechanical properties with hydraulic models. Soil Sci Soc Am J 68(1):57–65

Braudeau E, Frangi JP, Mohtar RH (2004) Characterizing nonrigid aggregated soil-water medium using its shrinkage curve. Soil Sci Soc Am J 68:359–370

Brooks RH, Corey AT (1964) Hydraulic properties of porous media. Colorado: Hydrology Paper, No. 3, Colorado State University, Fort Collins

Cerato AB, Lutenegger AJ (2006) Shrinkage of clays. In: Proceedings of the fourth international conference on unsaturated soils, 2–6 Apr 2006, Carefree, AZ, vol 2, pp 1097–1108

Chertkov VY (2000) Modeling the pore structure and shrinkage curve of soil clay matrix. Geoderma 95:215–246

Chertkov VY (2008) The physical effects of an intra-aggregate structure on soil shrinkage. Geoderma 146:147–156

Chertkov VY (2007) The reference shrinkage curve at higher than critical soil clay content. Soil Sci Soc Am J 71:641–655

Chertkov VY (2003) Modelling the shrinkage curve of soil clay pastes. Geoderma 112:71–95

Chertkov VY (2004) A physically based model for the water retention curve of clay pastes. J Hydrol 286:203–226

Chertkov VY (2012a) Physical modeling of the soil swelling curve vs. the shrinkage curve. Adv Water Resour 44:66–84

Chertkov VY (2012b) An integrated approach to soil structure, shrinkage, and cracking in samples and layers. Geoderma 173–174:258–273

Cornelis WM, Corluy J, Medina H, Hartmann R, Van Meirvenne M, Ruiz ME (2006) A simplified parametric model to describe the magnitude and geometry of soil shrinkage. Eur J Soil Sci 57:258–268

Crescimanno G, Provenazano G (1999) Soil shrinkage characteristic curve in clay soils: measurement and prediction. Soil Sci Soc Am J 63:25–32

Cuisinier S, Laloui L (2004) Fabric evolution during hydromechanical loading of a compacted silt. Int J Numer Anal Meth Geomech 28:483–499

Ferber V, Airol JC, Cui YJ, Magnan JP (2009) On the swelling potential of compacted high plasticity clays. Eng Geol 104:200–210

Fredlund MD, Wilson GW, Fredlund DG (2002) Representation and estimation of the shrinkage curve. In: Proceedings of the third international conference on unsaturated soils, UNSAT 2002, Brazil, pp 145–149

Fredlund DG, Xing AQ (1994) Equations for the soil-water characteristic curve. Can Geotech J 31: 521-532

Gregory AS, Bird NRA, Whalley W, Matthewsb P (2010a) Estimating relative hydraulic conductivity from the water release characteristic of a shrinking clay soil. Soil Sci Soc Am J 74:1113–1118

Gregory AS, Bird NRA, Whalley W, Matthewsb P, Young M (2010b) Deformation and shrinkage effects on the soil water release characteristic. Soil Sci Soc Am J 74:1104–1112

Groenevelt PH, Grant CD (2001) Re-evaluation of the structural properties of some British swelling soils. Eur J Soil Sci 52:469–477

Groenevelt PH, Grant CD (2002) Curvature of shrinkage lines in relation to the consistency and structure of a Norwegian clay soil. Geoderma 106:235–245

Huang SY, Barbour SL, Fredlund DG (1998) Development and verification of a coefficient of permeability function for a deformable unsaturated soil. Can Geotech J 35:411–425

Krisdani H, Rahardjo H, Leong EC (2008) Effects of different drying rates on shrinkage characteristics of a residual soil and soil mixtures. Eng Geol 102:31–37

Lan MT, Wang DL, Yang Q, Li PY (2008) Experimental study on drying shrinkage of unsaturated compacted soils. Chin J Geotech Eng 30(1):118–122

Lapierre C, Leroueil S, Locat J (1990) Mercury intrusion and permeability of Louisville clay. Can Geotech J 27:761–773

Li X (2009) Dual-porosity structure and bimodal hydraulic property functions of coarse granular soils. PhD thesis, The Hong Kong University of Science and Technology, Hong Kong

Li X, Zhang LM (2009) Characterization of dual-structure pore-size distribution of soil. Can Geotech J 46(2):129–141

Li X, Zhang LM, Li JH (2009) Development of a modified axis translation technique for measuring SWCCs for gravel soils at very low suctions. Geotech Test J 32(6):478–488

Lu N, Likos WJ (2004) Unsaturated soil mechanics. Wiley

Lloret A, Villar MV, Sanchez M, Gens A, Pintado X, Alonso EE (2003) Mechanical behavior of heavily compacted bentonite under high suction changes. Geotechnique 53:27–40

Markgraf W, Horn R (2006) Scanning electron microscopy-energy dispersive scan analyses and rheological investigations of south-Brazilian soils. Soil Sci Soc Am J 71(3):851–859

Masad E, Jandhyala VK, Dasguptan N, Somadevan N, Shashidhar N (2002) Characterization of air void distribution in asphalt mixes using X-ray computed tomography. J Mater Civil Eng 14(2):122–129

Mitchell AR (1992) Shrinkage terminology: escape from "normalcy." Soil Sci Soc Am J 56:993–994

Ng CWW, Pang YW (2000) Influence of stress state on soil-water characteristics and slope stability. J Geotech Geoenviron Eng 126(2):157–166

Olsen PA, Haugen LE (1998) New model of the shrinkage characteristic applied to some Norwegian soils. Geoderma 83:67–81

Peng X, Dorner J, Zhao Y, Horn R (2009) Shrinkage behavior of transiently- and constantly-loaded soils and its consequences for soil moisture release. Eur J Soil Sci 60:681–694

Peng X, Horn R (2005) Modeling soil shrinkage curve across a wide range of soil types. Soil Sci Soc Am J 69:584–592

Peng X, Horn R (2013) Identifying six types of soil shrinkage curves from a large set of experimental data. Soil Sci Soc Am J 77(2):372–381

Penumadu D, Dean J (2000) Compressibility effect in evaluating the pore-size distribution of kaolin clay using mercury intrusion porosimetry. Can Geotech J 37:393–405

Prost R, Koutit T, Benchara A, Huard E (1998) State and location of water adsorbed on clay minerals: consequences of the hydration and swelling-shrinkage phenomena. Clays Clay Min 46(2):117–131

Reddi LN, Poduri R (1997) Use of liquid limit state to generalize water retention properties of fine-grained soils. Geotechnique 47(5):1043–1047

Richard G, Cousin I, Sillon JF, Bruand A, Guerif J (2001) Effect of compaction on the porosity of a silty soil: influence on unsaturated hydraulic properties. Eur J Soil Sci 52:49–58

Saiyouri N, Hicher Y, Tessier D (2000) Microstructural approach and transfer water modeling in compacted unsaturated swelling clays. Mech Cohesive-Frictional Mater 5:41–60

Simms P, Yanful E (2001) Measurement and estimation of pore shrinkage and pore distribution in a clayey till during soil-water characteristic curve tests. Can Geotech J 38:741–754

Salager S, Nuth M, Ferrari A, Laloui L (2013) Investigation into water retention behavior of deformable soils. Can Geotech J 50:200–208

Salager S, Youssoufi MSE, Saix C (2010) Definition and experimental determination of a soil-water retention surface. Can Geotech J 47(6):609–622

Stange CF, Horn R (2005) Modeling the soil water retention curve for conditions of variable porosity. Vadose Zone J 4:602–613

Stewart RD, Abou N, Majdi R, Rupp DE (2012) An image-based method for determining bulk density and the soil shrinkage curve. Soil Sci Soc Am J 76:1217–1221

Thom R, Sivakumar R, Sivakumar V, Murray EJ, Mackinnon P (2007) Pore size distribtution of unsaturated compacted kaolin: initial states and final states following saturation. Geotechnique 57(5):469–474

Tripathy S, Rao KS, Fredlund DG (2002) Water content—void ratio well-shrink paths of compacted expansive soils. Can Geotech J 39:938–959

Tuller M, Or D, Dudley LM (1999) Adsorption and capillary condensation in porous media: liquid retention and interfacial configurations in angular pores. Water Resour Res 35(7):1949–1964

Willson C, Lu N, Likos WJ (2012) X-ray computer-aid-tomography for unsaturated sand. Geotech Test J 35(6):1–13

Yang H, Rahardjo H, Leong EC, Fredlund DG (2004) Factors affecting drying and wetting soil-water characteristic curves of sandy soils. Can Geotech J 41(5):908–920

Zhang LM, Li X (2010) Micro-porosity structure of coarse granular soils. J Geotech Geoenviron Eng ASCE 136(10):1425–1436

Chapter 7
Shearing Behavior of Widely-Graded Soils Under Low Confining Pressures

7.1 Introduction

Colluvial soils are widely distributed on natural terrains. Shallow-seated failures with sliding depths varying between 0.5 and 3.0 m are the main failure mode in colluvial soil deposits during rainstorms (e.g., Au 1998; Fannin et al. 2005; Cascini et al. 2010). The corresponding confining stresses are in a low range between 5 and 25 kPa. It is pertinent to study the behavior of loose, coarse, widely-graded colluvial soils under very low confining pressures to provide better understanding of shallow-seated failures in coarse-grained soils.

As discussed in Chap. 2, colluvial soils have a wide range of coarse contents as the soils have been subject to transportation and particle segregation. As discussed in Chaps. 3 and 5, the coarse content significantly affects the microporosity and hydraulic properties of coarse widely-graded soils. Several studies have been performed to investigate the effect of coarse content on the void ratio (e.g., Lade et al. 1998), collapsibility (e.g., Pitman et al. 1994), shear strength (e.g., Vallejo and Mawby 2000), stress–strain relationships (e.g., Dash and Sitharam 2011), and critical-state characteristics of soils (e.g., Wood and Maeda 2008; Bandini and Coop 2011). The saturated clean sands often exhibit dilative behavior at low confining pressures (Fukushima and Tatsuoka 1984; Bolton 1986; Fannin et al. 2005; Chakraborty and Salgado 2010). When a certain fine is added, some saturated loose silty sands show contractive behavior under low confining pressures (Yamamuro and Lade 1998; Yamamuro et al. 2008; Zhao 2013; Zhao et al. 2013; Zhao and Zhang 2013). However, colluvial deposits are often unsaturated. In this chapter, the effect of coarse content on the compressibility and shearing behavior of saturated and unsaturated coarse and widely-graded soils under very low confining pressures are investigated. The critical-state characteristics of saturated and unsaturated granular soils are also investigated.

X. Li et al., *Fundamental Behavior of Unsaturated Widely-Graded Soil*,
https://doi.org/10.1007/978-981-19-3402-5_7

7.2 Experimental Techniques

7.2.1 Experimental Apparatuses

7.2.1.1 Triaxial Pressure Plate System

A triaxial pressure plate system was used to conduct the SWCC tests and isotropic consolidation tests on unsaturated soils. Figure 7.1a shows a photograph of the setup of the triaxial pressure plate system. The system consists of five main components, namely a net normal stress control, a suction control, a flushing system, a measuring system for water volume change, and a new measuring system for total volume change. The cell pressure is applied from the top of the outer cell. The negative water column technique and axis-translation technique was adopted for control suction in this system. The total volume change of the test specimen was measured by adopting

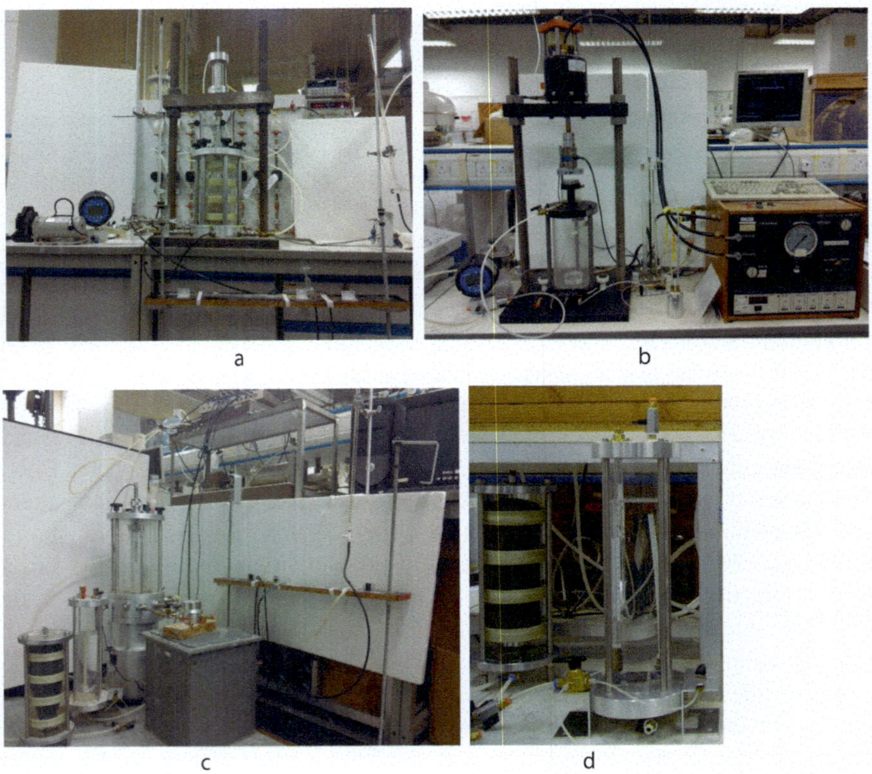

Fig. 7.1 Setup of the testing system: **a** triaxial pressure plate system; **b** CKC triaxial cell; **c** CKC triaxial cell; **d** photograph of the flushing system

a double-cell volume measuring system within the triaxial apparatus. A pump was used to flush the system efficiently.

7.2.1.2 CKC Triaxial Testing System

The automatic CKC triaxial system (Li et al. 1988) was used to carry out the triaxial tests on saturated specimens. Figure 7.1b shows a photograph of the setup of a CKC triaxial testing system. The system composes a loading frame (with a loading piston mounted on the top), a triaxial cell, a controllable pneumatic loading unit, a signal conditioner, a process interface, and a computer for system control and data collection. The pneumatic loading unit is used for regulating cell pressure (σ_r), water pressure (u_w), and pressure difference in the loading piston for controlling axial stress (σ_a). By adjusting the pressure difference between the upper and lower chambers in the loading piston, axial force can be applied and controlled. Five transducers are used during testing. These include an external load cell for measuring and controlling the axial force that is connected to the top cap with a frictionless loading rod; a linear variable differential transformer (LVDT) mounted on the top of the triaxial cell and fixed to the loading rod to measure the external axial displacement; two pressure transducers for monitoring the cell pressure as well as the pressure difference between the cell and pore-water pressure, respectively; and a volume-change measuring device which is mounted with the two pressure transducers. The five transducers are connected to five channels in the signal conditioner. The axial stress, cell pressure, and water pressure are applied through the pneumatic loading units. At the same time, the pressures are measured by the load cell and the pressure transducers and fed back to the computer. Based on the measurements, the computer adjusts the pneumatic loading unit automatically for accurate pressure control. The CKC triaxial testing system is capable of performing both stress- and strain-controlled tests in triaxial space. Both triaxial compression and extension tests can be performed.

All the tests were conducted under very low confining pressures (i.e., 5, 15, and 25 kPa). In order to improve the accuracy, an air pressure gauge with a resolution of 0.05 kPa was used to control and monitor the air pressure. During the tests, the fluctuations of the air pressure were within 0.5 kPa. The maximum possible errors of confining pressure were 10%, 3%, and 2% at confining pressures of 5, 15, and 25 kPa, respectively. A high-accuracy load cell with a low capacity of 0.45 kN was used to measure the axial force. Some repeatability tests were conducted, and the test results were repeatable under the low confining pressures.

7.2.1.3 GDS Triaxial Testing System

A GDS triaxial testing system was used to conduct triaxial tests on unsaturated soils. The general layout of the GDS triaxial testing system is shown in Fig. 7.1c. Axial force and axial strain are applied to each specimen by a loading piston which is fixed to the base pedestal of the triaxial cell. This loading piston is actuated hydraulically from an integral lower chamber in the base of the cell which contains pressurized de-aired water. An LVDT is mounted on the cross-arm to measure the axial displacement of the piston, which is equal to the external axial displacement of a specimen. When the loading piston moves vertically up and down, the base pedestal moves together. The top cap of each specimen is kept in contact with the loading ram which is fixed on the top of the cell during testing. An internal load cell is attached to the loading ram inside the cell to measure the axial force applied to the specimen directly. More details of the GDS triaxial testing system are given by Zhan (2003).

Diffused Air Volume Indicator

Bishop and Donald (1961) stated that although saturated ceramic disk resists the passage of free air at a pressure below air entry value of ceramic disk, dissolved air may diffuse through ceramic disk. Air diffusion becomes more prominent for long-lasting unsaturated soil testing due to low water hydraulic conductivity of an unsaturated soil. Diffused air may accumulate beneath the high air entry value ceramic disk and break the continuity of the water phase if no technique is employed to remove it. This may impose obvious influence on the reliability of experimental results, no matter undrained or drained tests (Fredlund 1975). Accumulated air induces an error with the measurement of water content change in drained tests. In order to overcome, the problems caused by diffused air during unsaturated soil testing. Based on the technique described by Fredlund (1975), a diffused air volume indicator was developed by Chen (2007) and adopted in this study. Figure 7.1d shows a photograph of the flushing system for measuring the volume of diffused air. Under an applied pressure gradient, de-aired water was introduced from one air–water interface through the flushing grooves beneath the ceramic disk in the base pedestal and finally to the diffused air volume indicator, which was kept at a slightly lower pressure than the interface. By using a four-way valve, similar to that by Sharma (1998), the flow direction for flushing diffused air can be reversed, so that any remaining air bubbles can be removed effectively. Based on the ideal gas law, Fredlund and Rahardjo (1993) presented a detailed procedure for computing the volume of diffused air. It is assumed that the air pressure equals the water pressure underneath ceramic disk. The initial and final water levels in the burette are measured. The volume of diffused air can be calculated as:

$$v_{fa} = \frac{\mu_{a2} V_{a2} - \mu_{a1} V_{a1}}{\mu_{fa}} \tag{7.1}$$

where V_{fa} is the volume of diffused air, μ_{a1} and μ_{a2} are initial and final absolute water pressures in the burette, respectively; V_{a1} and V_{a2} are initial and final volumes of air in the burette, respectively; μ_{fa} is the absolute pressure of accumulated diffused air under ceramic disk before flushing. During testing in this study, diffused air was flushed every day. The diffused air volume was deducted from the measured volume of water flow out of soil specimen. At low suctions, the measured diffused air volume was negligible.

Calibrations

The devices that required calibration included the internal load cells, the LVDTs, a cell pressure transducer, the pore-air and pore-water transducers, and the total volume change measuring system. Both the external load cell in CKC triaxial testing system and the internal load cell in GDS triaxial testing system were calibrated by a 3 kN proving ring with high accuracy. The LVDTs were calibrated by a micrometer with a resolution of 1 mm. All the cell pressure, pore-air pressure, and pore water pressure transducers were calibrated by a pressure gauge with a resolution of 0.05 kPa. The transducers were calibrated for at least two full cycles.

The setup of the calibration for the total volume change measuring system simulates the formal test except by replacing the soil sample with a rigid aluminum dummy sample. For the calibration of the DPT, a water volume change in the inner cell was realized through inflow/outflow of a known amount of de-aired water to/from the inner cell because the dummy sample cannot produce volume change by itself, whereas the water level in the reference tube was kept unchanged. Simultaneously, outputs of the DPT were recorded correspondingly as long as they were stabilized. Figure 7.2a shows the calibration results of the DPT for water inflow/outflow of the inner cell under a cell pressure of 100 kPa, respectively. The calibration results under other cell pressures show similar linear relation with slightly different sensitivities (Fig. 7.2b), respectively. The calibration result of the apparent volume change of the relative movement of the loading ram to the inner cell wall is shown in Fig. 7.2c. The calibration results show that linear calibration curves were obtained, and no significant hysteresis was observed. Refer to Zhan (2003) for a detailed description of the calibration procedures for the total volume change measuring system.

Control of Suction

The negative water column was connected to the system to control the suction values lower than 10 kPa, and axis-translation technique was adopted for control suction larger suction.

Negative water column technique for controlling very low suctions (Fig. 4.5): The air entry values of coarse granular soils are very small. Rahardjo et al. (2008) reported that the air-entry values of sand are around 1 kPa. Such soils experience a significant change of water content at relatively small suctions (i.e., 2 kPa). The triaxial apparatuses for testing unsaturated soils often use the axial translation technique or osmotic technique to control the suction. Those methods show low accuracy for controlling very low suctions (i.e., smaller than 10 kPa). As the coarse granular soils are the primary concern in this book, it is necessary to adopt a proper device to control the suction at very low suctions. Li and Zhang (2009) developed a device for SWCC measurement at very low suctions (as described in Chap. 4). The water supply and pressure control system were connected to the GDS and pressure plate triaxial system for controlling suctions smaller than 10 kPa (i.e., water heads up to -1000 mm).

Axis-translation technique for high suctions: The axis-translation technique was used to control the matric suction when the suction value is larger than 10 kPa. The

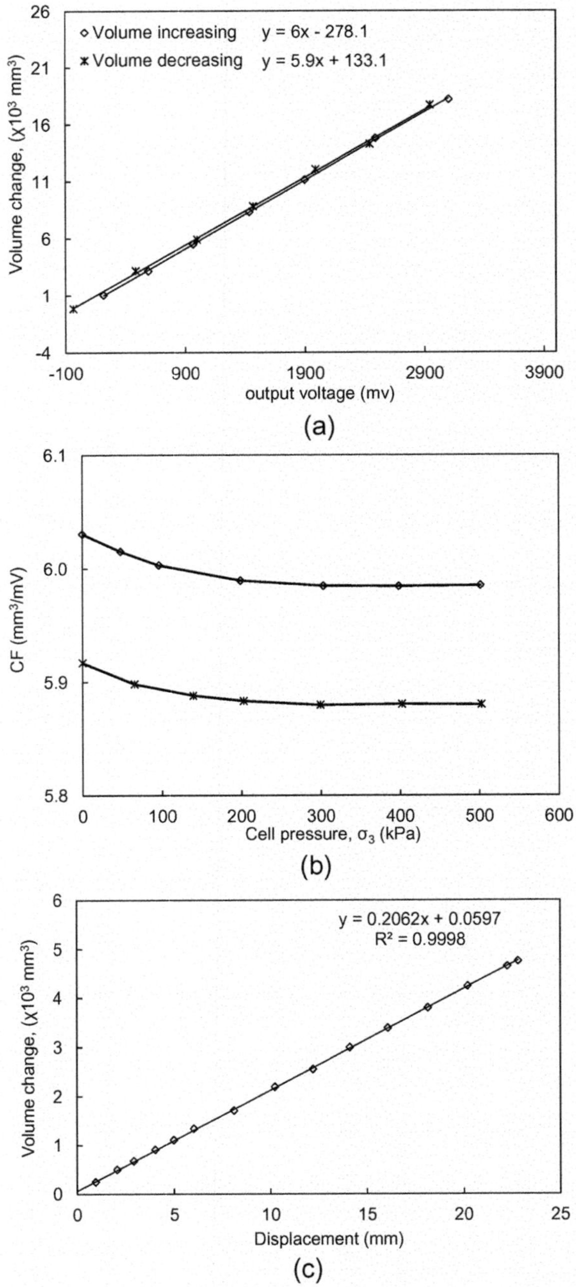

Fig. 7.2 Calibration of GDS triaxial system: **a** Relationship between water volume change in the inner cell and output voltage of DPT. **b** Bilinear relationship between calibration factor and cell pressure. **c** Relationship between the volume change and the relative displacement between the loading ram and the inner cell

axis-translation technique controls matric suction by controlling the pore-air pressure and the pore water pressure at positive values independently. The pore-air pressure is controlled through a coarse porous stone placed on top of each soil specimen. A filter paper is placed between the coarse corundum disk and the specimen, so that soil particles will not block the void of the coarse corundum disk. Then, the air phase is ensured to be continuous in the specimen and the air pressure controller. The pore water pressure is controlled through a saturated high air entry value ceramic disk sealed to the base pedestal of the triaxial apparatus. The high air entry value ceramic disk allows the passage of water but prevents the flow of free air from the specimen to the water control and drainage system underneath it. Thus, the matric suction is controlled as the difference between the pore-air pressure and the pore water pressure. It should be noted that the matric suction in each soil specimen must not exceed the air entry value of the ceramic disk. Otherwise, air will enter the ceramic disk and hence causes an error in the control and measurement of pore water pressure.

7.2.2 *Experimental Program and Stress Paths*

Four series of triaxial tests were conducted in this study and the stress paths of these tests are shown in Fig. 7.3. The detailed test plans are summarized in Tables 7.1, 7.2, 7.3, and 7.4. The tests consist of isotropic consolidation tests on saturated and unsaturated specimens (stress path 1), drained triaxial tests on saturated and unsaturated soils (stress path 2), undrained triaxial tests on saturated soils (stress path 3), and wetting tests under constant deviator stress and mean net stress on unsaturated specimens (stress path 4).

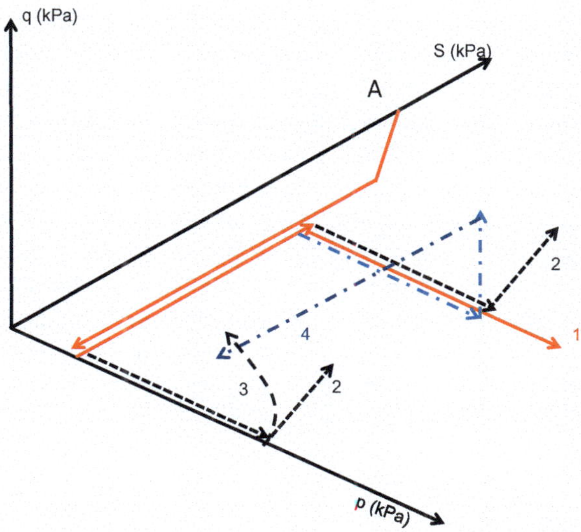

Fig. 7.3 Stress paths for triaxial tests on three types of soil with $f_c = 50$, 68, and 90%

Table 7.1 Plan of isotropic consolidation tests

Soil type (%)	Suction s (kPa)	Void ratio e	Objectives
$f_c = 20$	0	0.62	To investigate the variation of compressibility of coarse granular soils with coarse content
$f_c = 50$	0	0.62	
$f_c = 68$	0	0.62	
$f_c = 75$	0	0.62	
$f_c = 90$	0	0.62	
$f_c = 50$	0	0.52	To investigate the variation of the compression behavior of coarse granular soils with void ratio ($e = 0.52$, 0.62 and 0.87)
$f_c = 50$	0	0.87	
$f_c = 68$	0	0.52	
$f_c = 68$	0	0.87	
$f_c = 90$	0	0.52	
$f_c = 90$	0	0.87	
$f_c = 90$	0.5	0.62	To investigate the effect of suction, s, on the compression behavior of coarse granular soils with bimodal SWCCs ($s = 0$, 0.5, 10 and 80 kPa)
$f_c = 90$	10	0.62	
$f_c = 90$	80	0.62	

The plan of the isotropic consolidation tests is listed in Table 7.1. Three series of tests were conducted. In the first series, isotropic consolidation tests were conducted on five types of soils with different coarse contents ($f_c = 20$, 50, 68, 75, and 90%) under the same initial void ratio $e = 0.62$ to investigate the variation of compressibility of granular soils with different coarse contents. In the second series, the compression behavior of the three types of soils ($f_c = 50$, 68, and 90%) with different void ratios (e = 0.52, 0.62 and 0.87) was compared. In the third series, the isotopic consolidation tests were conducted on the $f_c = 90\%$ specimens under different suctions ($s = 0$, 0.5, 10, and 80 kPa). The objective of this series of tests is to investigate the effect of suction on the compression behavior of coarse granular soils with bimodal SWCCs.

Table 7.2 summarizes the plan for triaxial shearing tests. In the first series, 13 drained and undrained triaxial tests were conducted on saturated coarse granular soils ($f_c = 50$, 68, and 90%) with the same void ratio of $e = 0.62$. The aim is to investigate the effect of coarse content on the shearing behavior of the soils under low confining pressures. In the second series of tests, drained and undrained triaxial tests were conducted on three types of soils ($f_c = 50$, 68, and 90%) under different void ratios ($e = 0.52$, 0.62, and 0.87). The objective of this series of tests is to investigate the instability behavior of saturated granular soils. Drained triaxial tests were conducted on unsaturated soils in the third series to investigate the shearing behavior and critical-state characteristics of unsaturated coarse granular soils with different coarse contents.

Table 7.2 Plan of triaxial shearing tests

Soil type (%)	Suction (kPa)	Void ratio	Drainage condition		Confining pressure (kPa)		Objectives
			CD	CU	CD	CU	
$f_c = 50$	0	0.62	√	√	5, 15, 25	15	To study the effect of coarse content on the shearing behavior under very low confining pressures
$f_c = 68$	0	0.62	√	√	5, 15, 25	15, 25	
$f_c = 90$	0	0.62	√	√	5, 15, 25	15	
$f_c = 50$	0	0.52	√	√	15	15	To study the instability behavior of saturated granular soils
$f_c = 50$	0	0.87	√	√	15	15	
$f_c = 68$	0	0.52	√	√	5, 15, 25	15	
$f_c = 68$	0	0.87	√	√	5, 15, 25	15	
$f_c = 90$	0	0.52	√	√	5, 15, 25	15	
$f_c = 90$	0	0.87	√	√	5, 15, 25	15	
$f_c = 50$	80	0.62	√		5, 15, 25		To investigate the shearing behavior and critical-state characteristics of unsaturated coarse granular soils with different coarse contents
$f_c = 50$	280	0.62	√		15, 25		
$f_c = 68$	10	0.62	√		5, 15, 25		
$f_c = 68$	80	0.62	√		5, 15, 25		
$f_c = 68$	280	0.62	√		15, 25		
$f_c = 90$	0.5	0.62	√		5, 15, 25		
$f_c = 90$	10	0.62	√		5, 15, 25		
$f_c = 90$	80	0.62	√		5, 15, 25		

Notes
CD consolidated and drained tests
CU consolidated and undrained tests

Table 7.3 Plan of wetting tests

Soil type (%)	Confining pressures (kPa)	Stress ratio η	Objective
$f_c = 68$	15	0	To investigate the instability behavior of unsaturated soils subjected to different stress ratios
$f_c = 68$	15	1.0	
$f_c = 68$	15	1.5	
$f_c = 68$	15	1.87	
$f_c = 68$	5	1.5	To investigate the instability behavior of unsaturated soils subjected to different confining pressures
$f_c = 68$	25	1.5	

Table 7.4 State properties of the specimens in the drained tests

Coarse content (%)	Confining pressure (kPa)	Initial void ratio e_0	Critical void ratio e_c	State parameter ψ
50	5	1.623	1.680	−0.057
50	15	1.620	1.630	−0.010
50	25	1.613	1.608	0.005
68	5	1.615	1.635	−0.020
68	15	1.586	1.563	0.023
68	25	1.572	1.535	0.037
90	5	1.618	1.708	−0.090
90	15	1.612	1.666	−0.054
90	25	1.610	1.645	−0.035
Toyoura Sand (Fukushima and Tatsuoka 1984)	4.9	0.681	0.791	−0.11

7.2.3 Preparation of Soil Specimens

The soil was oven-dried at 105 °C for 24 h and then cooled inside a desiccator. Then the soil was mixed according to designated particle-size distributions (Chap. 2). Based on the targeted optimum water content, the required amount of de-aired water was added to the oven-dried soils sample step by step using a sprayer to avoid any possible generation of soil lump. Thereafter, the well-mixed wet soil samples were kept inside a plastic bag for approximately 7 days in a humidity- and temperature-controlled room for moisture homogenization. Before compacting the sample, any aggregations of particles were broken down by hand.

A 70 mm diameter three-way split compaction mold was used for specimen compaction. The static compaction was conducted on a WF loading frame, in which the axial displacement rate can be controlled. A proving ring was attached to monitor and control the static pressure applied to the soil specimen.

Figure 7.4 shows the setup of specimen compaction. To produce a uniform specimen, the compaction of a 140 mm high triaxial specimen was conducted in 10 layers. The soil sample was then statically compacted to the desired dry density at a rate of 0.25 mm/min. The amount of soil for each layer was the same. To prevent any disturbance to the $f_c = 90\%$ specimen or to avoid collapse of the granular soil specimen during the installation, a layer of membrane was wrapped inside the mold by applying a small vacuum during compaction. The specimens produced by this method were relatively uniform, and the interfaces between layers were barely apparent. The repeatability of the preparation method was also found to be excellent. The specimen was wrapped by another layer of membrane after compaction to avoid punching of the inner membrane during compaction.

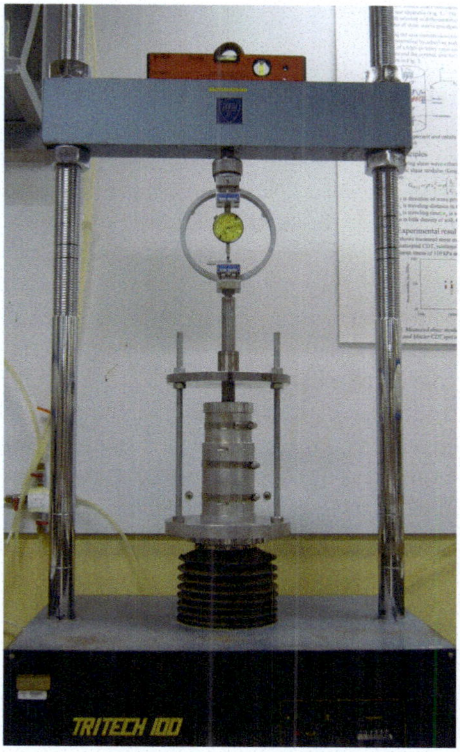

Fig. 7.4 Photograph of the static compaction procedure for sample preparation

After disassembling from the compaction mold, each compacted specimen was left in an airtight container for approximately 30 min to allow for elastic swelling due to the release of lateral pressure. The initial dimensions of the specimen were measured by a caliper (readable to 0.01 mm) for height and by a PI tape for direct measurement of diameter (readable to 0.01 mm, manufactured by Escondido, CA). The initial weight of the specimen was measured by a digital balance (readable to 0.01 g). The water content of the soil left in the plastic bag was measured to calculate the void ratio as follows:

$$\rho_{\mathrm{d}} = \frac{m}{V(1+w)} \tag{7.2}$$

$$e_i = \frac{G_{\mathrm{s}}}{\rho_{\mathrm{d}}} - 1 \tag{7.3}$$

where *D*, *H*, and *V* are, respectively, the diameter, height, and volume of the soil sample; m is the weight of the compacted specimen; w is the water content of the specimen; ρ_{d} is the sample's dry density; G_{s} is the specific gravity of soil; and e_i is the void ratio.

7.2.4 Test Procedures

7.2.4.1 Specimen Saturation

After the soil specimen was installed into the triaxial system, a low confining stress (i.e., 5 kPa) was applied to prevent collapse of the specimen and any leakage between the specimen and the membrane during the saturation step. First, carbon dioxide was slowly injected into the specimen from the bottom of the specimen to displace the air within the specimen. After approximately 2 h, the carbon dioxide was replaced by de-aired water. The de-aired water was slowly introduced into the specimen by sucking the base water into the sample under vacuum, with most of it flowing into the vacuum chamber connected to the upper drainage and a very small amount being kept in the sample to saturate it. After all the visible air bubbles were removed from the specimen, the triaxial cell was connected to an automatic control system. After which, back pressure was applied to saturate the specimen until a minimum pore-water pressure coefficient, B, of 0.96 was achieved.

7.2.4.2 Isotropic Consolidation Tests on Saturated and Unsaturated Soils

Figure 7.5 shows the stress path, plotted as suction *s* against mean net stress *p* for the equalization and isotropic consolidation stages. On initial setting up in the triaxial cell, each sample is at point A, with the cell pressure and pore-air pressure both zero (resulting in $p = 0$) and a negative value of pore-water pressure produced by the compaction process. Then, a small confining pressure (i.e., 5 kPa) was applied to the

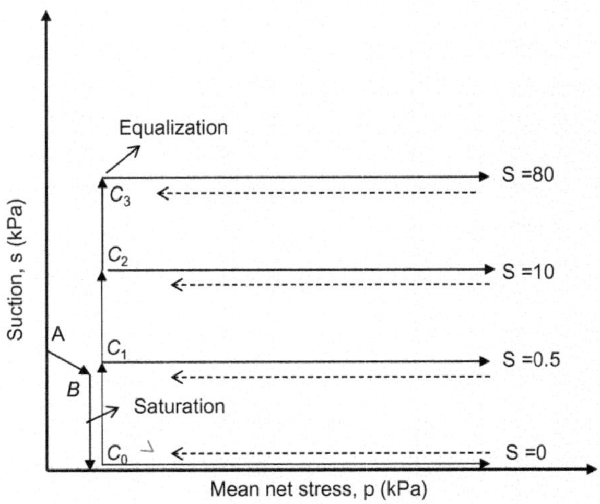

Fig. 7.5 Stress paths for isotropic consolidation tests on saturated and unsaturated soils

samples to move the soil state to point B to saturate the sample. After the specimens were saturated, the specimens were equalized to the desired suctions. During the subsequent equalization stage the soil state moves to point C_0 ($s = 0$ kPa), C_1 ($s =$ 0.5 kPa), C_2 ($s = 10$ kPa), and C_3 ($s = 80$ kPa). Then isotropic consolidation tests were conducted under selected matric suctions. Two methods were used to control the suction. The negative water column technique was adopted when the suction values were smaller than 10 kPa, and the axis-translation technique was used for suctions larger than 10 kPa.

Isotropic consolidation of saturated soil was achieved by increasing the cell pressure while keeping the back pressure unchanged. Continuously increasing cell pressure was applied during the process with loading rates ranging from 15 to 30 kPa/h depending on the effective confining pressure. During loading at a low effective confining pressure (i.e., smaller than 300 kPa), a relatively slow loading rate was used to ensure the full dissipation of excess pore-water pressure and complete drainage. During loading at a high effective confining pressure (i.e., larger than 300 kPa) as well as unloading, a relatively fast loading rate could be used. The specimens were consolidated to a high effective stress in several steps. In each step, the test was terminated if the volume change was smaller than 5 mm^3. The axial displacement and volume change of the specimen were monitored during the process. As a result, a continuous isotropic consolidation curve in the e-logp' space can be obtained. In this study, effective consolidation pressures ranging from 15 to 1000 kPa were applied.

Isotropic consolidation tests on unsaturated soil with $f_c = 90\%$ were conducted in a triaxial pressure plate system designed for measuring the volume change of unsaturated soil samples. Pore-water pressure, u_w, was applied or measured at the base of the sample through a ceramic disk. Pore-air pressure u_a was applied at the top of the sample. The cell pressure, σ_3, and pore-air pressure, u_a, were each controlled independently by regulators. In order to accurately control the pressures, two air pressure gauges with accuracy 0.5 kPa were used in the pressure control system, instead of the original pressure gauges with accuracy of 5 kPa. The isotropic consolidation tests on unsaturated specimens were achieved by increasing the cell pressure but keeping the pore water pressure and pore-air pressure unchanged. The rate of increase of cell pressure was 0.6 kPa/h to allow dissipation of the excess pore-water pressure throughout the sample. Several incremental stages were used to attain the target value of p, a rest long period (10–12 days) was allowed to ensure complete dissipation of any excess pore-water pressure. Finally, the unloading tests were conducted. The water volume changes and total volume changes were recorded in each step.

7.2.4.3 Triaxial Tests on Saturated and Unsaturated Soil Specimens

The initial suction in the prepared specimen was generally different from the specified suction value, under which the compression and the shearing tests were conducted. The first step after setting up a specimen was to change the suction present in the soil specimen. In order to keep the suction path consistent, all the unsaturated specimens were saturated first and then dried to the desired suction.

For the triaxial tests conducted on saturated specimens, after saturation and isotropic consolidation to desired confining pressures (i.e., 5, 15, and 25 kPa), the saturated specimens were sheared under drained and undrained conditions according to the test plans. The triaxial shearing tests were rate controlled, the maximum shearing rates being determined according to ASTM D4767 (2002).

For the unsaturated specimen, after saturation, the specimens were dried to the desired suctions. The volume of water flowing in/out was continuously monitored during the equalization stage. The equalization was terminated once the drainage of water was smaller than 0.01% of the total dry soil weight until within 8 h. Once each specimen was equalized at a given suction, it was isotropically compressed at a constant suction to a preselected net mean stress, *p*. This was achieved by increasing the cell pressure at a low rate to the required value but keeping both the air and water back pressures unchanged. Then the specimens were sheared at a constant rate of axial strain. Ho and Fredlund (1982) suggested that the time required to fail an unsaturated soil specimen (t_{f}) could be estimated from the following equation, which was modified from the equation of Bishop and Henkel (1962):

$$t_{\mathrm{f}} = \frac{L^2}{(1 - u_{\mathrm{f}})\eta_m c_v^m} \tag{7.4}$$

where L is one half the height of the specimen; u_{f} is the average degree of dissipation of the excess pore-water pressure at failure; $c_{\mathrm{v}}{}^{\mathrm{w}}$ is the average coefficient of consolidation with respect to the water phase; η_m is a parameter defined by $0.75/(1 + 3/\omega)$ for single drainage; and ω is an impedance factor representing the impeding effect of the ceramic disk on water drainage:

$$w = \frac{k_{\mathrm{d}} L_{\mathrm{p}}}{k_{\mathrm{w}} L_{\mathrm{d}}} \tag{7.5}$$

where k_{d} is the coefficient of hydraulic conductivity of the high air entry value disk; k_{w} is the coefficient of hydraulic conductivity of the unsaturated soil to water; L_{p} is the length of drainage path in the soil; L_{d} is the thickness of the high air entry value disk.

7.3 Shearing Behavior of Saturated Widely-Graded Soils

7.3.1 Effect of Coarse Fraction on Compressibility

The compression behavior of five types of soil with coarse contents of 20, 50, 68, 75, and 90% was investigated through isotropic consolidation tests. The initial void ratio for all the soils was 0.62. The isotropic consolidation lines obtained are presented in Fig. 7.6 along with the isotropic consolidation lines for Sacramento River Sand (Lee

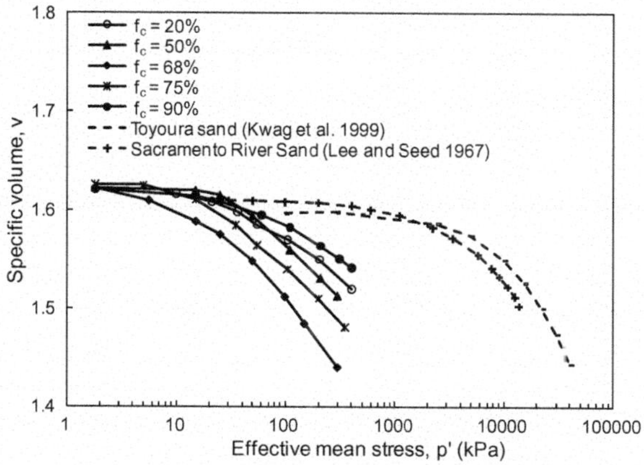

Fig. 7.6 Isotropic consolidation lines for saturated specimens

and Seed 1967) and Toyoura Sand (Kwag et al. 1999) that have the same void ratio of 0.62 for comparison. The Toyoura sand was a uniform dune sand with a mean diameter of $D_{50} = 1.6$ mm and a minimum void ratio of 0.61. The relative density was 95%, which means the sample was very densely compacted. The yield stresses of the five types of soil can be estimated using Casagrande's graphical method (Casagrande 1936). For example, a yield stress of about 60 kPa is estimated for the soil with a coarse content of 90%. Isotropic normal consolidation lines are identified within the range of the applied effective mean stress. To better compare the compressibility of these soils, the compression index, λ (=$\Delta V/\Delta \log p'$, where ΔV = volume change and p' = effective mean stress), is used. The changes in λ with coarse content are shown in Fig. 7.7. The parameter λ initially increases with coarse content and reaches its maximum at a coarse content of 68% and then decreases. As shown in Fig. 7.6, River Sand and Toyoura Sand have very low compressibility under low confining pressures,

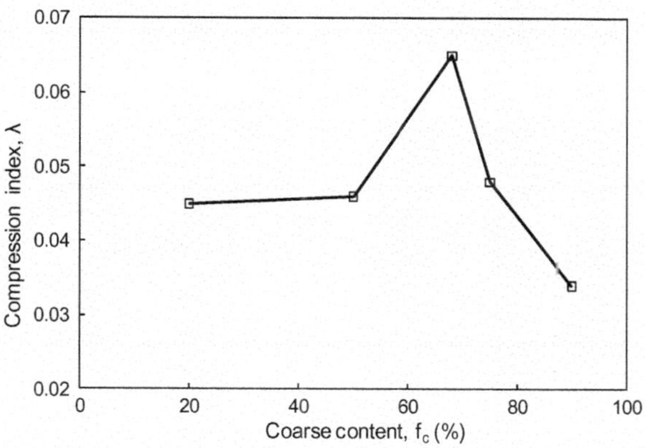

Fig. 7.7 Relationship between compression index λ and coarse content

and the compressibility increases with confining pressure. The soil compressibility will affect the volumetric behavior in drained tests or excess pore-water pressures in undrained tests. A saturated soil with high compressibility tends to contract during drained loading and generate positive pore-water pressures during undrained loading.

Based on the results of the isotropic consolidation tests in Fig. 7.7, three representative test soils with coarse contents of 50, 68, and 90% were chosen to investigate their detailed microstructures and shear behavior under drained and undrained conditions. The three types of soil are noted as sandy CL ($f_c = 50\%$), SM with gravel ($f_c = 68\%$), and SW-SM with gravel ($f_c = 90\%$), respectively.

Microstructures are used to explain the compressibility of the specimens with $f_c = 50\%$, 68%, and 90%. The specimen with $f_c = 68\%$ has the highest value of $\lambda = 0.065$ (Fig. 7.7). A high λ value means high compressibility and low stiffness of the skeleton. As Fig. 7.8b shows, the skeleton of the $f_c = 68\%$ specimen is supported

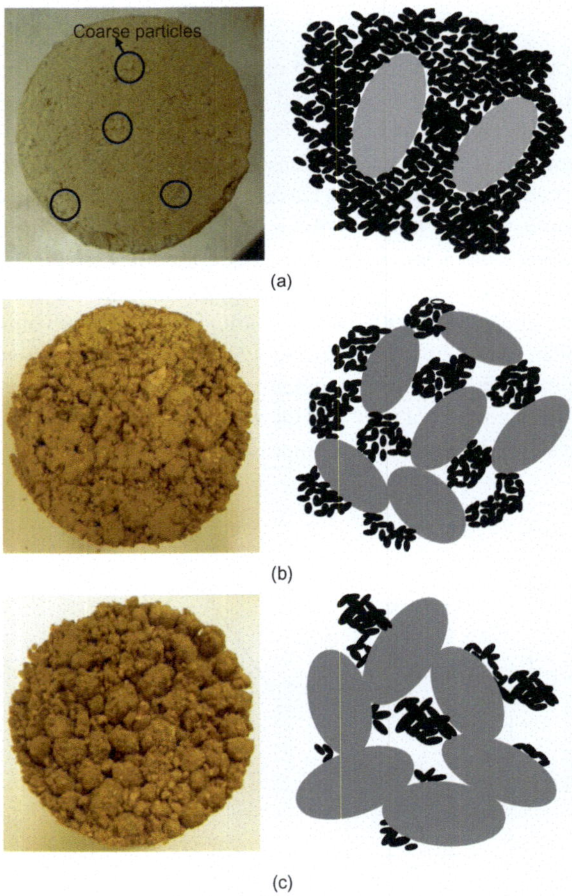

Fig. 7.8 Photographs and schematic microstructures of three soils: **a** $f_c = 50\%$; **b** $f_c = 68\%$; **c** $f_c = 90\%$

by both clay aggregates and coarse particles, which form inter-aggregate pores with the pore sizes ranging from 2 to 75 μm. Upon isotropic consolidation, the large inter-aggregate pores are compressed, which causes high compressibility of the soil. The compaction curve of the $f_c = 68\%$ soil is flatter than those of the other two soils as shown in Fig. 7.6. It means that the $f_c = 68\%$ soil can be easily compacted to a large dry density in a wide range of water contents. This also supports that the $f_c = 68\%$ soil is more compressible than the other two soils. Clay aggregates form the skeleton of the $f_c = 50\%$ soil and small intra-aggregate pores are present. The intra-aggregate pores are more difficult to compress. This is the reason for the relatively low compressibility of the $f_c = 50\%$ soil ($\lambda = 0.037$). The specimen with $f_c = 90\%$ has the lowest compressibility even though it also contains large inter-particle pores. The coarse particles form a high stiffness skeleton as shown in Fig. 7.8c, which is the reason for the low compressibility of the soil. Similar to the specimen with $f_c = 90\%$, the rigid skeleton of the dense Toyoura sand leads to the low compressibility of the sand at low confining pressures.

7.3.2 *Effect of Coarse Content on Shearing Behavior*

7.3.2.1 Effect of Coarse Content on Drained Shearing Behavior

Figures 7.9, 7.10, and 7.11 show the stress–strain relationships from the drained triaxial compression tests on the saturated $f_c = 50\%$, $f_c = 68\%$, and $f_c = 90\%$ specimens, respectively, under effective confining stresses of 5, 15, and 25 kPa. As the confining pressure increases, the volumetric behavior becomes more contractive. Figure 7.9b shows that the true critical state has not been achieved at an axial strain of 35%, as the volumetric strain continues to change even at such a large strain. However, the tests do show that the rate of increase of volumetric strain decreases toward the end of the tests. In order to compare the test results later, the state at an axial strain of 35% is taken as the critical state. This may overestimate the critical void ratio and the limitation should be noted. The stress–strain behavior of the $f_c = 68\%$ specimens under these confining stresses is all contractive as shown in Fig. 7.10b. The rate of volume change decreases as the axial strain increases, and the state at an axial strain of 35% is used to define the critical state as well. Strong strain-softening behavior is observed for the $f_c = 90\%$ specimens as shown in Fig. 7.11b. The specimens compress first and then swell until reaching the critical state. With increasing confining pressure, the strain at the peak state increases and the tendency of dilation becomes weaker. The stress–strain relationships of Toyoura Sand at an effective confining pressure of 4.9 kPa (Fukushima and Tatsuoka 1984) are also plotted in Fig. 7.11 for comparison. The void ratio of the Toyoura sand specimen is 0.68 that is slightly larger than the specimens in this study. Nevertheless, very strong dilative behavior is observed.

The failure modes of the three types of soil are shown in Fig. 7.12. The geometry is generally symmetrical, with bulging at large axial strain levels. The soil structure

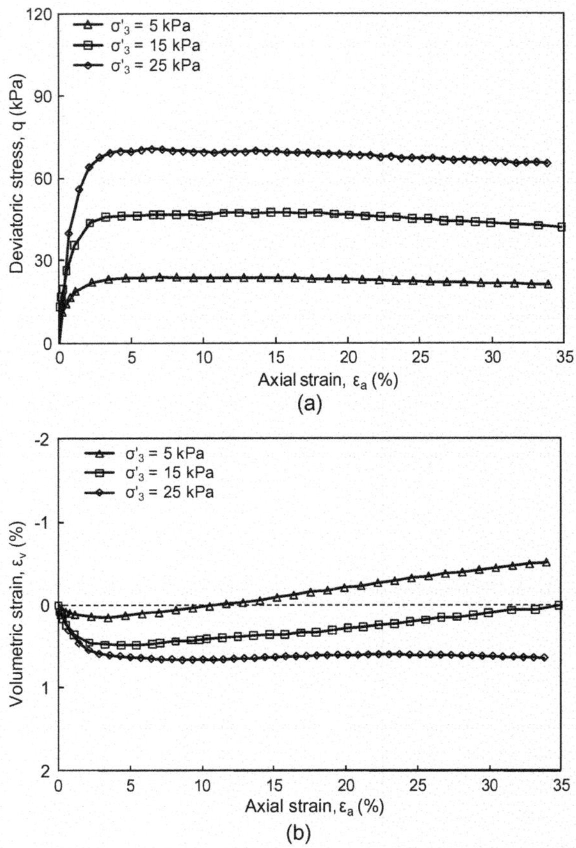

Fig. 7.9 **a** Stress–strain relationship and **b** relationship between volumetric strain and axial strain for the $f_c = 50\%$ specimen

in the maximum bulging section of each specimen is examined. During shearing, the saturated clay aggregates in the $f_c = 68\%$ specimens collapse into the inter-aggregate pores, causing contraction of the soil skeleton. After shearing, as Fig. 7.12 shows the inter-aggregate pores in the $f_c = 68\%$ specimens are filled by fines. Compared with Fig. 7.12, the particle arrangement of the $f_c = 90\%$ specimens becomes more random.

The soil compressibility affects its volumetric behavior in drained tests. The $f_c = 68\%$ specimen shows the highest compressibility upon loading, producing large contractive volumetric strains. The $f_c = 90\%$ specimen has lower compressibility due to the higher stiffness of its skeleton, causing dilation under drained loading. Compared with the soils used in this study, Toyoura Sand has the lowest compressibility under low confining pressures (Fig. 7.6), and in turn the largest dilative volumetric strain as shown in Fig. 7.11b.

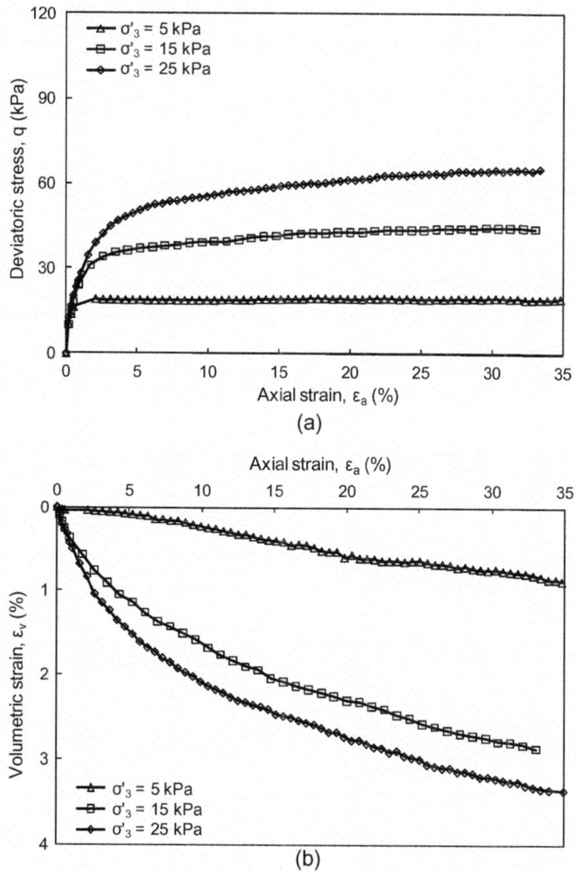

Fig. 7.10 **a** Stress–strain relationship and **b** relationship between volumetric strain and axial strain for the $f_c = 68\%$ specimen

The critical-state lines and isotropic consolidation lines in the $v - \log p'$ space are shown in Fig. 7.13. As the shearing tests were not conducted at high confining stresses, the critical-state lines were extrapolated to higher stresses assuming that they are approximately parallel to the respective isotropic consolidation lines. The values of λ for the critical-state lines are 0.046, 0.065, and 0.034 for the soils with $f_c = 50\%$, 68%, and 90%, respectively. A specimen exhibits contractive behavior when the initial state is located above its critical-state line, whereas a specimen initially located below the critical-state line shows dilative behavior. Been and Jefferies (1985) used a state parameter, ψ, to describe the state of the soil which accounts for both the void ratio and stress level. Table 7.4 summarizes the state properties of the five types of test soil and Toyoura sand. The soil response is consistent with the value of the state parameter. Strong dilation is apparent for a high negative value of state parameter (Toyoura Sand). The soil with a coarse content of 68% exhibits strong contractive

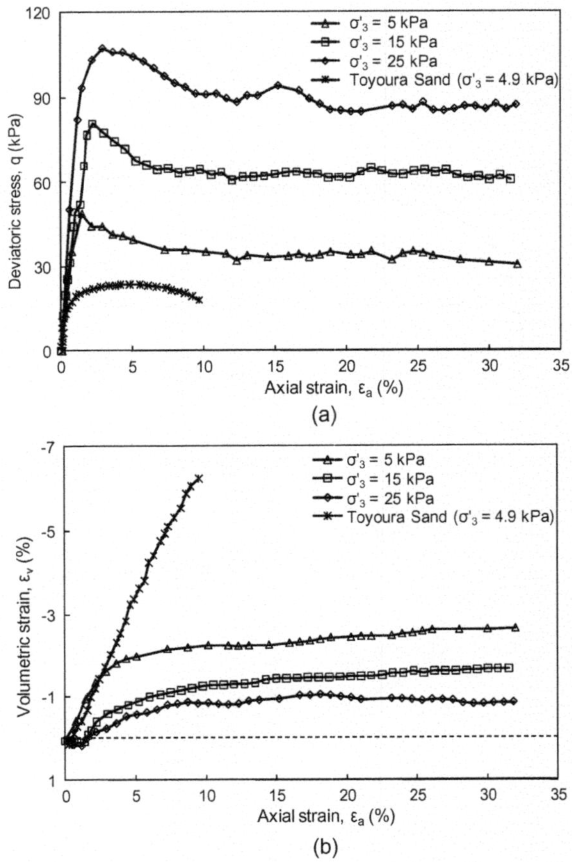

Fig. 7.11 **a** Stress–strain relationship and **b** relationship between volumetric strain and axial strain for the $f_c = 90\%$ specimen and Toyoura Sand. The Toyoura sand test results are taken from Fukushima and Tatsuoka (1984)

behavior when the state parameter is positive. However, the state parameter for the $f_c = 68\%$ specimen at an effective confining pressure of 5 kPa is slightly negative ($\psi = -0.020$) but the volumetric behavior is slightly contractive. The main reason may be the overestimation of the critical void ratio at a strain of 35% (Fig. 7.10b).

7.3.2.2 Effect of Coarse Content on Undrained Shearing Behavior

To investigate the effect of coarse content on undrained shearing behavior, undrained triaxial tests were conducted on the specimens with $f_c = 50$, 68, and 90% under an effective confining pressure of 15 kPa. Figure 7.14a shows the stress–strain relationships. The relationships between the excess pore-water pressure and the axial strain are shown in Fig. 7.14b. Strain hardening is observed for the $f_c = 50\%$ specimen,

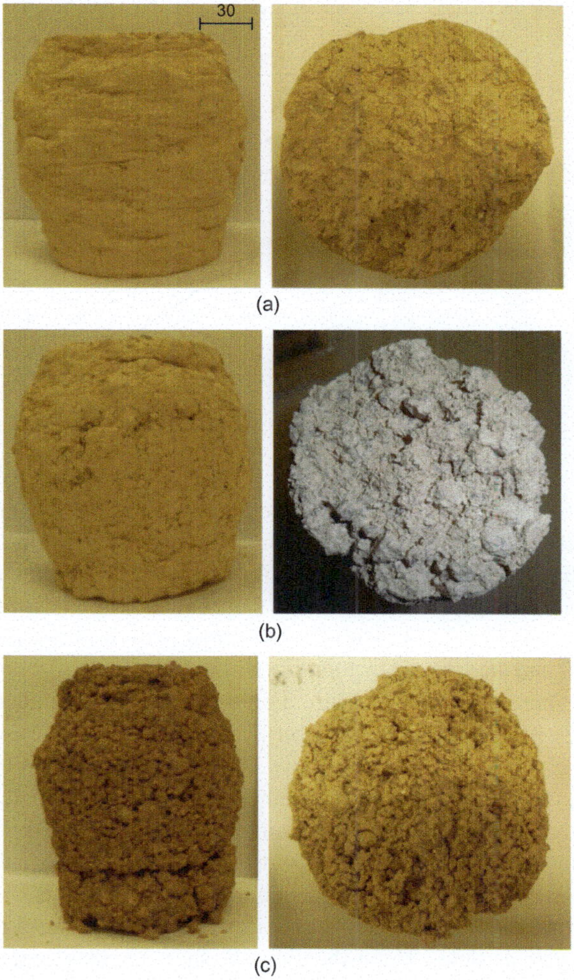

Fig. 7.12 Failure modes and cross-sectional photographs: **a** $f_c = 50\%$; b $f_c = 68\%$; **c** $f_c = 90\%$

whereas strain softening is observed for the $f_c = 68\%$ specimen. The difference in the shear strength at the end of the test and the peak strength is 23% for the $f_c = 68\%$ specimen, and positive excess pore-water pressure is recorded during shearing. For the $f_c = 90\%$ specimen, negative excess pore-water pressures develop and the shear strength is much higher than those of the other two soils.

Figure 7.15 shows the effective stress paths for the undrained tests and the critical-state lines for the three types of soil on the $p' - q$ space, along with the total stress paths for reference. The total and effective stress paths originate at the stress points along the horizontal axis corresponding to the isotropic consolidation pressures. The total stress path is at a slope of 3:1. The pore-water pressure can be determined from the horizontal distance between the total stress path and the effective stress path.

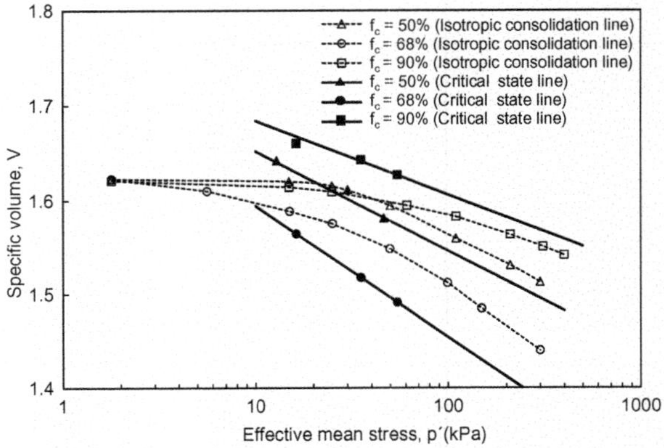

Fig. 7.13 Critical-state lines and isotropic consolidation lines on the $v - \log p'$ space

For the $f_c = 50\%$ specimen in Fig. 7.15a, the pore-water pressure increases initially and decreases after it reaches a turning point. The turning point indicates that the soil behavior changes from contractive to dilative. Finally, the effective stress path reaches the critical-state line. The $f_c = 68\%$ specimen exhibits strong contractive tendency as shown in Fig. 7.15b. The effective stress path reaches a peak. After that, the soil collapses and the stress path goes down toward the critical-state line, resulting in continuously increasing pore pressure. For the $f_c = 90\%$ specimen (Fig. 7.15c), after the effective stress path crosses the critical-state line, the pore-water pressure increment changes from positive to negative, which is an indication of dilation.

In summary, the behavior of the $f_c = 50\%$ specimen is governed by preconsolidated clay aggregates. Its behavior is similar to that of slightly overconsolidated clays. Dilative tendency was observed in undrained shearing under a confining pressure of 15 kPa. For the $f_c = 68\%$ specimen, a large volume of 2–75 μm inter-aggregate pores are formed between the clay aggregates and between the clay aggregates and the coarse particles. Upon shearing, the saturated clay aggregates collapse into the inter-aggregate pores. Therefore, strong contractive tendency and positive pore-water pressure are observed under undrained shearing. Coarse particles form the skeleton of the $f_c = 90\%$ specimen. Due to the high stiffness of its skeleton, the inter-particle pores are difficult to compress. Upon undrained shearing, the coarse particles rearrange themselves and show highly dilative tendency.

7.3.2.3 Comparison of Drained and Undrained Behavior Under Different Coarse Contents

Figure 7.16a, c, and e shows the critical-state lines on the $p' - q$ space based on the combined data from the drained tests under confining pressures of 5, 15, and 25 kPa and the undrained tests under the confining pressure of 15 kPa. As shown in the figures, the drained and undrained tests reach the same critical-state line on the $p' -$

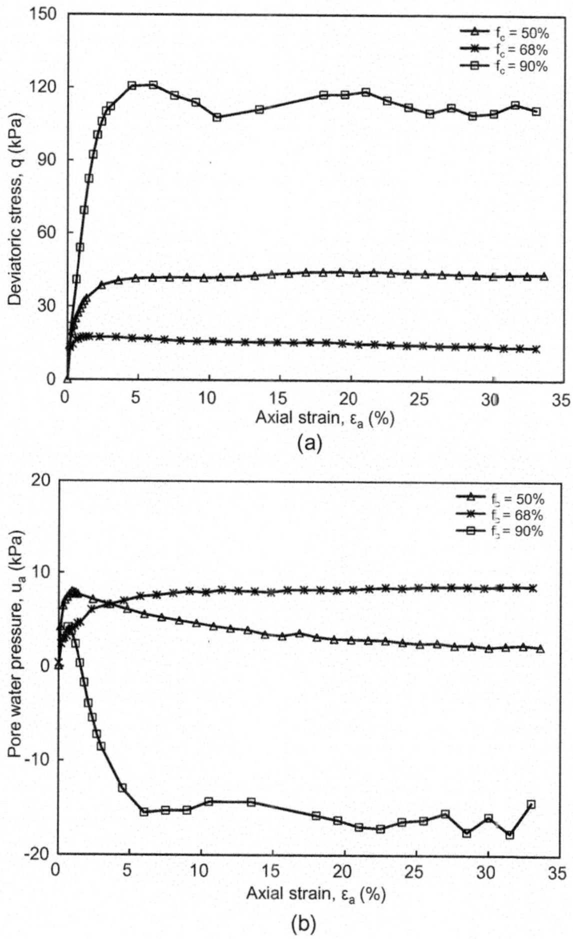

Fig. 7.14 **a** Stress–strain relationship and **b** relationship between excess pore-water pressure and axial strain under a confining pressure of 15 kPa

q space for each test soil. Figure 7.16b, d, and f shows the stress paths on the $v - \log p'$ space for the drained and undrained tests under a confining pressure of 15 kPa. The initial states for each soil in Fig. 7.16b, d, and f are almost identical in the drained and undrained tests. The initial states for the $f_c = 50\%$ specimen lie on the dry side of the critical-state line (Fig. 7.16b). Under the undrained condition, the void ratio remains constant and the stress state reaches the critical state eventually. Under the drained condition, the void ratio decreases at the beginning of the test due to contraction and then increases due to dilation. The critical-state void ratio values are almost the same in the drained and undrained tests, so are the critical deviatoric stresses in Fig. 7.17. As shown in Fig. 7.16d, the initial states of the $f_c = 68\%$ specimen are on the wet side of the critical-state line. The void ratio in the drained test keeps decreasing due to

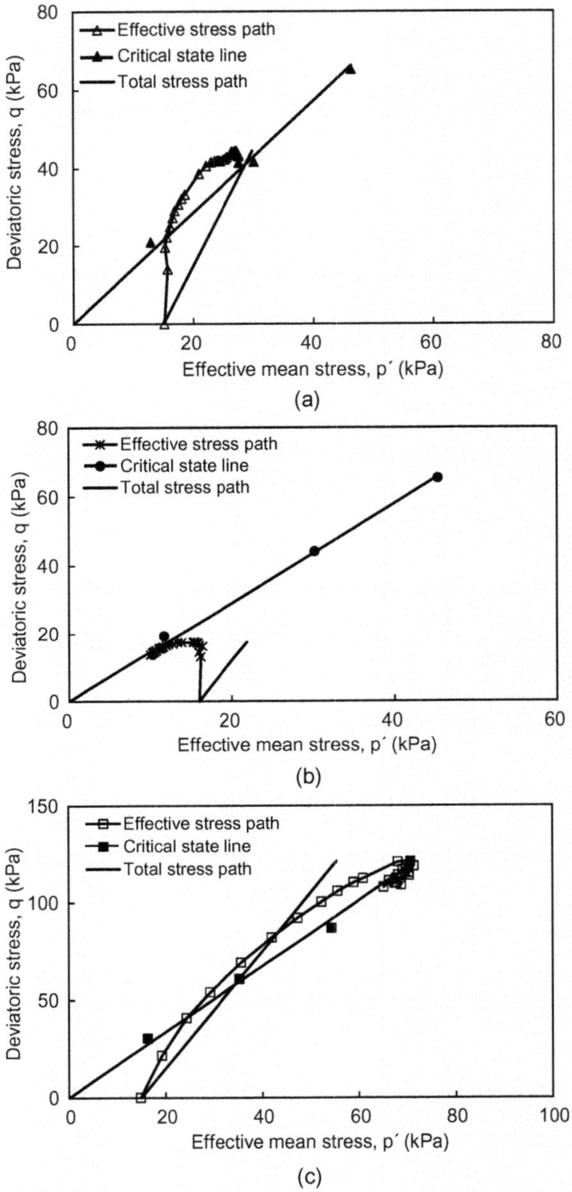

Fig. 7.15 Effective stress paths in undrained tests under a confining pressure of 15 kPa: **a** f_c = 50%; **b** f_c = 68%; **c** f_c = 90%

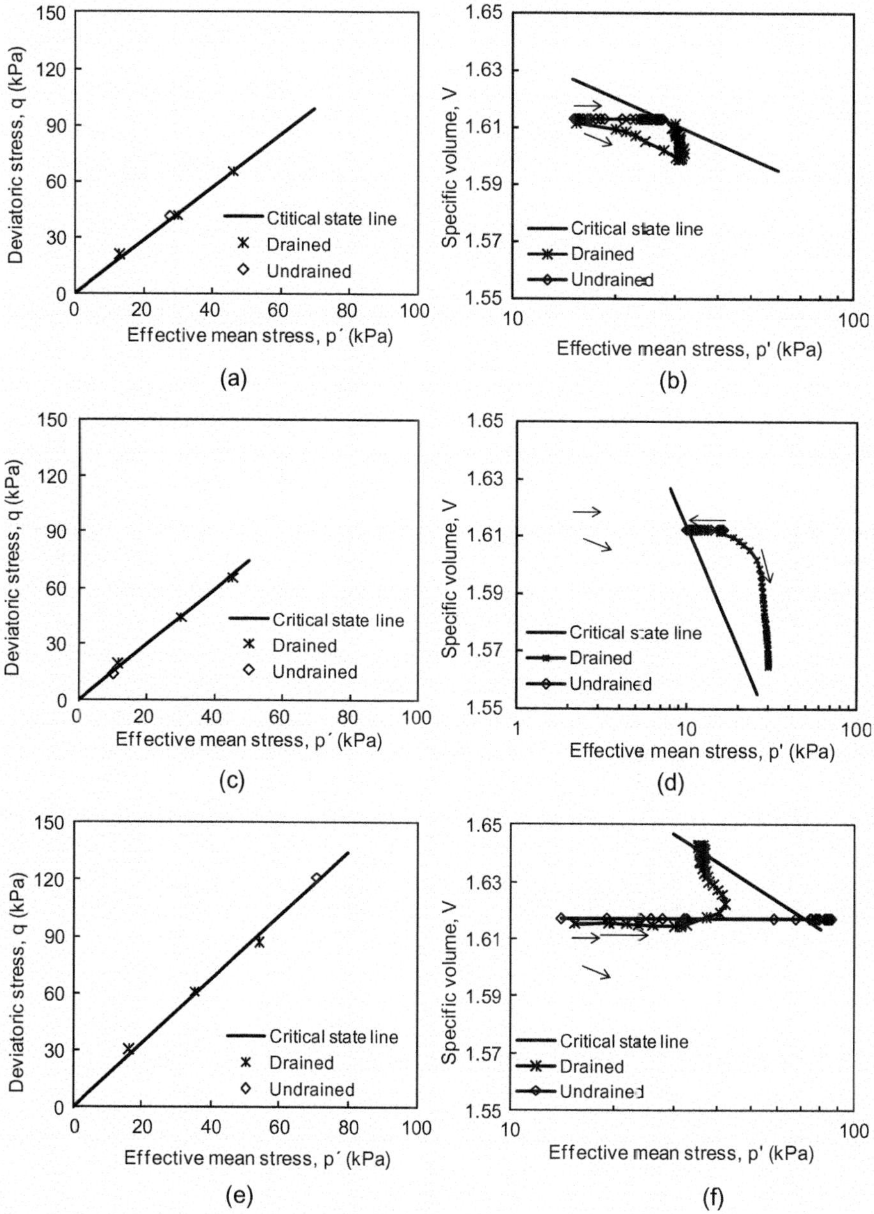

Fig. 7.16 Critical-state lines and effective stress paths on the $v - \log p'$ space under drained and undrained conditions: **a**, **b** $f_c = 50\%$; **c**, **d** $f_c = 68\%$; **e**, **f** $f_c = 90\%$

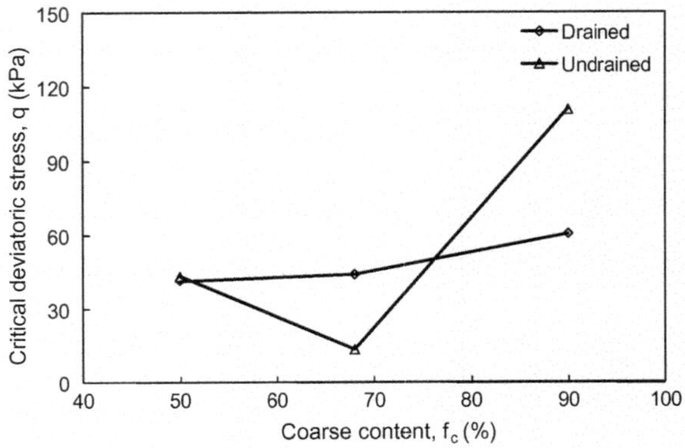

Fig. 7.17 Relationships between critical deviatoric stress and coarse content under a confining pressure of 15 kPa

the collapse of the skeleton. Hence, the critical-state void ratio is much smaller than that under the undrained condition. For the $f_c = 90\%$ specimen, the coarse particles form a stiff skeleton. When the soil is subjected to drained shearing from the dry side, the particles rearrange and the void ratio increases as shown in Fig. 7.16f.

The critical deviatoric stress increases with the coarse content under an effective confining stress of 15 kPa in the drained condition (Fig. 7.17). However, under the undrained condition, the critical deviatoric stress decreases when the coarse content increases from 50 to 68%, which is in accordance with the fact that the stiffness of the skeleton becomes smaller as the coarse content increases from 50 to 68%. The $f_c = 68\%$ specimen develops the largest positive pore-water pressure during undrained shearing due to the high compressibility of the soil. Hence, its critical deviatoric stress is the smallest (Fig. 7.17). When the coarse content is increased to 90%, a stiff skeleton is formed by coarse particles. The soil then shows dilative tendency during undrained loading and the undrained critical deviatoric stress is high.

7.3.3 Effect of Coarse Content on Critical Friction Angle

The slope of the critical-state line, M_{cs}, is obtained by fitting the test data on the $p' - q$ space as shown in Fig. 7.19a, c, and e. The values of M_{cs} are 1.41, 1.48, and 1.68 for the $f_c = 50\%$, 68%, and 90% specimens, respectively. The corresponding critical friction angles, ϕ'_{cs}, are 34.9°, 36.4°, and 41.1°, respectively. As shown in Fig. 7.18, the critical friction angle increases with f_c. Grain-size analysis shows that the values of the mean particle diameter, D_{50}, are 0.06, 1.2, and 3 mm for the $f_c = 50\%$, $f_c = 68\%$, and $f_c = 90\%$ specimens, respectively. Hence, ϕ'_{cs} increases with increasing D_{50}. The mass fraction of coarse content, f_c, is an important parameter for a coarse-grained soil. A low f_c value means the coarse particles are not in contact, and the soil behavior is primarily governed by the fines matrix. At a high value of f_c, the coarse particles are in contact and the soil behavior is governed by these coarse particles. This is in agreement with findings from studies on sand mixtures (Irfan and Tang 1993; Vallejo and Mawby 2000).

The critical-state friction angle is also affected by particle mineralogy and angularity. The same type of soil with similar mineralogy and angularity was used in

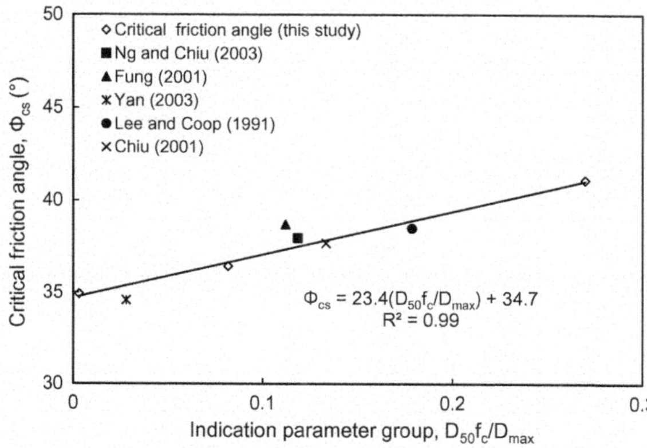

Fig. 7.18 Relationship between critical friction angle and particle-size distribution indication parameters

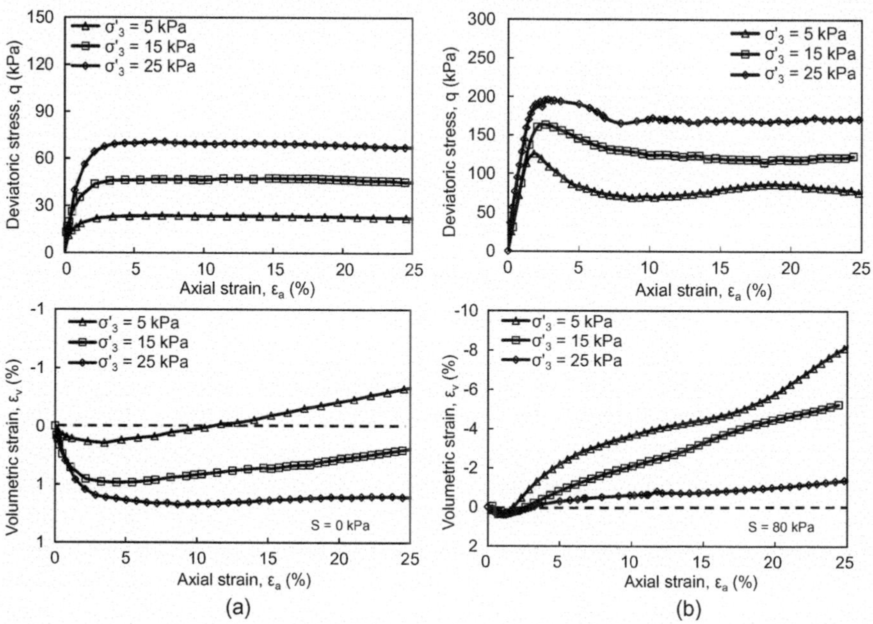

Fig. 7.19 Stress–strain relationships for the soil with $f_c = 50\%$ under **a** $S = 0$ kPa; **b** $S = 80$ kPa

Table 7.5 Particle-size properties and values of critical-state friction angle of CDG soils

Group symbols	Particle-size distribution properties			Indication parameter group, $D_{50} f_c/D_{max}$	Critical friction angle (°)
	Mean diameter, D_{50} (mm)	Maximum diameter, D_{max} (mm)	Coarse content, f_c		
Sandy CL	0.06	10	0.5	0.003	34.9
SM with gravel	1.2	10	0.68	0.0816	36.4
SW-SM with gravel	3.0	10	0.9	0.27	41.1
Ng and Chiu (2003)	0.8	5	0.74	0.1184	38.0
Fung (2001)	0.7	5	0.8	0.112	38.7
Yan (2003)	0.2	5	0.7	0.028	34.6
Lee and Coop (1995)	1.9	10	0.94	0.1786	38.5
Chiu (1991)	0.9	5	0.74	0.1332	37.7

this study. Hence, the grain-size distribution is the main factor affecting the shearing behavior of these test soils.

Table 7.5 summarizes the results of tests on the same kind of soil with information of f_c and D_{50}. An empirical equation is sought to express the relation between ϕ'_{cs} and f_c and D_{50}. The maximum particle size, D_{max}, is also used to make the influence factor group, $D_{50} f_c/D_{max}$, dimensionless. As indicated in Fig. 7.19, the following linear regression fits the relationship between ϕ'_{cs} and $D_{50} f_c/D_{max}$ very well with a correlation coefficient of $R^2 = 0.99$:

$$\phi'_{cs} = 23.4 D_{50}\, f_c/D_{max} + 34.7 \tag{7.6}$$

7.3.4 Implication to Engineering Practice

Many natural colluvial soil slopes consist of widely-graded coarse soils represented by SM with gravel ($f_c = 68\%$). The results of this study show that under the same initial void ratio of 0.62, SM with gravel with a coarse content of about 70% is highly compressible and exhibits strong contractive behavior even under very low confining pressures of 5–25 kPa. This is in contrast to the dilative behavior of clean or dune sand with the same initial void ratio under low confining pressures. The contractive behavior of the $f_c = 68\%$ specimen implies that soil deposits of this nature are susceptible to instability during rainfall infiltration since large positive pore-water pressures could develop during undrained shearing. Frequent rain-induced failures of recently deposited colluvial slopes have indeed been observed (e.g., Zhang et al.

2011). Geotechnical engineers can use the results from this study in slope stability analysis and landslide prediction. The critical friction angle is a function of coarse content, f_c, and mean particle diameter, D_{50}. The critical friction angles at sampling sites can be estimated using Eq. (7.6). Rapid evaluation of landslide hazards might be conducted by comparing the slope angles with the critical friction angles of the soils. The shear strength parameters obtained in this study can be applied to limit equilibrium analysis of slope stability. The soil behavior (dilative or contractive) can also be evaluated through the analysis of the coarse contents of the colluvial soils. For example, under very small effective confining pressures, loose colluvial soils may exhibit contractive behavior if the coarse content is smaller than 70% so that the probability of failure under rainfall conditions may be high.

7.4 Shearing Behavior of Unsaturated Widely-Graded Soils

7.4.1 *Drained Shearing Behavior of Unsaturated Soils*

A total of 31 suction-controlled drained triaxial shear tests (refers to the stress path 2 in Fig. 7.3 and drained test plans in Table 7.4) were performed on the samples with $f_c = 50\%$, 68%, and 90%. These tests were conducted under very low net confining pressures of 5, 15, and 25 kPa. Five suctions (0, 0.5, 10, 80, and 280 kPa) were selected.

Figures 7.19, 7.20, 7.21, 7.22, 7.23, and 7.24 show the variation of the deviator stress, q, and the volumetric strain, ε_v, with the axial strain, ε_a, at different suctions. At the saturated state, the $f_c = 68\%$ specimens show strain-hardening behavior; the $f_c = 90\%$ specimens show distinct strain-softening behavior; and the $f_c = 50\%$ specimens exhibit slightly strain-softening behavior. This is because the saturated clay aggregates in the $f_c = 68\%$ specimens tend to collapse into the large inter-aggregate pores and contract. In the unsaturated state, most of the stress–strain curves exhibit strain-softening behavior. The strain-softening behavior is more pronounced when the specimen is subject to high suctions. The failure mode changes from ductile

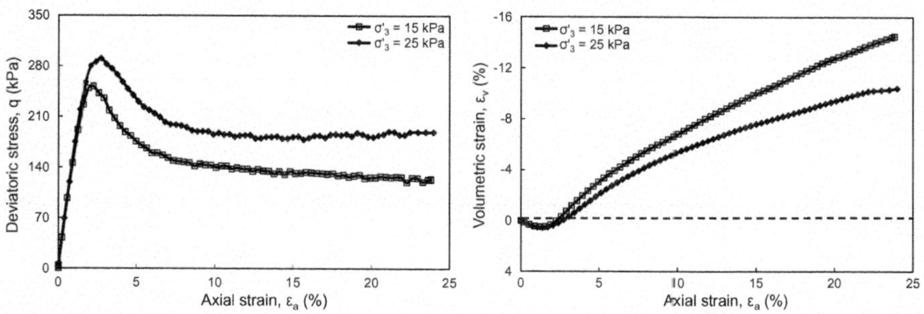

Fig. 7.20 Stress–strain relationships for the soil with $f_c = 50\%$ under $S = 280$ kPa

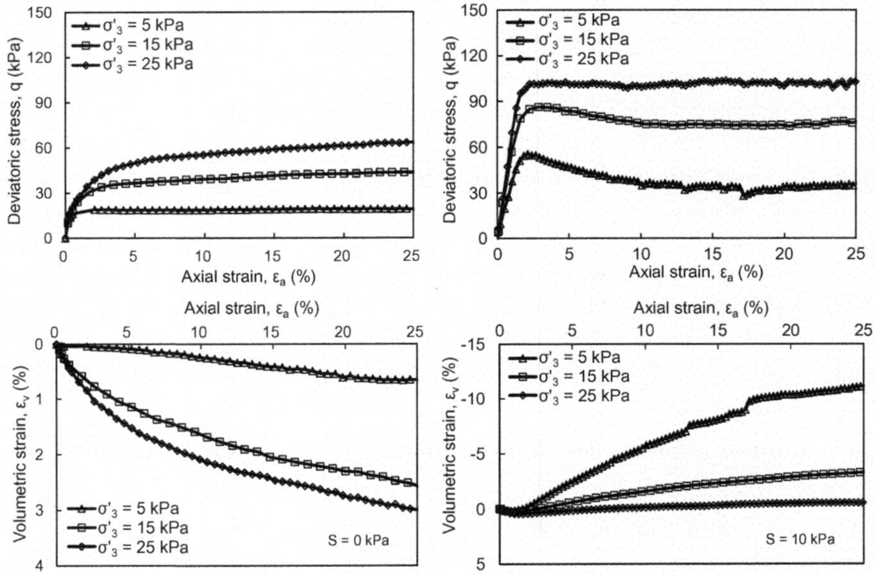

Fig. 7.21 Stress–strain relationships for the soil with $f_c = 68\%$ under $S = 0$ kPa and $S = 10$ kPa

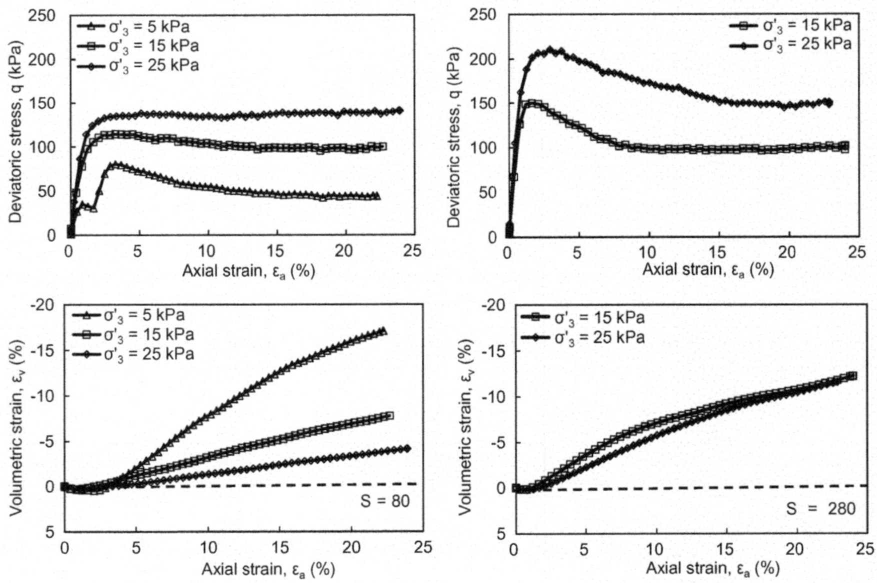

Fig. 7.22 Stress–strain relationships for the soil with $f_c = 68\%$ under $S = 80$ kPa and $S = 280$ kPa

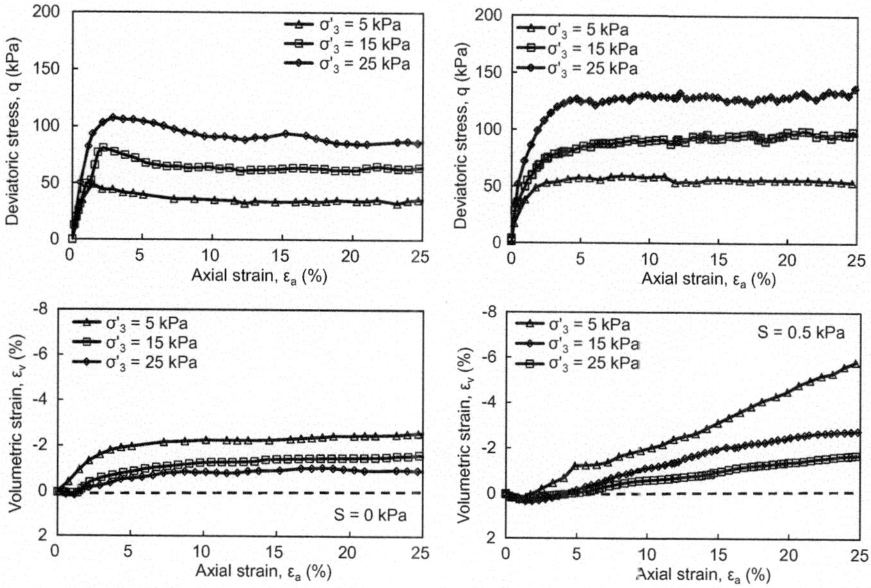

Fig. 7.23 Stress–strain relationships for the soil with $f_c = 90\%$ under $S = 0$ kPa and $S = 0.5$ kPa

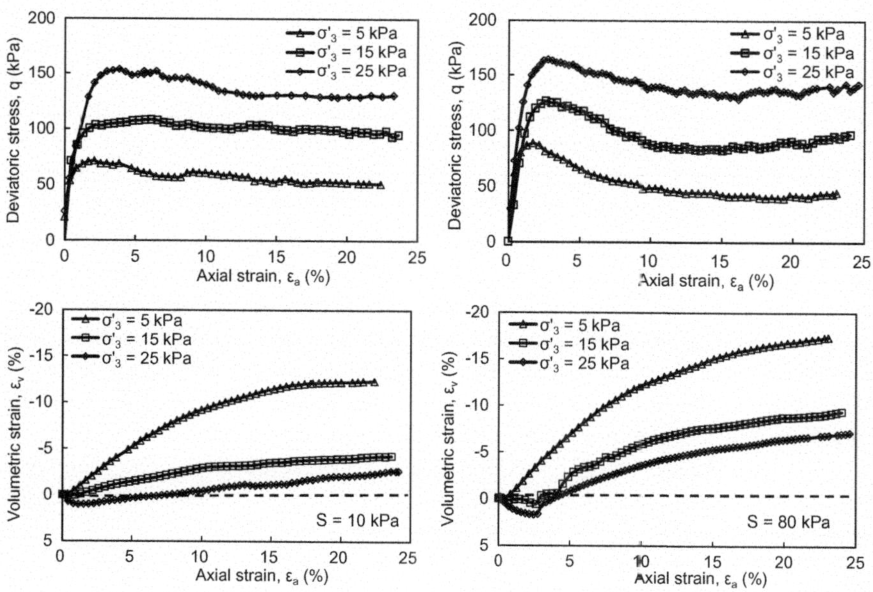

Fig. 7.24 Stress–strain relationships for the soil with $f_c = 90\%$ under $S = 10$ kPa and $S = 80$ kPa

Fig. 7.25 Failure modes of the soil with $f_c = 50\%$ under an effective confining pressure of 25 kPa: **a** ductile failure under $S = 0$ kPa; **b** brittle failure under $S = 280$ kPa

failure to brittle failure with increasing suction. Take the soil with $f_c = 50\%$, for example, under the saturated condition, the failure mode is ductile with bulging at large axial strain levels. When the specimen is subject to a high suction value of 280 kPa, a shear band is observed (Fig. 7.24). The ductility of the specimens increases with increasing net confining stress. The lower the net confining stress is, the more distinct the dilatant behavior is, and the smaller the axial strain required for the specimen to reach the maximum deviator stress (Fig. 7.25).

The volumetric change is compressive at first, then dilative as the axial strain increases during shearing. It is observed that, for a given net normal stress, the degree of dilation increases with suction. This is because a higher suction results in more tightly bound and rigid aggregates of soil particles. The aggregates do not slip easily but rather tend to ride up over each other as the shear stress is increased. The tendency to dilate is most pronounced at low net normal stress values and high suctions. The fabric plays a significant role in the behavior of the unsaturated compacted soil. When the three types of soil are subject to the same net normal stress of 15 kPa and the same suction of 80 kPa, the volumetric strains at the critical state are −5.3, −7.8, and −9.4% for the soils with $f_c = 50\%$, 68%, and 90%, respectively. The volumetric strain at the critical state increases with increasing coarse content. Therefore, the shear-induced volume change is related to the SWCCs of the soils.

7.4.2 *Variation of Peak Shear Strength of Unsaturated Soils with Coarse Content*

7.4.2.1 Peak Shear Strength Envelopes

Figure 7.26 shows the failure envelopes for the three types of unsaturated soil by taking the shear strength measured at the saturated condition as a reference. Matric suction contributes to the peak shear strength via two mechanisms: the contribution

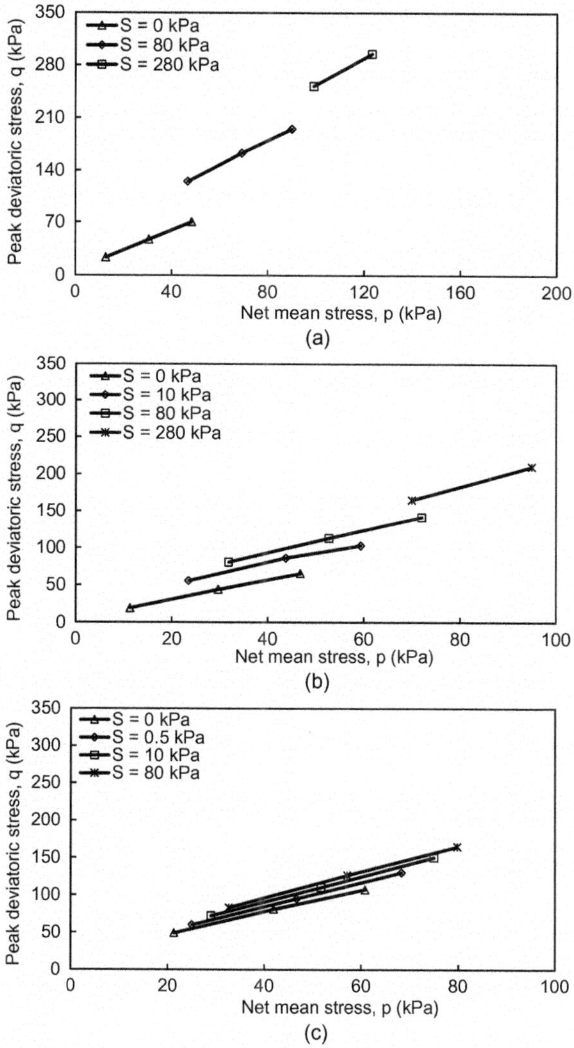

Fig. 7.26 Peak state lines in the $p - q$ space: **a** $f_c = 50\%$; **b** $f_c = 68\%$; **c** $f_c = 90\%$

of capillary force to inter-particle normal stresses and the effect of suction on soil dilatancy. The effects decrease with increasing coarse content. For example, under an effective confining pressure of 5 kPa, when the suction increases from 0 to 80 kPa, the increments in the peak shear strength are 101, 61, and 34 kPa for the soils with $f_c = 50\%$, 68%, and 90%, respectively.

As shown in Fig. 7.26, the failure envelopes for the three types of soil are almost linear. The strength envelopes at different suctions are not exactly parallel. Let $M_{peak}(s)$ denote the slope of the failure envelopes and $\mu_{peak}(s)$ denote the intercept of the failure envelope. The differences in $M_{peak}(0)$ and $\mu_{peak}(0)$ of the corresponding saturated soils are due to the initial fabric. As shown in Fig. 7.27a, the soil with $f_c =$ 90% has the highest value of $M_{peak}(0)$. This is because of the high stiffness skeleton formed by coarse particles. Upon drained shearing, the coarse particles rearrange themselves and gain higher $M_{peak}(0)$ through dilation. The soil with $f_c = 68\%$ has the lowest $M_{peak}(0)$ as it exhibits strong contractive behavior. The soil with $f_c = 50\%$

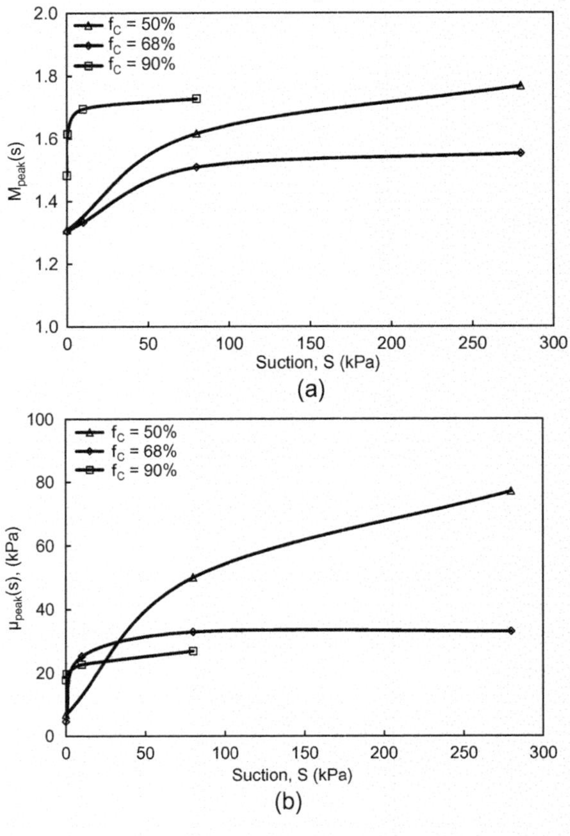

Fig. 7.27 Friction slope M_{peak}(s) and cohesion intercept $\mu_{peak}(s)$ against suction: **a** $M_{peak}(s)$; **b** $\mu_{peak}(s)$

gives intermediate $M_{\text{peak}}(0)$ values as its skeleton is governed by preconsolidated clay aggregates.

Values of friction slope $M_{\text{peak}}(s)$ and cohesion intercept $\mu_{\text{peak}}(s)$ as functions of suction are shown in Fig. 7.27. Both $M_{\text{peak}}(s)$ and $\mu_{\text{peak}}(s)$ increase nonlinearly with suction until a specific suction is reached, after which they remain nearly constant. The suction to reach the peak decreases with coarse content. These suction values lie in the first residual stage of the bimodal SWCCs, and therefore, the effect of suction has reached its ultimate in the first drainage region. For the $f_{\text{c}} = 50\%$ specimen, the residual suction has not been reached in the considered suction range, so $M_{\text{peak}}(s)$ does not reach a steady value. The increase of the cohesion intercept, $\mu_{\text{peak}}(s)$, with suction has been reported by Escario et al. (1989). But the increase of $M_{\text{peak}}(s)$ with suction has not been reported. The increase of $M_{\text{peak}}(s)$ of unsaturated soil is governed by the change in the soil "structure" associated with the inter-particle capillary stress generated by water menisci. In the first drainage region, as the water drains from the inter-aggregate pores, the capillary stress induced by the water meniscus surrounding the contact points of soil particles gradually increases. The capillary stress acts as an internal stress at every contact point and stiffens the soil skeleton. Thus, $M_{\text{peak}}(s)$ increases continuously in this region. In the second drainage region, the desaturation rate is much smaller than that in the first region since the intra-aggregate pores are dominant. So the contribution of suction to parameters $M_{\text{peak}}(s)$ and $\mu_{\text{peak}}(s)$ is limited. The effects of suction on $M_{\text{peak}}(s)$ and $\mu_{\text{peak}}(s)$ decrease with increasing coarse content.

7.4.2.2 The Maximum Dilatancy, *D*

The maximum rate of dilatancy coincides with the peak state in the stress–strain relationship. The dilatancy of a soil can be examined by studying $\mathrm{d}\varepsilon_v^p/\mathrm{d}\varepsilon_q^p$, where $\mathrm{d}\varepsilon_v^p$ and $\mathrm{d}\varepsilon_q^p$ are the plastic volumetric and deviatoric strain increments, respectively. In the triaxial condition with axial and volumetric strain measurements, one has $\varepsilon_q = \varepsilon_a - \varepsilon_v/3$. It is assumed that the elastic components of the volumetric and shear strain increments are relatively small and negligible. The dilatancy of the soil can thus be approximated by $\mathrm{d}\varepsilon_v/\mathrm{d}\varepsilon_q$.

Figure 7.28 shows the variation of the maximum dilatancy, *D*, with the degree of saturation under different confining pressures. The contribution of suction to the maximum dilatancy depends on the coarse content. The initial fabric dominates the behavior of the saturated soil. The maximum dilatancy is the largest for the soil with $f_{\text{c}} = 90\%$ due to the high stiffness of the skeleton. The soil with $f_{\text{c}} = 68\%$ shows contractive behavior attributed to its highly compressible skeleton. As shown in Fig. 7.28, the dilatancy increases with a decrease in the degree of saturation for the three types of soil. When the specimens become unsaturated, the suction-induced fabric difference dominates the variation. The effect of degree of saturation on dilatancy is related to the grading of the soil. As shown in Fig. 7.3, the pores are expected to shrink as the water meniscus progresses from large pores to smaller pores when the suction is increased. The dilatancy of the soil with $f_{\text{c}} = 50\%$ increases

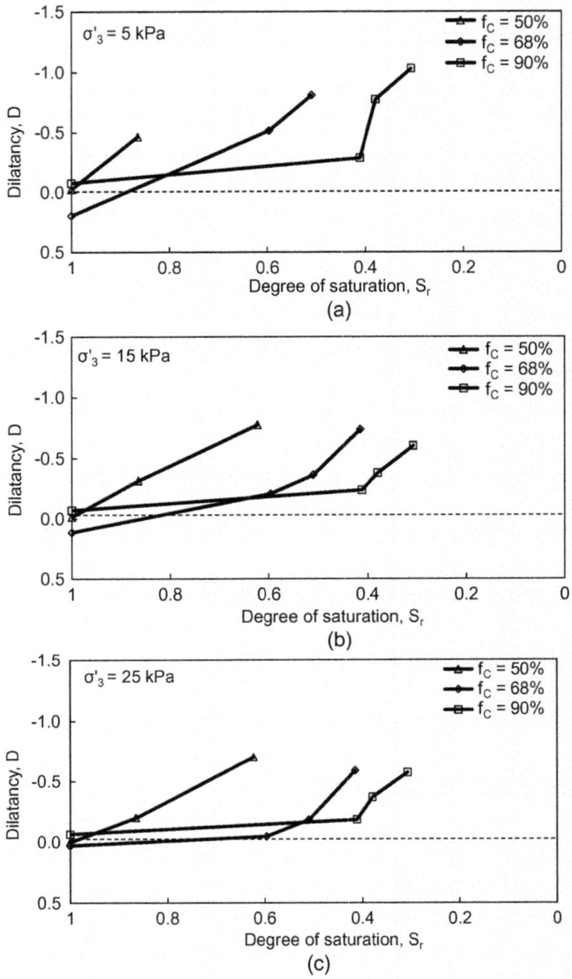

Fig. 7.28 Variation of the maximum dilatancy, *D*, with degree of saturation for the three types of soil: **a** $\sigma'_3 = 5$ kPa; **b** $\sigma'_3 = 15$ kPa; and **c** $\sigma'_3 = 25$ kPa

monotonously with the degree of saturation in the considered suction range (Fig. 7.28) because the fabric change is from the gradual drainage of water from the intra-aggregate pores.

For the soils with a bimodal SWCC ($f_c = 68$ and 90%), the pore water in the two-pore series contributes to dilatancy differently during the desaturation process. In stage *b*, the bulk water drains from the inter-aggregate pores and the water phase is continuous. During the drying process, the capillary stress increases gradually and the inter-particle contacts become more stable, which increases the dilatancy. In stage *c*, the water phase becomes discontinuous except for the water bounded between the particles and the clay aggregates in the form of water bridges. In this stage, the radii of menisci are small and the capillary stresses are high, which increases the resistance to particle slippage and the dilatancy. In the second drainage region, the clay aggregates play a major role as water mainly drains from the intra-aggregate pores. The stiffness of the clay aggregates increases after shrinkage of the clay aggregates. The aggregates tend to ride up over each other, making the sample more dilative.

7.4.3 *Variation of the Critical-State Strength of Unsaturated Soils with Coarse Content*

The critical-state framework for unsaturated soil has been proposed based on saturated soil mechanics. Wheeler and Sivakumar (1995) suggested critical-state relationships for q and specific volume v as

$$q = M(s)p + \mu(s) \tag{7.7}$$

$$v = \Gamma(s) - \psi(s)\ln\frac{p}{p_{\mathrm{at}}} \tag{7.8}$$

where M(s) is the slope of the projection of the critical-state line in the q: p space; μ(s) is the apparent cohesion intercept caused by suction; $\Gamma(s)$ is the intercept at p = 1 kPa; p_{at} is the atmospheric pressure; and $\psi(s)$ is the slope of the projection of the critical-state line in the v – log$p$ space. All the four parameters are functions of suction.

The stress–strain relationships (Figs. 7.21, 7.22, 7.23, and 7.24) of the soils under different suctions show that the true critical state has not been achieved, as the volumetric strain continues to change even at very large strains. However, the tests do show that the rate of increase of volumetric strain decreases toward the end of the tests. In order to compare the test results later, the last point of each test at an axial strain of approximately 25% is taken as the critical state.

7.4.3.1 Critical-State Lines in the $p - q$ Space

Figure 7.29 summarizes the critical-state data in the p:q space from the drained triaxial tests on the soils with f_c = 50, 68, and 90% at different suctions. The critical-state shear strength is meant to eliminate the contribution of soil dilatancy to the shear strength. The critical-state values of q and p at each suction level appear to fall on a straight line. The critical-state lines at different suctions are again not parallel. Equation (7.8) is used to obtain parameters M(s) and μ(s).

M(0) under the saturated state is an intrinsic material property characterizing the intrinsic friction coefficient of the particle assembly for a given initial soil fabric, but M(s) depends on suction. Figure 7.30a shows the variation of M(s) with suction,

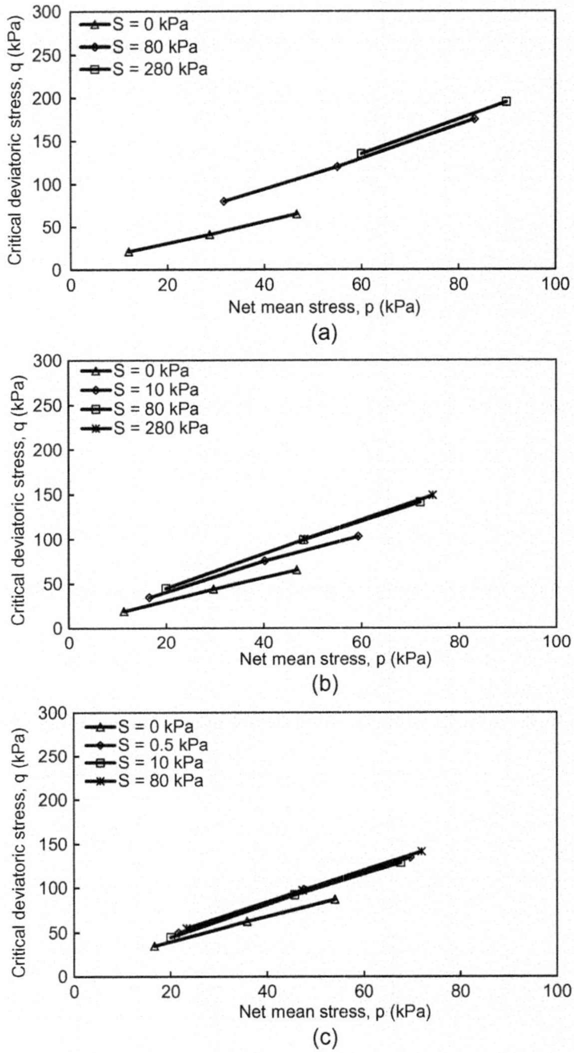

Fig. 7.29 Critical-state lines in the $p - q$ space: **a** $f_c = 50\%$; **b** $f_c = 68\%$; **c** $f_c = 90\%$

which has the same trend as $M_{\text{peak}}(\text{s})$. Values of $M(\text{s})$ of the soils with $f_c = 68\%$ and $f_c = 90\%$ increase with suction until the first residual stage (stage *c*) of each of the soils has been reached. However, the $M(\text{s})$ of the $f_c = 50\%$ soil increases monotonously since the residual suction has not been reached. It can also be observed from Fig. 7.29 that the critical-state lines lie close to each other for suctions larger than 10 kPa and 80 kPa for the soils with $f_c = 90\%$ and $f_c = 68\%$, respectively. Thus, for the soils with $f_c = 90\%$ and $f_c = 68\%$, the effect of suction has reached its maximum value within the first drainage region.

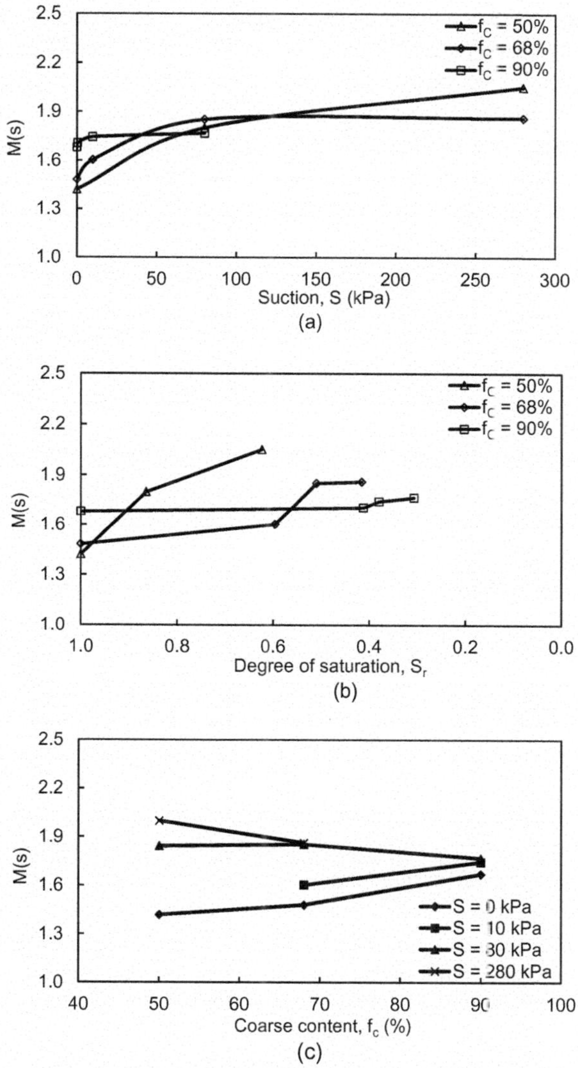

Fig. 7.30 Variation of M(s) of the three types of soil at critical state: **a** with suction; **b** with degree of saturation; **c** with coarse content

The variation of M(s) can also be expressed as a function of degree of saturation as shown in Fig. 7.30b. The change in M(s) with degree of saturation can be related to the different roles of the pore water. The value of M(s) for the $f_c = 50\%$ specimen increases almost monotonously as water drains from the intra-aggregate pores gradually. The increases of M(s) for the soils with $f_c = 68\%$ and 90% exhibit highly nonlinear features, which is related to the drainage process of the bimodal SWCC (Fig. 5.6). In stage *b*, the pore water drains sharply from the inter-aggregate pores at low suctions. The capillary stress generates a resistance to the slippage at the contacts between the particles (or aggregates) which affect $M(\mathrm{s})$. In stage *c*, the meniscus water is bounded between the particles and the clay aggregates in the form of a bridge. The capillary stress increases as the meniscus radius decreases. The generated capillary force increases the friction between particles. Therefore, $M(\mathrm{s})$ increases with decreasing degree of saturation. In the second drainage region, the stiffness of the clay aggregates increases as the clay aggregates become unsaturated and provide a high friction resistance.

The variation of $M(\mathrm{s})$ with coarse content under different suctions is shown in Fig. 7.30c. $M(\mathrm{s})$ increases with coarse content. The increases of $M(0)$ under the saturation state are due to the initial fabric. Under the unsaturated condition, both the soil fabric and suction affect $M(\mathrm{s})$. $M(\mathrm{s})$ can be described as a sum of $M(0)$ and an incremental component due to suction, $\Delta M(s)$. The effect of suction on $\Delta M(s)$ decreases with increasing coarse content. For example, when the suction increases from 0 to 80 kPa, the values of $M(\mathrm{s})$ for the soils with $f_c = 50\%$, 68%, and 90% increase by 0.581 (from 1.419 to 2), 0.381 (from 1.48 to 1.861) and 0.222 (from 1.67 to 1.892), respectively.

$\mu(\mathrm{s})$ quantifies the effect of suction on the shear strength which represents a similar effect of ϕ^b proposed by Fredlund et al (1978) and M_b proposed by Toll (1990). Vanapalli et al. (1996) pointed out that the area of the pore water in contact with soil particles is responsible for the increase in the shear strength of soils. So the value of $\mu(\mathrm{s})$ is related to the SWCC of the soil. The log scale is used in this plot to reflect the changes in $\mu(\mathrm{s})$ under very low suctions. A value of 0.01 kPa, which is smaller than the smallest AEV (0.07 kPa), is used to represent the saturated state. As shown in Fig. 7.31, the $\mu(\mathrm{s})$ of the soil with $f_c = 50\%$ increases with suction monotonically. However, the values of $\mu(\mathrm{s})$ for the soils with $f_c = 68\%$ and 90% show bimodal features which are related to their bimodal SWCCs. In the first drainage region (stage *b*), $\mu(\mathrm{s})$ increases with increasing matric suction and decreases near the first residual state (stage *c*). In the second drainage region, $\mu(\mathrm{s})$ increases monotonically with increasing suction. The $\mu(\mathrm{s})$ of the soil with a bimodal SWCC is similar to that of sand in the first drainage region, but similar to that of clay in the

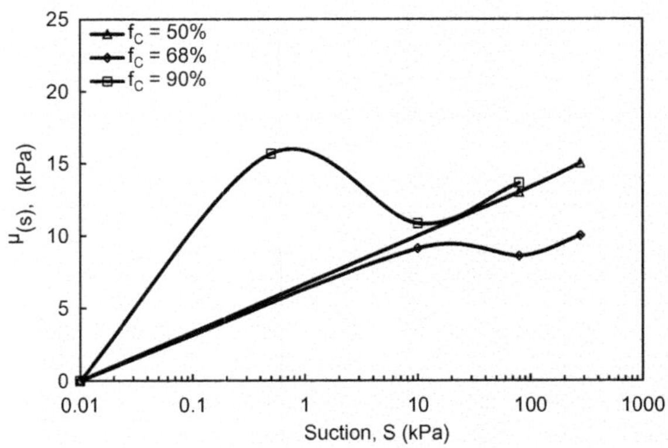

Fig. 7.31 Variation of $\mu(\mathrm{s})$ with suction

second drainage region. In stage *b*, with increasing suction, the additional bonding force induced by suction increases monotonically and μ(s) increases accordingly. When the soil enters stage *c* which is the residual stage of the first drainage region, the water menisci vanish and may not transmit suction effectively to the soil particles or aggregate contact points. Thus, μ(s) decreases with increasing suction. In the second drainage region, the clay aggregates control the unsaturated behavior of the soil. μ(s) increases as water drains from intra-aggregate pores. Note that this is different from the peak cohesion intercept $\mu_{\text{peak}}(s)$, which keeps increasing at high suctions attributed to the suction-induced dilatancy.

7.4.3.2 Critical-State Lines in the $v - \log p$ Space

Figure 7.32a–c plots the relationships between specific volume v (i.e., $v = 1 + e$) and the logarithm of net mean stress for the soils with $f_{\text{c}} = 50\%$, 68%, and 90%, respectively. There appears a linear relationship at a constant matric suction. The position and the slope of the critical-state line at zero suction are very different from those of the critical-state lines in unsaturated conditions. This further confirms the significant influence of suction on the soil fabric. Equation (7.8) is used to obtain parameters $\Gamma(s)$ and $\psi(s)$.

Figures 7.33 and 7.34 plot the variations of $\psi(s)$ and $\Gamma(s)$ with suction or degree of saturation. For the saturated soil, the soil with $f_{\text{c}} = 68\%$ has the largest value of $\psi(0)$. This is because the $f_{\text{c}} = 68\%$ specimen has the largest volume of inter-aggregate pores (4–10 μm) and the saturated clay aggregates tend to collapse into the large inter-aggregate pores and contract. As shown in Fig. 7.33a, the values of $\psi(s)$ increase with suction. Similar results are also observed by other researchers (Wheeler and Sivakumar 1995; Toll and Ong 2003). However, the values of $\psi(s)$ and $\Gamma(s)$ of the $f_{\text{c}} = 68\%$ specimen and $f_{\text{c}} = 90\%$ specimen reach a peak at the first residual stages of the soils and maintain the value thereafter. This is because the desaturation rate and the contribution of suction in the second drainage region are very small as the intra-aggregate pores dominate. For the $f_{\text{c}} = 50\%$ specimen, the residual suction has not been reached in the considered suction range; hence, $\psi(s)$ and $\Gamma(s)$ do not reach a steady value. The suction values at the peak state decrease with coarse content. Figures 7.33b and 7.34b plot the variation of $\psi(s)$ and $\Gamma(s)$ with the degree of saturation. Similar to the variation of M(s), $\psi(s)$ and $\Gamma(s)$ increase monotonously with the degree of saturation for the $f_{\text{c}} = 50\%$ specimen as the pore water drains from the intra-aggregate pore gradually. The increase of parameters $\psi(s)$ and $\Gamma(s)$ for the soils with $f_{\text{c}} = 68$ and 90% is related to their bimodal SWCCs. In stage *b*, the capillary stress increases as bulk water drains from the inter-aggregate pores. In stage *c*, the capillary stress also increases as the radius of the meniscus water decreases. The generated capillary stress acts as an internal stress and generates a resistance to the slippage at the contacts between particles (or aggregates), which increase the dilatancy and $\Gamma(s)$. In stage c, the stiffness of the aggregates increases as the clay aggregates become unsaturated. The coarse particles and aggregates tend to be more dilative, resulting in higher values of $\Gamma(s)$.

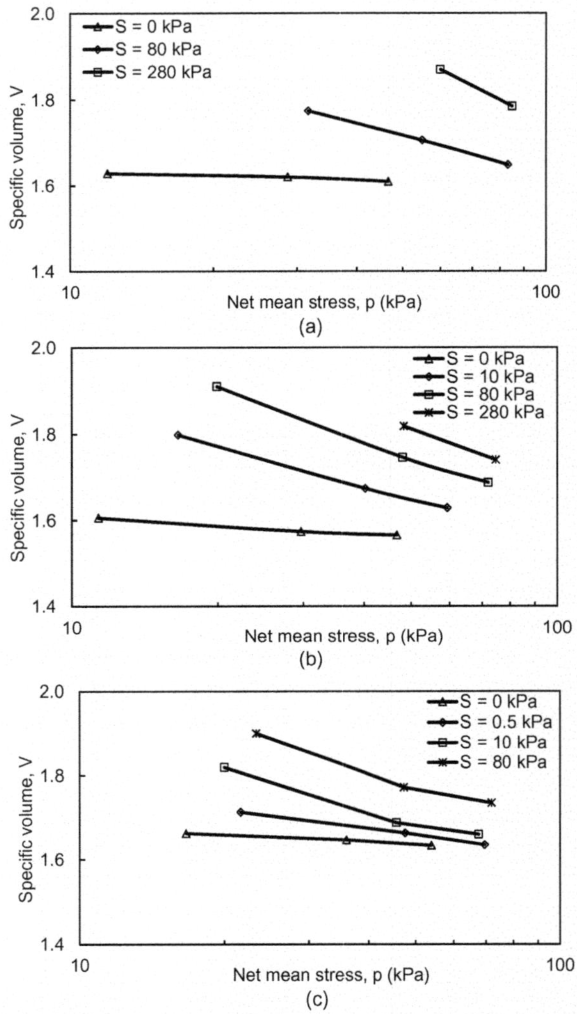

Fig. 7.32 Critical-state lines in the $v - \log p$ space: **a** $f_c = 50\%$; **b** $f_c = 68\%$; **c** $f_c = 90\%$

The aggregate particles significantly influence the shear-induced volumetric behavior and the critical state. The shrinkage ability increases with increasing fines content as shown in Fig. 7.33. The role of clay aggregates is more marked in the finer soil. Figures 7.33c and 7.34c show that the contribution of suction on $\psi(s)$ and $\Gamma(s)$ decreases as increasing coarse content. The experimental data on critical-state lines for unsaturated soils reported in this study show a clear pattern of the variations of $\psi(s)$ and $\Gamma(s)$ with suction, which may be useful for the development of constitutive models for unsaturated soils, especially bimodal soils.

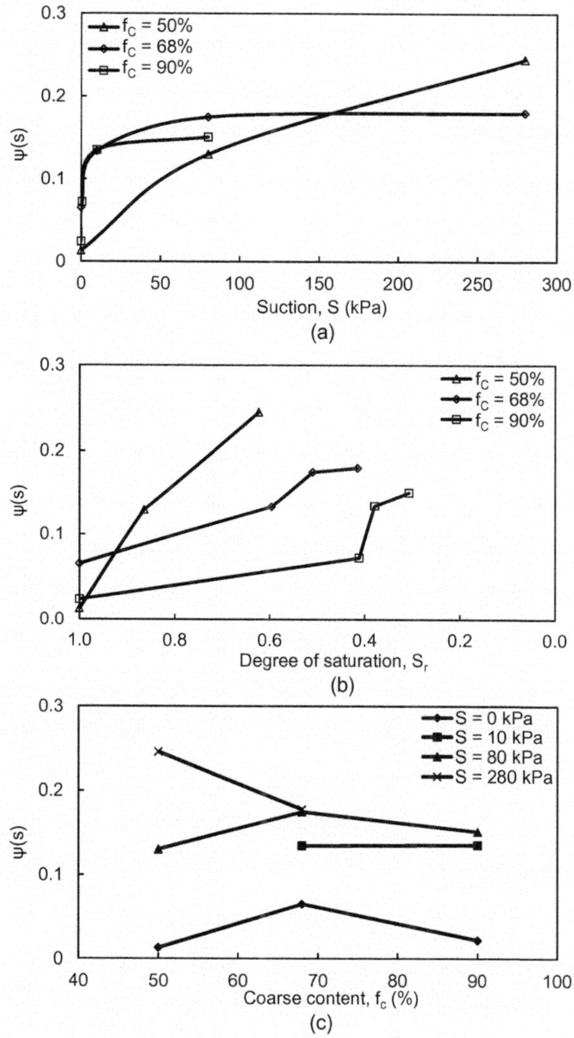

Fig. 7.33 Variation of ψ(s): **a** with suction; **b** with degree of saturation; **c** with coarse content

7.5 Summary

In this chapter, the effect of coarse content on the compressibility and shear strength of saturated and unsaturated coarse soils under very low confining pressures are investigated. The main findings are as follows:

The soil compression initially increases with coarse content and reaches the maximum at a coarse content of 68%. Thereafter the trend is reversed with further increases of coarse content. The specimen with $f_c = 68\%$ has the highest compressibility due to the presence of large inter-aggregate pores in the soil. The specimens with $f_c = 90\%$ have the lowest compressibility due to the high stiffness of the skeleton of the soil. Small intra-aggregate pores are dominant in the $f_c = 50\%$ specimen, which is difficult to compress and result in low compressibility of the soil.

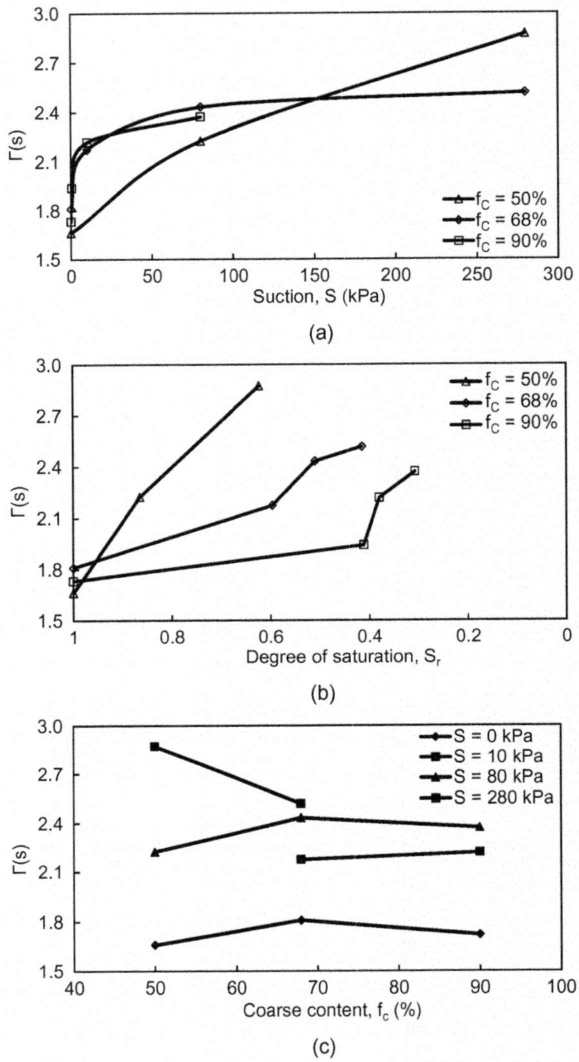

Fig. 7.34 Variation of Γ(s): **a** with suction; **b** with degree of saturation; **c** with coarse content

The coarse content affects the shear behavior significantly. Under the same initial void ratio of 0.62 and low confining pressure of 15 kPa, a soil with a coarse content near the critical value of 70% shows strong contractive behavior during drained loading and generates high positive pore-water pressure during undrained loading. The saturated clay aggregates in the $f_c = 68\%$ specimens tend to collapse into the large inter-aggregate pores and contract. Soils with lower or higher coarse contents than the critical value tend to be more dilative.

The peak shear strength increases with suction due to the contribution of capillary force to inter-particle normal stresses and the effect of suction on soil dilatancy. The effect of suction on the friction slope $M_{\text{peak}}(s)$ and cohesion intercept $\mu_{\text{peak}}(s)$ of a bimodal soil reaches its maximum within the first drainage region. The contribution of suction to the peak shear strength decreases with increasing coarse content.

The maximum dilatancy, *D*, decreases with degree of saturation due to suction-induced fabric changes. For the bimodal soils with $f_c = 68$ and 90%, the pore water in the two-pore series contributes to dilatancy differently during the desaturation process.

Critical-state parameters $M(s)$, $\psi(s)$, and $\Gamma(s)$ increase with suction until the residual or first residual stage of the soil has been reached. The contribution of suction to $\mu(s)$ changes from unimodal to bimodal features with increasing coarse content.

References

Au SWC (1998) Rain-induced slope instability in Hong Kong. Eng Geol 51(1):1–36

ASTM D4767-02. (2002) Standard Test Method for Consolidated Undrained Triaxial Compression Test for Cohesive Soils. Annual Book of ASTM Standards, American Society for Testing and Materials, Philadelphia, Pa

Bandini V, Coop MR (2011) The influence of particle breakage of the location of the critical state line of sands. Soils Found 51(4):591–600

Been K, Jefferies MG (1985) A state parameter for sands. Geotechnique 35(2):99–112

Bishop AW, Donald B (1961) The experimental study of partly saturated soil in the triaxial apparatus. In: 5th international conference on geotechnical engineering and soil mechanics, vol 1, pp 13–21

Bishop AW, Henkel DJ (1962) The measurement of soil properties in the triaxial test. E. Arnold, London

Bolton MD (1986) The shear strength and dilatancy of sands. Geotechnique 36(1):65–78

Casagrande A (1936) The determination of the preconsolidation load and its practical significance. In: Proceedings of the international conference on soil mechanics and foundation engineering, vol 3. Cambridge, MA, pp 60–64

Cascini L, Cuomo S, Pastor M, Sorbino G (2010) Modeling of rainfall-induced shallow landslides of the flow-type. J Geotech Geoenviron Eng 136(1):85–98

Chakraborty T, Salgado R (2010) Dilatancy and shear strength of sand at low confining pressures. J Geotech Geoenviron Eng 136(3):527–532

Chen R (2007) Experimental study and constitutive modelling of stress-dependent coupled hydraulic hysteresis and mechanical behavior of an unsaturated soil. Ph.D. thesis. The Hong Kong University of Science and Technology

Chiu CF (1991) "Mechanical behavior of compacted decomposed granite soil." Ph.D. thesis, The Hong Kong University of Science and Technology, Hong Kong

Dash HK, Sitharam TG (2011) Undrianed monotonic response of the sand-silt mixtures: effect of nonplastic fines. Geomech Geoeng 6(1):47–58

Escario V, Juca JFT, Coppe MS (1989) Strength and deformation of partly saturated soils. In: Congrès international de mécanique des sols et des travaux de fondations, vol 12, pp 43–46

Fannin RJ, Eliadorani A, Wilkinson JMT (2005) Shear strength of cohesionless soils at low stress. Geotechnique 55(6):467–478

Fredlund DG (1975) A diffused air volume indicator for unsaturated soils. Can Geotech J 12:533–539

Fredlund DG, Morgenstern NR, Widger RA (1978) Shear strength of unsaturated soils. Can Geotech J 15(3):313–321

Fukushima S, Tatsuoka F (1984) Strength and deformation characteristics of saturated sand at extremely low pressure. Soils Found 24(4):30–48

Fredlund DG, Rahardjo H (1993) Soil mechanics for unsaturated soils. John Wiley & Sons, Inc, New York

Fung W T (2001) "Experimental study and centrifuge modeling of loose fill slope." M.Phil. thesis, The Hong Kong Univ of Science and Technology Hong Kong
Ho DYF, Fredlund DG (1982) Strain rates for unsaturated soil shear strength testing. In Proceedings 7th South-east Asian Geotechnical Conference. Hong Kong
Irfan TY, Tang KY (1993) Effect of coarse fractions on the shear strength of colluviums. GEO report no. 23. Geotechnical Engineering Office, Hong Kong
Kwag JM, Ochiai H, Yasufuku N (1999) Yielding stress characteristics of carbonate sand in relation to individual particle fragmentation strength. In: AI-Shafei (ed) Engineering for calcareous sediments, vol 90, pp 79–86
Lade PV, Liggio CD Jr, Yamamuro JA (1998) Effects of non-plastic fines on minimum and maximum void ratios of sand. Geotech Test J 21(4):336–347
Lee KL, Seed HB (1967) Drained strength characteristics of sands. J Soil Mech Found Div ASCE 9(6):117–141
Lee IK, Coop MR (1995) "The intrinsic behavior of a decomposed granite soil." Geotechnique 45(1):117-130
Li XS, Chan CK, Shen CK (1988) An automatic triaxial testing system. In: Advanced triaxial testing of soil and rock, ASTM STP977, pp 95–106
Li X, Zhang LM, Li JH (2009) Development of a modified axis translation technique for measuring SWCCs for gravel soils at very low suctions. Geotech Test J 32(6):1–11
Ng CWW, Chiu CF (2003) "Laboratory study of loose saturated and unsaturated decomposed granitic soil." J Geotech Geoenviron Eng 129(6):550-559
Pitman TD, Robertson PK, Sego DC (1994) Influence of fines on the collapse of loose sands. Can Geotech J 31(5):728–739
Rahardjo H, Leong EC, Rezaur RB (2008) Effect of antecedent rainfall on pore-water pressure distribution characteristics in residual soil slopes under tropical rainfall. Hydrolog Process 22(4):506–523
Sharma RS (1998) Mechanical behavior of unsaturated highly expansive clays. Ph.D. thesis. Oxford University
Toll DG (1990) A framework for unsaturated soil behavior. Geotechnique 40(1):31–44
Toll DG, Ong BH (2003) Critical state parameters for an unsaturated residual sandy clay. Geotechnique 53(1):93–103
Vallejo LE, Mawby R (2000) Porosity influence on the shear strength of granular material-clay mixtures. Eng Geol 58(2):125–136
Vanapalli SK, Fredlund DG, Pufahl DE, Clifton AW (1996) Model for the prediction of shear strength with respect to soil suction. Can Geotech J 33:379–392
Wheeler SJ, Sivakumar V (1995) An elasto-plastic critical state framework for unsaturated soil. Geotechnique 45(1):35–53
Wood DM, Maeda K (2008) Changing grading of soil: effect on critical states. Acta Geotech 3:3–14
Yamamuro JA, Lade PV (1998) Steady-state concepts and static liquefaction of silty sands. J Geotech Geoenviron Eng 124(9):868–877
Yamamuro JA, Wood FM, Lade PV (2008) Effect of depositional method on the microstructure of silty sand. Can Geotech J 45(11):1538–1555
Yan W M (2003) "Experimental study and constitutive modeling of recompacted completely decomposed granite." Ph.D. thesis, The Hong Kong Univ. of Science and Technology, Hong Kong
Zhan LT (2003) Field and laboratory study of an unsaturated expansive soil associated with rain-induced slope instability. Ph.D. thesis. The Hong Kong University of Science and Technology
Zhang LL, Zhang J, Zhang LM, Tang WH (2011) Stability analysis of rainfall-induced slope failure: a review. Geotech Eng 164(5):299–316
Zhao HF (2013) Behavior of saturated and unsaturated coarse widely graded soils under low confining pressures. Ph.D. thesis. The Hong Kong University of Science and Technology
Zhao HF, Zhang LM (2013) Effect of coarse content on the shear behavior of unsaturated widely graded soils. Can Geotech J 51:1371–1383
Zhao HF, Zhang LM, Chang DS (2013) Behavior of coarse widely graded soil under low confining pressures. J Geotech Geoenviron Eng 139(1):35–48

Chapter 8
Bimodal Features for Widely-Graded Soils

8.1 Introduction

The microporosity structure of coarse granular soils with various coarse/fines contents has been introduced in Chap. 3. The skeletons of five types of compacted and saturated specimens change from clay aggregates to coarse particles gradually. The soils with coarse contents greater than 70% have a coarse-controlled microstructure and the soils with coarse contents lower than 70% have a fines-controlled microstructure. The coarse-controlled soils, such as GW-GM with sand and SM with gravel, dual-porosity structure are formed during the soil compaction and show limit changes after wetting and drying. As presented in Chap. 5, Figs. 5.6, 5.7, and 5.32, the SWCCs and SHCFs for colluvial soils with high coarse fractions show a bimodal feature with two air entry values and is essentially relates to a dual-porosity structure.

Extensive laboratory investigations have been conducted on the shear strength behavior of fine-grained soils. Numerous equations have also been published for estimating the shear strength of unsaturated soils. Most of these equations use the SWCC as a controlling parameter function together with the saturated shear strength parameters to predict the unsaturated shear strength of soil (Fredlund et al. 1978; Vanapalli et al. 1996; Oberg and Sallfors 1997; Lee et al. 2005). The typical SWCC for a fine-grained soil is a sigmoid. Kim and Borden (2011), Sheng et al. (2011) and Guan et al. (2010) evaluated some of shear strength equations. These equations provide a reasonable prediction tool for the shear strength of fine-grained soils. However, the existing shear strength prediction equations may evaluate rather low values of shear strength under low suction conditions. The air entry value (AEV) of the coarse-grained soil is typically very low (e.g., smaller 2 kPa) and the soil desaturates with modest increases in matric suction. The shear strength envelopes exhibit nonlinear behavior, showing a peak near the AEV and then decreasing with further increases in matric suction (e.g., Houston et al. 2008; Likos et al. 2012). Therefore, those equations should be used with caution for coarse-grained soils.

In this chapter, the bimodal features for coarse widely-graded soil are summarized. A new bimodal SWCC model is proposed based on the mathematical description

X. Li et al., *Fundamental Behavior of Unsaturated Widely-Graded Soil*,
https://doi.org/10.1007/978-981-19-3402-5_8

of the dual-porosity geometry introduced in Chap. 3. Then, a bimodal hydraulic conductivity function model and bimodal shear strength model are proposed based on a dual-porosity structure and related to the bimodal SWCC model proposed.

8.2 Bimodal Pore Structure of Coarse Widely-Graded Soil

8.2.1 Pore-Size Distribution and Soil Fabric

The colluvial soils with a coarse content larger than 70% are characterized as coarse-controlled soils, such as GW-GM with sand and SM with gravel. Figure 8.1a shows a digital image of compacted SW-SM with gravel. The soil forms a skeleton of grain-to-grain contacts, with the finer particles partially filling the large inter-particle

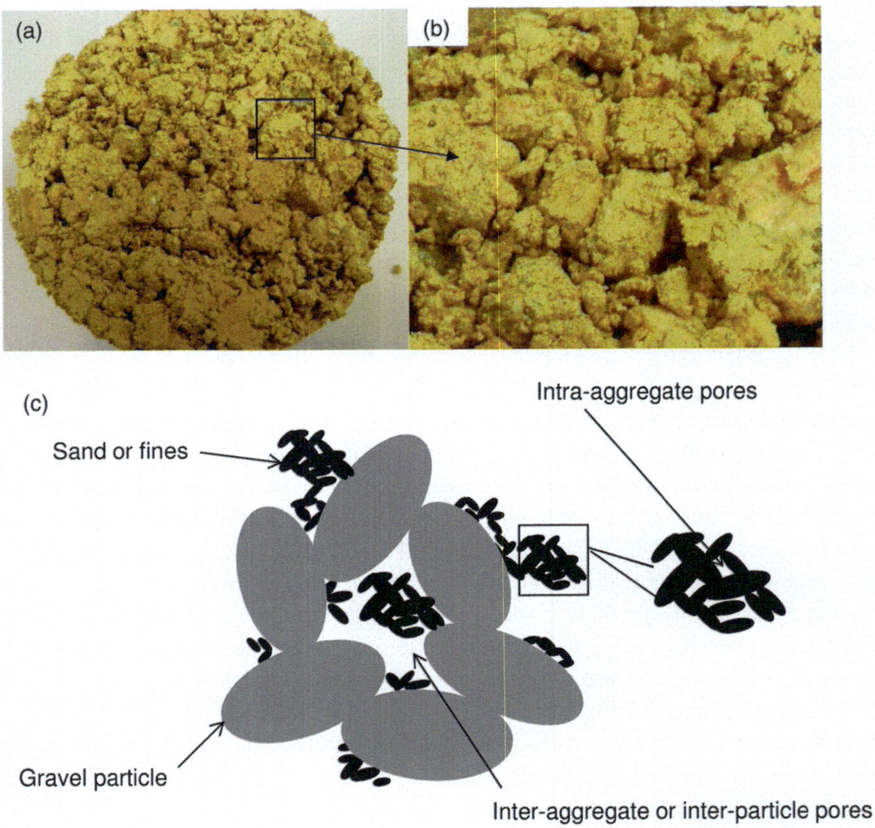

Fig. 8.1 Fabric of SW-SM with gravel: **a** and **b** digital image of the compacted specimen; **c** schematic microstructure

pores. The fines in the compacted specimen are grouped in assemblages. Figure 8.1b indicates that in the SW-SM with gravel specimen, sand and fines adhere to large gravel particles and sometimes form aggregates. Three types of fabric features can be identified, i.e., gravel particles, aggregates of fines, and pore spaces. The schematic microstructure of SW-SM with gravel is given in Fig. 8.1c, where coarse particles form the skeleton and leave large inter-particle pores between the coarse particles and intra-aggregate pores are present in the fines aggregates (Zhao et al. 2013).

In Chap. 3, the microporosity of the five types of soils has been reported in detail. A dual-porosity structure is formed for the coarse-controlled soils as shown in Chap. 3. (Figs. 3.18 and 3.19). The pore-size density function of the compacted SW-SM with gravel samples under saturated (i.e., samples compacted at their optimum water contents and then saturated) and saturated and oven-dried conditions are plotted in Fig. 8.2a. The pore-size distributions of SW-SM with gravel under saturated and dried states show bimodal features with a radius boundary of 2 μm. The specimens are considered to have a dual-porosity structure with both intra-aggregate pores (i.e., pores with radii smaller than 2 μm) and inter-aggregate pores (i.e., pores with radii larger than 2 μm). The dominant pore radii (peaks) are 1.04 and 28.1 μm, and an intermediate peak occurs at a pore radius of 3.0 μm at the saturated state as shown in Fig. 8.2a. After the specimen was oven dried, the peak on the left (1.04 μm) corresponding to the most dominant pore size decreased to 0.52 μm. The peak on the right increased from 28.1 to 40 μm, which indicates that the pore size in the specimen increased. The pores in the radius range of 0.3–2 μm were smaller. Such decreases in volume resulted from shrinkage of the clay aggregates. The coarse particles constitute 90% of the soil and form a skeleton structure. When the clayey aggregates shrink upon drying, the soil skeleton structure supported by the coarse particles remains stable and does not show significant deformation. The shrinkage of the clay aggregates leads to some increase in the volume of the inter-aggregate pores. The mathematical description of dual-porosity geometry has been introduced in Chap. 3.

Figure 8.2b shows the typical pore-size distribution of kaolin clay (Wang and Siu 2006) and clean Ottawa sand (Juang and Holtz 1986). Different from the bimodal pore-size distribution of SW-SM with gravel, both the kaolin clay and clean sand show a unimodal distribution. For Ottawa sand, inter-particle pores dominate; a single mode is characteristic, which occurs at around 200 μm. The kaolin clay also exhibits a unimodal distribution essentially consisting of intra-aggregate pores. The predominant pore radius of the kaolin clay falls in the range of 0.01–10 μm, with a significant volume of pores having radii of 0.1–0.7 μm.

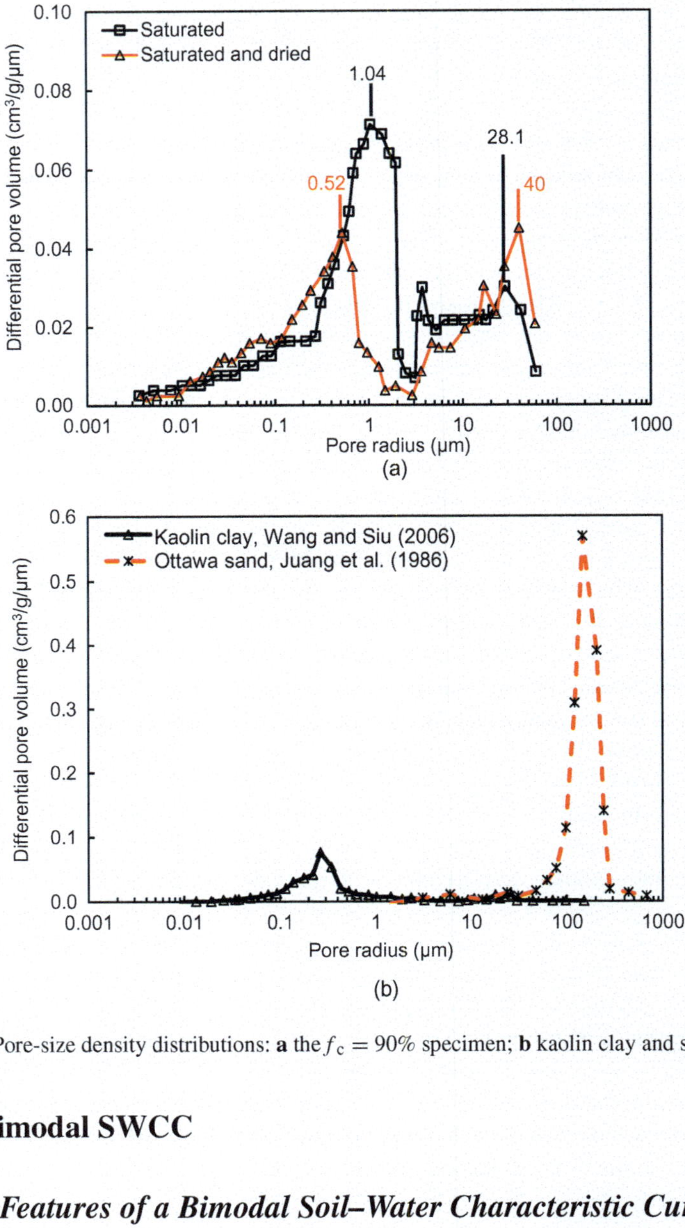

Fig. 8.2 Pore-size density distributions: **a** the $f_c = 90\%$ specimen; **b** kaolin clay and sand

8.3 Bimodal SWCC

8.3.1 Features of a Bimodal Soil–Water Characteristic Curve

As reported in Chap. 5, the SWCC for the $f_c = 90\%$ (SW-SM with gravel) and $f_c = 68\%$ (SM with gravel) shows strong bimodal distribution features. The bimodal curve of the $f_c = 90\%$ specimen is adopted to illustrate the feature of bimodal SWCC. As shown in Fig. 8.3, the water content decreases sharply in the first drainage period (i.e., smaller than 1 kPa) and then decreases gradually in the second drainage period.

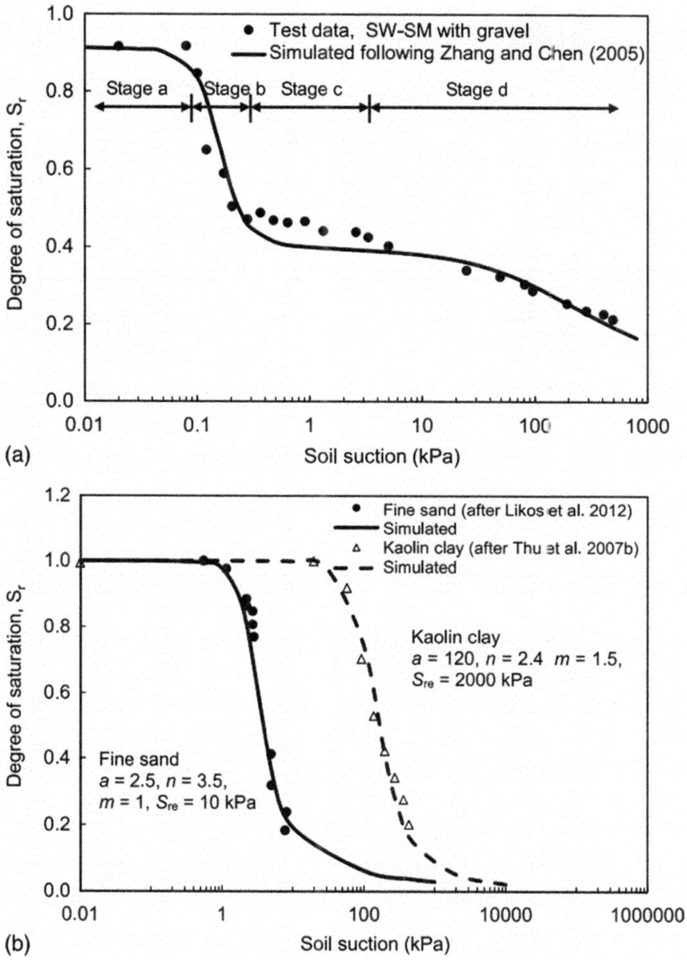

Fig. 8.3 Soil water characteristic curves: **a** the $f_c = 90\%$ specimen; **b** kaolin clay and sand

This phenomenon is related to the pore structure of the soil specimen because water is stored in the soil pores and flows through an interconnected chain of pores. The SW-SM with gravel shows a dual-porosity structure including two major pore series. The large inter-aggregate pores are located among the coarser particles and between the coarser particles and the clay aggregates. The intra-aggregate pores are found mainly within the clay aggregates. When air begins to enter the soil, water first drains from the largest pores. This is followed by drainage from the smaller pores. It is reasonable to expect that in the first drainage period only inter-aggregate pores are affected by desaturation while the intra-aggregate pores remain saturated. As matric suction increases, water gradually drains from the intra-aggregate pores, and the clay aggregates become unsaturated. In Fig. 8.3a, the degree of saturation starts at 0.93

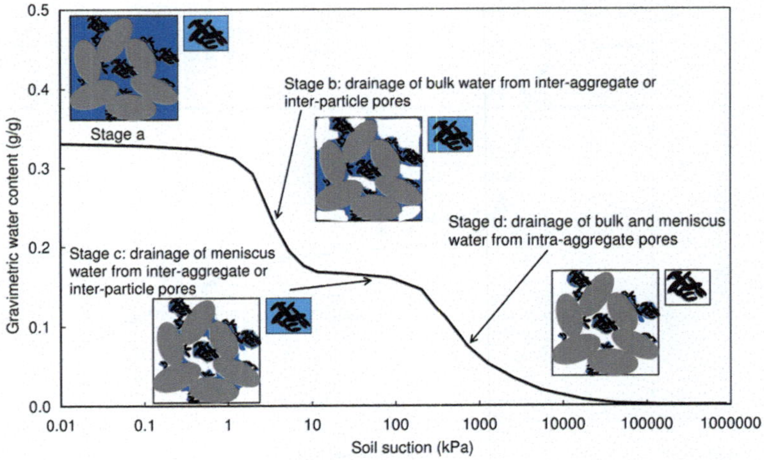

Fig. 8.4 General drainage process on a bimodal SWCC

because the specimen was not fully saturated initially as a result of trapped air in the specimen.

To better understand the bimodal feature of the SWCC for $f_c = 90\%$, the SWCCs for clay tested by Thu et al. (2007) and sand tested by Likos et al. (2010) are used for comparison. The clay was a mixture of kaolin clay (15%) and silts (85%). The maximum dry density was 1359 kg/m^3, and the optimum water content was 22%. The specific gravity of the soils was 2.65. The sand had more than 95% of the particles within the particle-size range of 0.1–0.5 mm. The mean grain diameter (d_{50}) was 0.25 mm. Figure 8.3b) shows the drying SWCCs for the clay (Thu et al. 2007) and sand (Likos et al. 2010), in which the SWCCs show a unimodal feature that is controlled by the unimodal pore-size structure. The predominant pore structure consisted of intra-aggregate pores for the clay and inter-particle pores for the sand (Fig. 8.1b]. Water drained gradually from the intra-aggregate pores or inter-particle pores; the drainage process had a single modal behavior. The AEV of sand is much smaller, and the desaturation rate is much faster than those for the clay. Vanapalli et al. (1996) developed a physical model for the unimodal SWCC. It was observed that in the boundary effect zone, the soil remained saturated. The soil started to desaturate in the transition zone and the water menisci in contact with the soil particles were continuous. The amount of water in the soil was small, and the water phase became discontinuous in the residual stage.

8.3.2 Physical Model for the Bimodal Soil–Water Characteristic Curve

Figure 8.4 shows the general drainage process in a dual-porosity structure soil. The entire drainage process can be divided into four stages. In the first drainage period, the stage before the AEV is defined as stage *a*. All the pores are filled with water in this stage. When the suction is larger than the AEV, air breaks into the inter-aggregate pores where the first drainage period commences. The bulk water stored in the inter-aggregate pores will drain during stage *b*. The soil mass becomes unsaturated. The water film forms a continuous phase during this stage. As soil suction increases, the drainage of the meniscus water in the inter-aggregate pores follows. A large increase in matric suction results in little changes in water content. This stage is defined as stage *c* and can be referred to as the first residual state. The pore water becomes discontinuous except for the water bounded between the particles and the clay aggregates in the form of water bridges. Further, in the second drainage period, the water stored in the intra-aggregate pores begins to drain (stage *d*). The clay aggregates subsequently become unsaturated.

The corresponding stages are also labeled in the drying SWCC of the $f_c = 90\%$ specimen in Fig. 8.3. The AEV is about 0.07 kPa. In stage *a*, the soil remains saturated under negative pore-water pressure. As matric suction exceeds the AEV, the soil becomes unsaturated. As shown in Fig. 8.3, in the saturated $f_c = 90\%$ specimen, the measured pore volume with radii above 73 mm is about 44%. When the soil suction increases to 0.5 kPa, over 50% of the water drains from the inter-aggregate pores in stage *b*. In the first residual state (stage *c*), the meniscus water drains from the inter-aggregate pores. The water content changes little as the suction increases from 0.5 to 10 kPa. In the following second drainage period, the clay aggregates begin to desaturate when the bulk water and meniscus water in the intra-aggregate pores start to drain (stage *d*). The desaturation rate in the second period is much smaller than that associated with the first period since the intra-aggregate pores are dominant.

8.3.3 New Bimodal SWCC Model

According to the capillary theory, a bimodal SWCC is associated with a bimodal pore-size distribution or a dual-porosity structure. To investigate the drainage process in soils with a dual-porosity structure, a brief review of the drainage process in uniform pores is helpful.

8.3.3.1 Bulk Water and Water-Pendular Rings

The soil–water characteristic curve for ideal uniform pores is shown in Fig. 8.5a, where the analytical solution of drainage curve in the simple cubic packing of

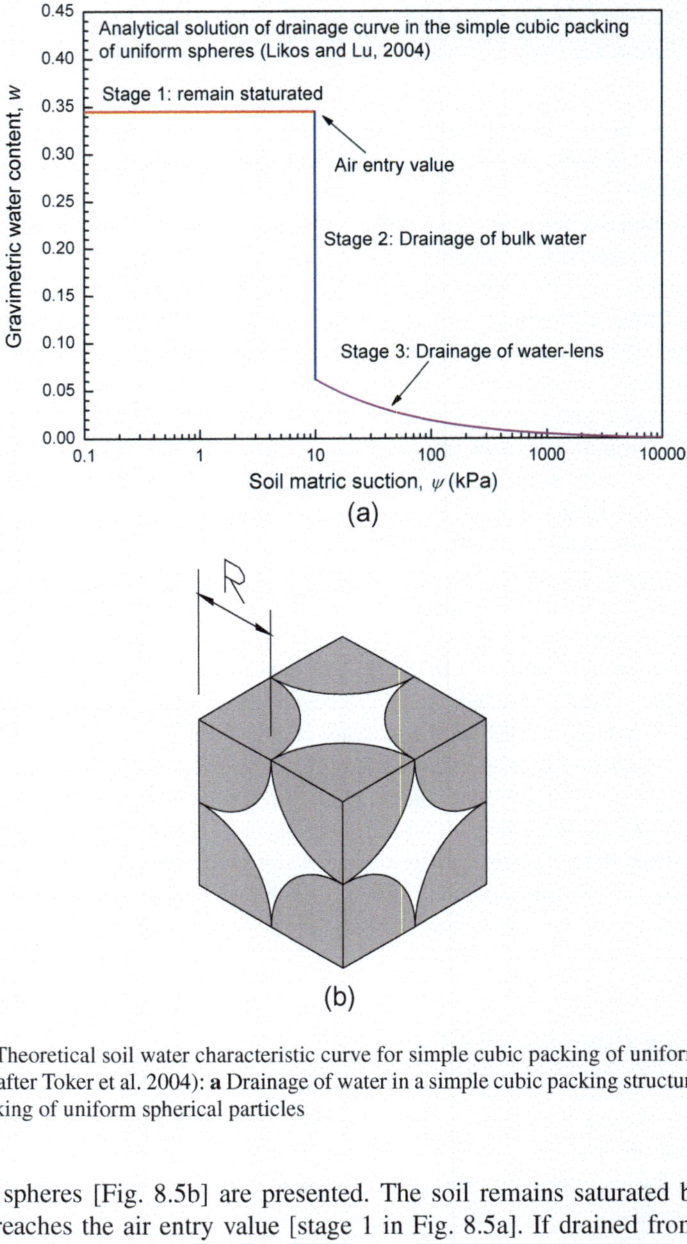

Fig. 8.5 Theoretical soil water characteristic curve for simple cubic packing of uniform spherical particles (after Toker et al. 2004): **a** Drainage of water in a simple cubic packing structure, **b** simple cubic packing of uniform spherical particles

uniform spheres [Fig. 8.5b] are presented. The soil remains saturated before the suction reaches the air entry value [stage 1 in Fig. 8.5a]. If drained from a slurry state, the soil will undergo remarkable volume change as suction changes from zero to the air entry value. So the water content will also change (while the degree of saturation remains 1.0). For a coarse soil or an overconsolidated soil, the volume change induced by suction changes from zero to the air entry value (AEV) is normally small because of its preconsolidation or small air entry value. In Fig. 8.5, the volume change of coarse soils induced by the suction change from 0 to a small AEV is ignored.

The air entry value is controlled by the routing radius (i.e., the radius of the largest sphere that can pass through the porous medium, which is also called "throat radius"). The routing radius is related to the soil particle packing structure and the particle diameter. When the soil suction reaches the air entry value, the "bulk" water starts to drain [stage 2 in Fig. 8.5a]. After that, the "bulk" water is replaced by air bubbles or other non-aqueous phase liquids and the remaining pore water is stored in the inter-particle pendular regime. The water stored in the inter-particle pendular regime is termed as "water lens." A meniscus interface will form between the water lens and the air phase (or any other non-aqueous phase liquid), controlled by the pressure difference between the water and air phases, i.e., the soil matric suction. Thus the water lens retreats slowly while the soil matric suction increases [stage 3 in Fig. 8.5a]. The water storage ability of water lens in stage 3 has been derived by Gvirtzman and Roberts (1991) and Likos and Lu (2004). Results show that the water stored in the water lens has a maximum volume of 6.3% of sample total volume in a simple cubic packing of uniform spheres and 3.2% in a tetrahedral packing of uniform spheres (Likos and Lu 2004).

In a random packing of soil particles, the bulk water is stored in pores with different routing radii. Then the drainage process of the bulk water [stage 2 in Fig. 8.5a] starts from the drainage of the largest bulk water spheres to the drainage of the smallest bulk water spheres. After the bulk water drains, the drainage of the water lens follows. The water lens drains slowly while the matric suction increases rapidly. Hence, the volume proportion of water lens determines the residual soil water content.

8.3.3.2 Simple SWCC Model with Meaningful Parameters

For a real soil, the pore routing radii are not unique, but distribute in a range. The drainage of soil bulk water will not happen at a single suction value, but in a suction range. After the bulk water drains, the drainage of the water lens follows. The water lens drains slowly while the matric suction increases rapidly. Hence, the volume proportion of water lens determines the residual soil water content.

As shown in Fig. 8.6a, SWCCs can be different from another for soils with different suction history. A SWCC describing the dehydration process of soil is termed as a drying curve and a SWCC describing the saturation process of soil is termed as a wetting curve. The resulting SWCCs for the drying path and wetting path exhibit hysteresis. Because of the hysteresis of soil water retention ability, SWCC depends on soil suction history and is normally located in the main drying curve and main wetting curve [i.e., the scanning curve in Fig. 8.6a]. In Fig. 8.6a, the SWCCs are in similar shapes, but with different characteristic values. By catching the shape, SWCC model is originated from a simple mathematical form (e.g., van Genuchten's model in Table 8.1) and some of the model parameters vary with soil wetting/drying history, which can reflect hysteresis phenomenon and often are determined by fitting to measured SWCC (Li 2005; Pedroso and Williams 2010 and many others). Tsiampousi et al. (2013) also offers a model for determining the scanning SWCC relying on given primary drying and wetting curve.

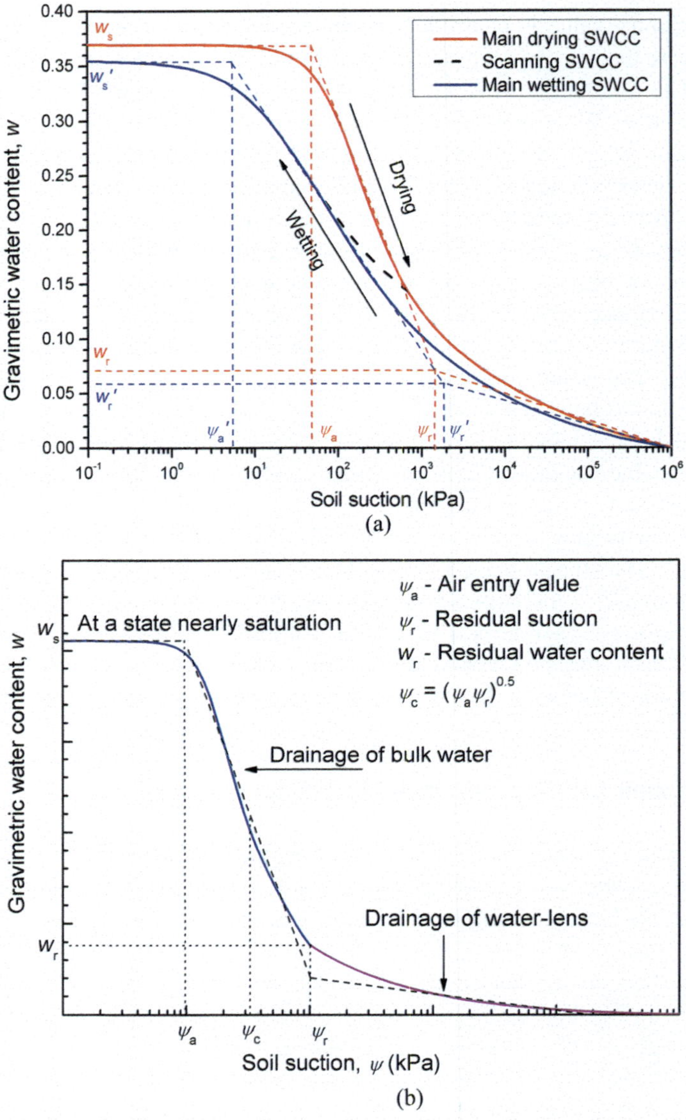

Fig. 8.6 Meaningful parameters used to describe the unimodal water retention curve: **a** Hysteresis phenomenon of water retention curve, **b** three-segment polyline used to approximate the water retention curve

As shown in Fig. 8.6, the drying SWCC for a soil with unimodal pore routing radii is a combination of a drainage curve of the bulk water and a drainage curve of the water lens. The SWCC in the primary draining stage, i.e., the draining curve for the bulk water, is often approximated by a straight line as shown in Fig. 8.6. Such approximation is proposed by Brooks and Corey (1964) and well accepted in the

Table 8.1 Summary of soil–water characteristic curve models

Model name	Expression	Parameters
Burdine(1953)	$S = 1/\left(1+(a\psi)^n\right)^{1-2/n}$	a, n = constant
Gardner (1956)	$S = 1/\left(1+a\psi^n\right)$	a, n = constant
Brooks and Corey (1964)	$\theta = (\psi_a/\psi)^\lambda$	λ = pore-size distribution index
Brutsaert (1966)	$S = 1/\left(1+(\psi/a)^n\right)$	a, n = constant
Campbell (1974)	$\frac{\psi}{\psi_a} = \left(\frac{\theta}{\theta_s}\right)^{-b}$	b = constant
Mualem (1976)	$S = \left(1+(a\psi)^n\right)^{-1+1/n}$	a, n = constant
van Genuchten (1980)	$S = \left(1+(a\psi)^n\right)^{-m}$	a, n, m = constant
Williams et al. (1983)	$\ln\psi = a + b\ln\theta$	a, b = constant
Mckee and Bumb (1987)	$S = \exp(-\psi/b)$	b = constant
Fredlund and Xing (1994)	$S = \left(\ln\left(e+(\psi/a)^n\right)\right)^{-m}$	a, n, m = constant
Kosugi (1994)	$S = 0.5erfc\left(\left[\ln\left(\frac{\psi_a-\psi}{\psi_a-\psi_{\text{mode}}}\right)-\sigma^2\right]/\sqrt{2\pi\sigma}\right)$	σ = constant; *erfc* = complimentary error function; Ψ_{mode} = mode related to pore-size distribution;
Gallipoli et al. (2003)	$e(p'',\xi) = \frac{e}{e_s}(\xi)e_s(p'')$	$e(p'',\xi)$ = the normal compression state surface; $e_s(p'')$ = the saturated normal compression line; ξ = bonding variable; p'' = isotropic average skeleton stress; e_s = the void ratio at saturated state
Li (2005)	$s^* = \alpha^* \pm \left(\left\|\overline{s}^*-\alpha^*\right\|^{\beta+1} - \left\|\overline{s}_0^*-\alpha^*\right\|^{\beta+1}\right)^{1/(\beta+1)}$	$S^* = \ln\psi$; $\alpha^* = \ln\alpha$; α = the impact of wetting/drying history; $\overline{s}^* = \ln\overline{s}$; β = material parameter; $\overline{s}_0^*$ = the value of $\overline{s}^*$ at the beginning of the scanning curve
Tarantino (2009)	$S = \left\{1+\left[\left(\frac{e}{a}\right)^{1/b}s\right]^n\right\}^{-b/n}$	a, b, n = *soil parameter*; e-void ratio

(continued)

Table 8.1 (continued)

Model name	Expression	Parameters
Pedroso and Williams (2010)	$S = -\lambda s + \frac{1}{\beta}\ln\left(c_3 + c_2 e^{c_1 s}\right)$	$s = \ln(1 + \psi)$; λ = tangent inclination for the reference curve; β = fitting coefficient; c_1,c_2,c_3 parameters depend only on the constitutive parameters and differs with wetting/drying history
Tsiampousi et al. (2013)	$S_{r,\text{pr}}^{\text{dr.wet}} = \frac{1-(1/s_0^*)s^*}{1+\alpha_{d,w}s^*}$ $S_{r,\text{scan}}^{\text{dr}} =$ $S_{r,A} - r_{\text{dr}} + \left[r_{\text{dr}}^2 + \left(\log s^* - \log s_A^*\right)^2\right]^{1/2}$	$S_{r,\text{pr}}^{\text{dr.wet}}$ = primary drying or wetting curve; $S_{r,\text{scan}}^{\text{dr}}$ = scanning drying curve; s^* = combined suction; s_A^* = combined suction at a retention point A; s_0^* = combined suction at degree of saturation of 0.0; α—fitting parameter; $s_{r,A}$ = degree of saturation at retention point A; r_{dr} = Radius of the circle

S is the degree of saturation; θ is the volumetric water content; ψ is the matric suction; ψ_a is the air entry value; θ_r is the residual water content; θ_s is the saturated water content

description of SWCC. Such a straight line can be used to determine the AEV, ψ_a, the residual suction, ψ_r and the residual water content, w_r together with another two tangent line (Brooks and Corey 1964). Clearly, ψ_a indicates the starting suction for the drainage of bulk water. Before the suction increases to ψ_a, water hardly drains and the water content remains at a value close to the saturated water content, w_s. ψ_r indicates the end suction for the drainage of bulk water and also the starting suction for the drainage of water lens. After the suction increases to ψ_r, the water content decreases to the residual water content, w_r and all the bulk water has drained. The gravimetric water content of bulk water is equal to $(w_s - w_r)$ and noted as w_b. The polyline with three segments shown in Fig. 8.6 can be used as an approximation to SWCC and fits many measured SWCCs well. For example, Fig. 8.7 shows the polylines fitted to the measured SWCCs for a clay and a residual soil, where the R^2, i.e., the coefficient of determination, in both cases are above 0.98.

As a combination of a drainage curve for the bulk water and a drainage curve for the water lens, the SWCC can be written as,

$$w = w_b f(\psi) + w_r g(\psi) \tag{8.1}$$

where $f(\psi)$ and $g(\psi)$ are functions used to describe the draining process of the bulk water and the water lens, respectively. Referring to Fig. 8.6 and Fig. 8.7, the drainage

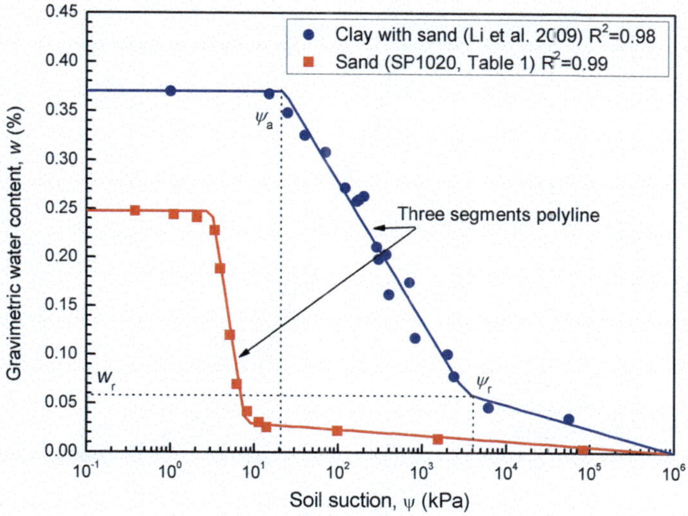

Fig. 8.7 Three-segment polyline for determination of ψ_a, ψ_r, w_r

function for the bulk water, $f(\psi)$, can be approximated by a straight line from [ψ_a, 1] to [ψ_r, 0]. A smooth approximation to such a line can be mathematically written as,

$$f(\psi) = \frac{\psi_c^i}{\psi^i + \psi_c^i} \tag{8.2}$$

where ψ_c is the suction corresponding to $f(\psi_c) = 0.5$ [as shown in Fig. 8.8a] and is equal to $(\psi_r\ \psi_a)^{0.5}$; ψ is matric suction; and i is a parameter that controls the shape of the drainage curve for the bulk water. Comparing Eq. 8.2 to the existing SWCC models in Table 8.2, Eq. 8.2 is a variation of the SWCC model presented by Brutsaert (1966).

Clearly, i varies with the ratio of ψ_r/ψ_a and can be expressed by,

$$i = \frac{n}{\log(\psi_r/\psi_a)} \tag{8.3}$$

where n is a coefficient. Now substituting Eqs. 8.3 into 8.2 yields

$$f(\psi) = \frac{\sqrt{\psi_a\psi_r}^{n/\log(\psi_r/\psi_a)}}{\psi^{n/\log(\psi_r/\psi_a)} + \sqrt{\psi_a\psi_r}^{n/\log(\psi_r/\psi_a)}} \tag{8.4}$$

Referring to Figs. 8.2 and 8.3, the drainage curve for the water lens also can be approximated by a straight line. Similarly, a smooth and round approximation of $g(\psi)$ can be written as,

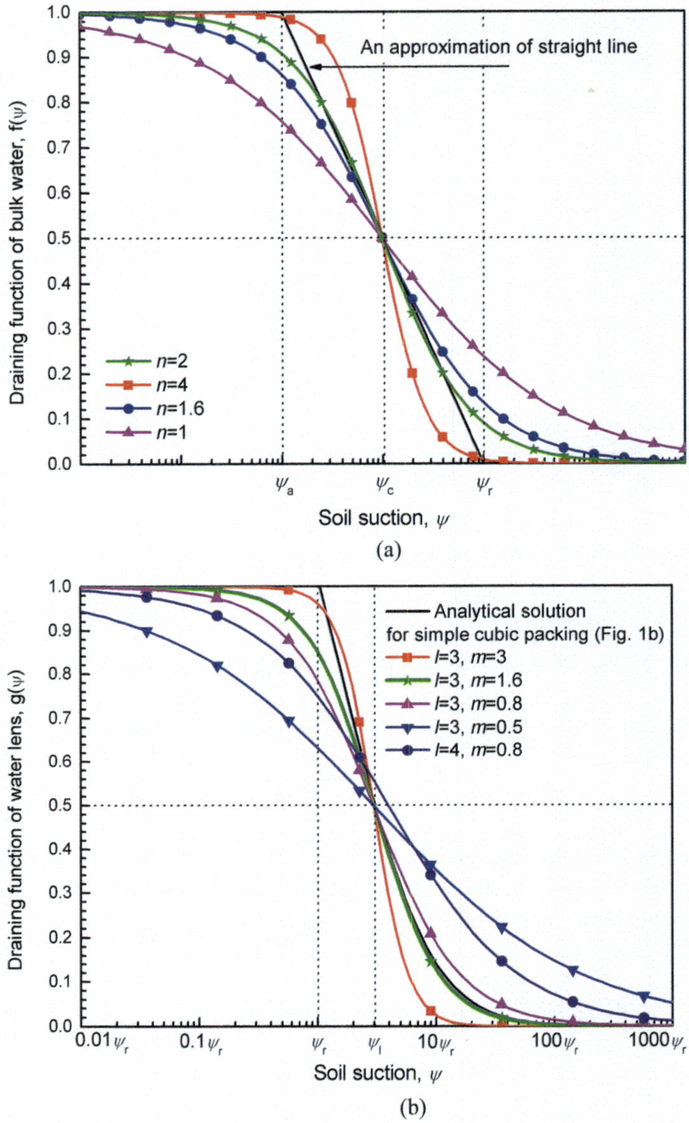

Fig. 8.8 Sensitivity studies of parameters involved in Eq. 9: **a** Draining function of bulk water, **b** draining function of water lens

$$g(\psi) = \frac{\psi_l^m}{\psi^m + \psi_l^m} \tag{8.5}$$

where ψ_1 is a suction corresponding to $g(\psi_1) = 0.5$ [as shown in Fig. 8.8b]; m is a parameter that controls the shape of the drainage curve for the water lens. If ψ_1 is expressed by the end drainage suction for the bulk water, ψ_r, as,

Table 8.2 Parameters estimated from soil GSD and void ratio for the bimodal SWCC model (Eqs. 13–17)

Soil type	Void ratio	d_{10} (mm)	d_{30} (mm)	d_{60} (mm)	w_s	ψ_a (kPa)	w_r	ψ_r (kPa)	ψ_t (kPa)	ψ_{a2} (kPa)	Fig. name	Reference
CL with sand	0.98	0.0019	0.0071	0.0302	0.370	21.0	0.058	4000	–	–	Figure 5	Li (2009)
GW-GM with sand	0.59	0.26	2.94	8.21	0.224	0.049	0.025	34.3	0.19	0.40	Figure 13d	Li (2009)
SM with gravel	0.63	0.0067	0.22	5.5	0.238	0.292	0.033	1016	4.65	20.7	Figure 13c	Li (2009)
Silt loam	1.29	0.0008	0.0199	0.0724	0.49	6.985	0.048	1412	72	116	Figure 12a	RS2000
Loam	0.73	0.00003	0.0293	0.0809	0.27	1.682	0.039	35,681	53	185	Figure 12b	SP1020
Sandy loam	0.61	0.0019	0.1397	0.4108	0.23	0.637	0.025	1550	6.85	44.5	Figure 12c	SP1020
Loamy sand	0.64	0.0569	0.1216	0.1835	0.24	2.114	0.022	32.7	7.49	6.07	Figure 12d	SP1020
Sand	0.66	0.1229	0.1738	0.2541	0.25	1.664	0.021	17.7	4.83	3.66	Figure 13a	SP1020
Well-graded gravel	0.51	1.2047	5.4907	11.8259	0.19	0.034	0.020	8.97	0.087	0.15	Figure 13b	SP1022

RS2000 and SP1020 are the dataset IDs in SoilVision database (1997)

$$\psi_1 = l\psi_r \tag{8.6}$$

where l is a coefficient to be determined. Equation 8.5 can then be modified as

$$g(\psi) = \frac{(l\psi_r)^m}{\psi^m + (l\psi_r)^m} \tag{8.7}$$

Substituting Eqs. 8.7 and 8.4 into Eq. 8.2, the SWCC for the unimodal random packing can be written as

$$w(\psi) = w_b \frac{\sqrt{\psi_a \psi_r}^{n/\log(\psi_r/\psi_a)}}{\psi^{n/\log(\psi_r/\psi_a)} + \sqrt{\psi_a \psi_r}^{n/\log(\psi_r/\psi_a)}} + w_r \frac{(l\psi_r)^m}{\psi^m + (l\psi_r)^m} \tag{8.8}$$

Equation 8.8 has three empirical coefficients, i.e., l, m, and n. To simplify the model, default values of l, m, n can be used. A sensitivity study of n is shown in Fig. 8.8(a). To simulate the drainage curve for the bulk water as an approximate straight line from (ψ_r, 0) to (ψ_a, 1), n is recommended as 2.0.

For the simple cubic packing of uniform spheres [Fig. 8.5b], an analytical solution of drainage curve for the water lens has been introduced by Likos and Lu (2004) and is shown in Fig. 8.8b. When l is equal to 3 and m is equal to 1.6, Eq. 8.7 fits the tail part of the drainage function for the water lens well for a simple cubic packing of uniform spheres [Fig. 8.8b]. For a real soil, the drainage curve for the water lens will span a much wider suction range (Fig. 8.9) because of the random packing of non-uniform particles. Fredlund and Xing (1994) indicated that the soil water content will approach to 0 as suction approaches 1,000,000 kPa, which is in good

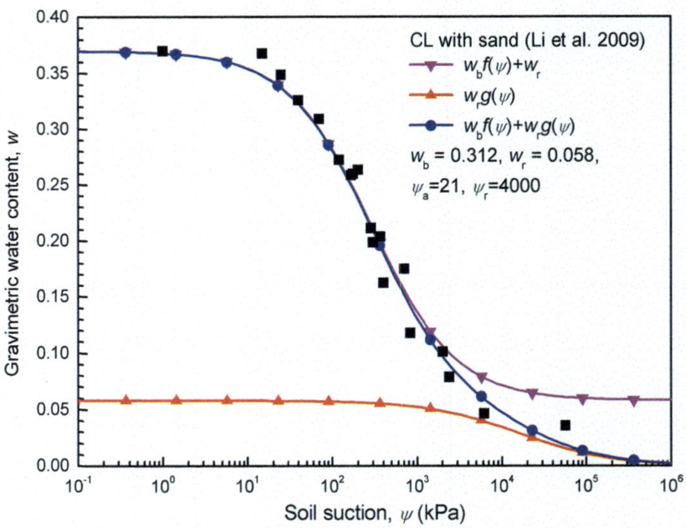

Fig. 8.9 SWCC fitted by Eq. 9 for CL with sand

agreement with many measured SWCCs. The draining function of water lens $g(\psi)$ is supposed to pass the point (1,000,000 kPa, 0). If ψ_r is taken as 3000 kPa, a value recommended by Fredlund and Xing (1994) and found to be proper in many SWCCs, $g(\psi)$ expressed as Eq. 8.7 will pass point (1,000,000 kPa, 0) when l is selected as 4 and m as 0.8. Such values of l and m lead to SWCCs that fit the measured data very well (referring to Fig. 8.9).

If l, m, and n are taken as the recommended values, i.e., 4, 0.8, and 2, Eq. 8.8 is simplified as,

$$w(\psi) = w_b \frac{\sqrt{\psi_a \psi_r}^{2/\log(\psi_r/\psi_a)}}{\psi^{2/\log(\psi_r/\psi_a)} + \sqrt{\psi_a \psi_r}^{2/\log(\psi_r/\psi_a)}} + w_r \frac{(4\psi_r)^{0.8}}{\psi^{0.8} + (4\psi_r)^{0.8}} \tag{8.9}$$

Some flexibility has been lost when employing default values of l, m, and n, but the complexity of Eq. 8.8 decreases by using default values of l, m, and n. In Eq. 8.9, all the involved parameters have clear physical meanings as listed in Table 8.3. Once the values of w_r, w_s, ψ_a, and ψ_r are determined using the tangent method (as shown in Figs. 8.6 and 8.7), a smooth SWCC with round corner can be determined by simply substituting these parameters into Eq. 8.9, as shown in Fig. 8.9. If typical values of w_r, w_s, ψ_a, and ψ_r for a silt and a clayey soil are used, the SWCCs predicted by Eq. 8.9 are shown in Fig. 8.10. The flexibility of Eq. 8.9 in fitting SWCC is good.

Table 8.3 Definition of frequently used symbols and model parameters

Quantity	Description
w_s	Saturated water content
w_r	Residual water content, i.e., mass content of water lens Mass content of water lens stored in micropores of soil
w_b	Mass content of bulky water
Λ	The ratio of the mass content of bulk water over that of water lens
ψ_a	Air entry value in unimodal SWCC; The first air entry value in bimodal or multimodal SWCC
ψ_r	Suction corresponding to residual water content in unimodal SWCC The second residual suction in bimodal SWCC
w_{b1}	Mass content of bulky water stored in macro-pores
w_{l1}	Mass content of water lens stored in macro-pores
w_{b2}	Mass content of bulky water stored in micropores
ψ_{a2}	The air entry value for water stored in micropores
ψ_t	The residual suction for water stored in macro-pores

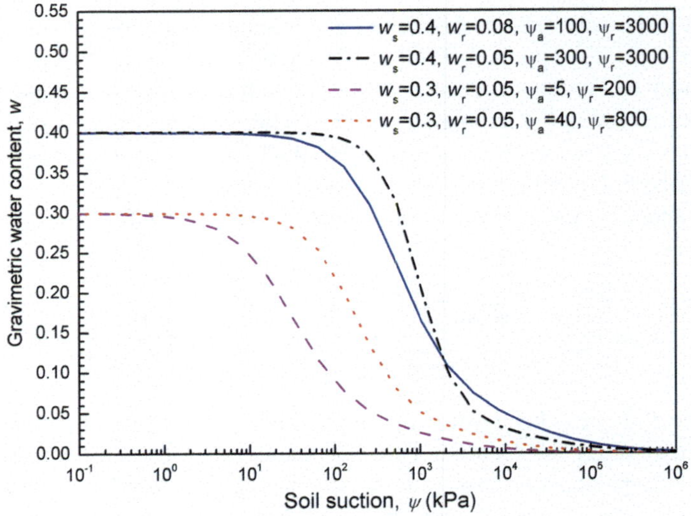

Fig. 8.10 Unimodal SWCC model using characteristic parameters

8.3.3.3 Bimodal SWCC Model Described by Meaningful Parameters

For a dual-porosity structure, there are two major pore series, i.e., macro-pores (i.e., inter-particle pores or inter-aggregate pores) and micropores (i.e., intra-aggregate pores). The drainage process in a dual-porosity structure is shown in Fig. 8.11. As the suction increases, the bulk water (BW) stored in the macro-pores will drain first. Then the drainage of water lens (WL) in the micropores follows. Further, when the suction increases to the starting drainage suction for the bulk water stored in the micropores, the bulk water stored in the micropores will drain. After the suction

Fig. 8.11 Conceptual bimodal soil water characteristic curve

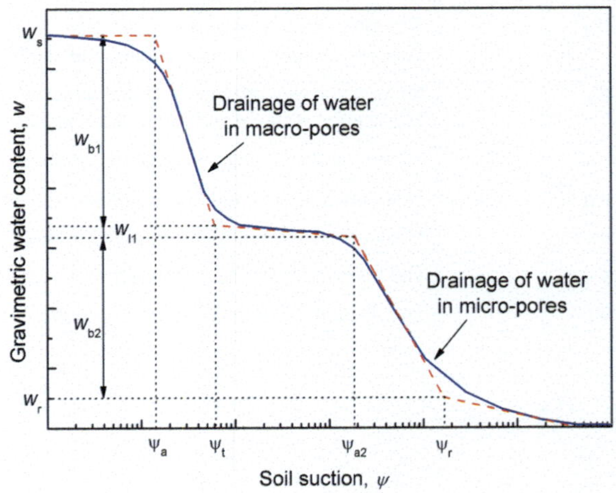

reaches the end suction for the drainage of the bulk water stored in the macro-pores, the water lens starts to drain.

The SWCC of the soil with a dual-porosity structure can be regarded as a superposition of two unimodal SWCCs, one for the water stored in the inter-particle pores and another for the water stored in the intra-particle pores. Then two sets of characteristic parameters of SWCC, i.e., ψ_a, ψ_r, w_r can be determined for the two unimodal SWCCs separately, as shown in Fig. 8.11. In the bimodal SWCC shown in Fig. 8.11, the first air entry value is regarded as the air entry value of the soil and termed as ψ_a. The second residual water content and suction are regarded as the residual water content and suction of the soil and termed as w_r and ψ_r, respectively. These parameters are listed in Table 8.3 together with their physical meanings. For the SWCC for the water stored in macro-pores, w_{b1}, w_{l1}, ψ_a, ψ_t refer to the mass content of bulk water in the macro-pores, the mass content of water lens in the macro-pores, the first air entry value, and the first residual suction, respectively. For the SWCC for the water stored in micropores, w_{b2}, w_r, ψ_{a2}, ψ_r refer to the mass content of bulk water in the micropores, the mass content of water lens in the micropores, the second air entry value, and the second residual suction, respectively. If λ is used to denote the ratio of the mass content of bulk water over that of water lens, we have $w_{b2} = \lambda w_r$ and $w_{b1} = \lambda w_{l1}$. Considering the relation $w_r + w_{b2} + w_{l1} + w_{b1} = w_s$, the SWCC for a dual-porosity structure, which is the superposition of the SWCC for the water stored in the macro-pores and that for the water stored in the micropores, can be written as

$$
\begin{aligned}
w(\psi) = {} & \lambda\left(\frac{w_s}{\lambda+1} - w_r\right)\frac{\sqrt{\psi_t\psi_a}^{\,n/\log(\psi_t/\psi_a)}}{\psi^{n/\log(\psi_t/\psi_a)} + \sqrt{\psi_t\psi_a}^{\,n/\log(\psi_t/\psi_a)}} \\
& + \left(\frac{w_s}{\lambda+1} - w_r\right)\frac{(l\psi_t)^m}{\psi^m + (l\psi_t)^m} \\
& + \lambda w_r\frac{\sqrt{\psi_r\psi_{a2}}^{\,n/\log(\psi_r/\psi_{a2})}}{\psi^{n/\log(\psi_r/\psi_{a2})} + \sqrt{\psi_r\psi_{a2}}^{\,n/\log(\psi_r/\psi_{a2})}} + w_r\frac{(l\psi_r)^m}{\psi^m + (l\psi_r)^m}
\end{aligned}
\tag{8.10}
$$

λ can be calculated as the ratio between $(w_s\text{-}w_r)/w_r$. Using the measured data (Li 2009; Soilvision 1997), w_s and w_r can be determined by the tangent method, leading to an average value of 3.12 for λ. Using default values of λ, l, m, and n of 3, 4, 0.8, and 2, respectively, Eq. 8.10 can be written as

$$
\begin{aligned}
w(\psi) = {} & \frac{(0.75w_s - 3w_r)\sqrt{\psi_a\psi_t}^{\,2/\log(\psi_t/\psi_a)}}{\psi^{2/\log(\psi_t/\psi_a)} + \sqrt{\psi_a\psi_t}^{\,2/\log(\psi_t/\psi_a)}} + \frac{(0.25w_s - w_r)(4\psi_t)^{0.8}}{\psi^{0.8} - (4\psi_t)^{0.8}} \\
& + \frac{3w_r\sqrt{\psi_{a2}\psi_r}^{\,2/\log(\psi_r/\psi_{a2})}}{\psi^{2/\log(\psi_r/\psi_{a2})} + \sqrt{\psi_{a2}\psi_r}^{\,2/\log(\psi_r/\psi_{a2})}} + \frac{w_r(4\psi_r)^{0.8}}{\psi^{0.8} + (4\psi_r)^{0.8}}
\end{aligned}
\tag{8.11}
$$

The bimodal SWCCs in Eq. 8.11 are shown in Fig. 8.12. If the 1st residual suction, ψ_t, is close to the 2nd AEV, ψ_{a2} (i.e., ψ_{a2}/ψ_t is smaller than 3), the bimodal SWCC

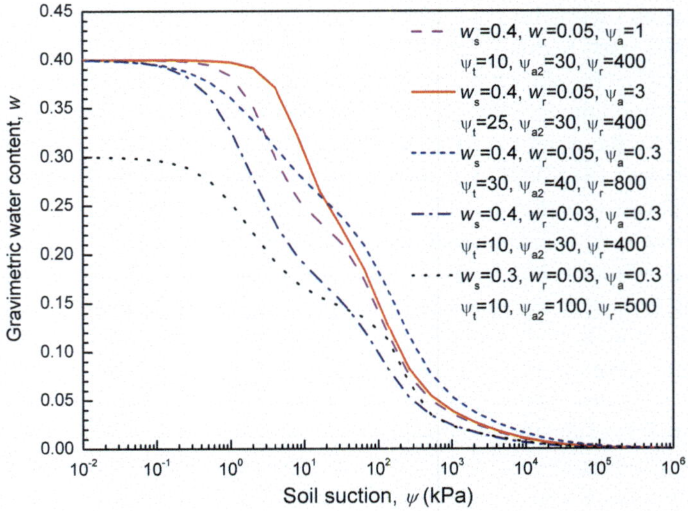

Fig. 8.12 SWCCs described by a bimodal SWCC model

model predicts SWCCs with weak bimodal features, which is similar to a unimodal SWCC. If ψ_{a2}/ψ_t is larger than 3, the model predicts SWCCs with distinct bimodal features.

By choosing proper parameters, the new bimodal SWCC model (Eq. 8.11) is fitted to the experimental data in Li et al. (2009). The fitted SWCCs are shown in Fig. 8.13. The bimodal SWCC model fits the measured bimodal SWCCs quite well. The measured SWCC shows the following features: (1) For the same soil, a denser sample shows a higher air entry value; (2) the parameters representing the water content stored in the micropores are similar for soils with different initial void ratios. That is to say, the micropores are not compressed much during compaction, which has been verified by pore-size distribution measurements Li and Zhang (2009).

8.3.4 SWCC Prediction Based on Grain-Size Distribution

In the absent of measured SWCC, e.g., in the preliminary design and analysis, it would be convenient to predict SWCC from GSD. Although such prediction is highly empirical and cannot substitute the measurement of SWCC, it still is helpful for implementing unsaturated soil mechanics in geotechnical and geo-environmental engineering. The soil–water characteristic curves reflect soil pore geometries, which are highly related to the soil grain-size distribution (Fredlund et al. 2002) and soil deformation history (such as compaction at an optimum water content, consolidation from slurry, or deposition in a natural landslide event etc.). Even for a particular soil sample, the soil pore geometries can change with stress history and wetting/drying

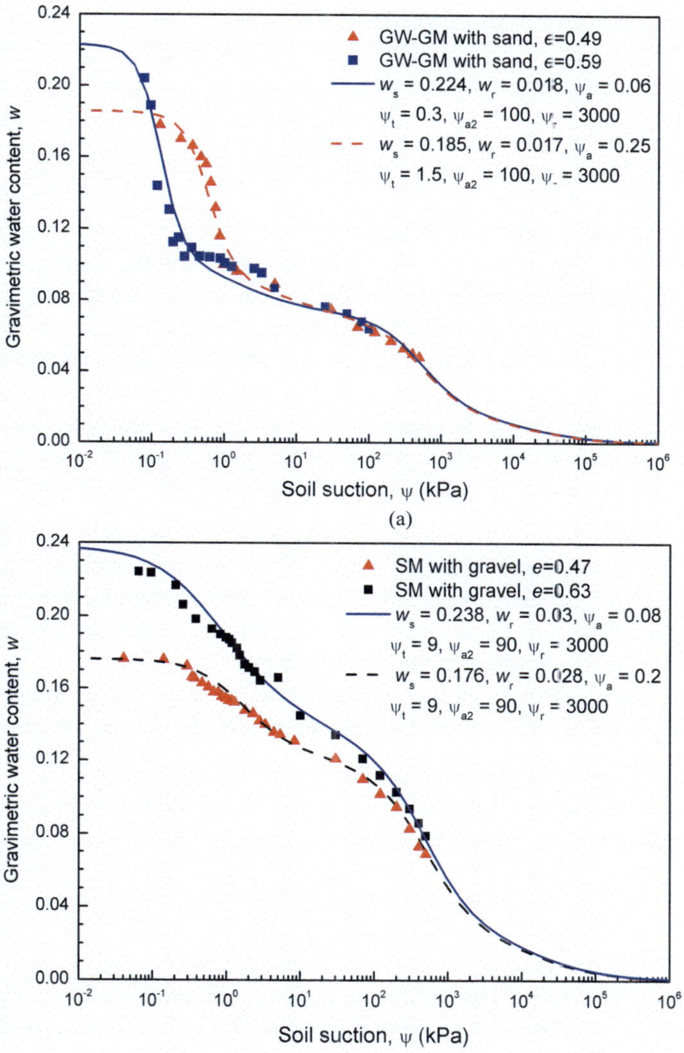

Fig. 8.13 Application of bimodal SWCC model: **a** Fitted to experimental SWCCs for GW-GM with sand, **b** fitted to experimental SWCCs for SM with gravel

cycles. Generally speaking, it is extremely difficult to predict the soil water characteristic curve for a soil sample accurately. However, some statistical relations can be derived from available experimental data and used to predict typical soil water characteristic curves.

8.3.4.1 Statistical Analysis of SWCC Parameter

The saturated water content, w_s, at a void ratio, e, is expressed as

$$w_s = \frac{e}{G_s} \tag{8.12}$$

where G_s is the specific density of solid in soil.

The controlling parameters such as air entry value, residual water content, and residual suction all vary with soil properties and can be calculated using empirical equations involving basic physical parameters, i.e., d_{10}, d_{30}, d_{50}, d_{60}, C_u etc. Here, d_{10}, d_{30}, d_{50}, and d_{60} are the particle diameters at 10%, 30%, 50%, and 60% passing in the sieving test, respectively; C_u is the coefficient of uniformity. ψ_a, ψ_r, and w_r are considered to have some relations with these soil physical properties. Regression analysis can be used to establish empirical equations for calculating ψ_a, ψ_r, and w_r based on soil physical properties. The regression procedure is as follows:

(1) The coefficient of correlation (COC) between ψ_a, ψ_r, w_r and d_{10}, d_{30}, d_{50}, d_{60}, C_u is calculated (Table 8.4). COC is also called the linear correlation coefficient, measures the strength and the direction of a linear relationship between two variables. The COC between ln(var1) and ln(var2) indicates the strength of a power relationship between the two variables.
(2) Regression analysis is carried out and leads to empirical relations between ψ_a, ψ_r, w_r and the variable, which has highest COC, i.e., strongest linear relationships, with ψ_a, ψ_r, and w_r.

Air entry value $\ln(\psi_a)$ is found to have high COCs with $\ln(d_{30})$ and e (Table 8.4). An empirical relation can be obtained as,

$$\psi_a = \frac{1.4^e}{3.6 d_{30} C_u^{0.25}} \tag{8.13}$$

where e is the void ratio. The estimated air entry values using Eq. 8.13 and the measured air entry values are plotted in Fig. 8.14a, where Eq. 8.13 has a coefficient of determination, R2, of 0.74.

The 2nd residual water content, w_r, shows high COCs with $\ln C_u$, and e (Table 8.4) and has an empirical relation [referring to Fig. 8.14b] for these 44 granular materials as follows,

$$w_r = 0.03e + 0.005 \log c_u \tag{8.14}$$

The 2nd residual suction $\ln\psi_r$ shows high COCs with $\ln C_u$ and $\ln d_{10}$ and yields [referring to Fig. 8.14c],

$$\psi_r = \frac{4 c_u^{0.4}}{d_{10}^{0.57}} \tag{8.15}$$

Table 8.4 Coefficient of correlation between characteristic parameters of SWCC and grain-size distribution

	D_{10}	D_{20}	D_{30}	D_{50}	D_{60}	e	C_u	C_c	$\ln(D_{10})$	$\ln(D_{20})$	$\ln(D_{30})$	$\ln(D_{50})$	$\ln(D_{60})$	$\ln(C_u)$	$\ln(C_c)$
ψ_a	−0.05	−0.07	−0.08	−0.08	−0.08	0.33	−0.08	−0.06	−0.33	−0.41	−0.52	−0.58	−0.60	0.11	0.00
$\ln(\psi_a)$	−0.29	−0.32	−0.42	−0.37	−0.34	0.49	−0.12	−0.19	−0.38	−0.50	−0.78	−0.73	−0.73	−0.18	−0.17
ψ_r	−0.11	−0.14	−0.15	−0.15	−0.15	0.13	0.11	−0.03	−0.39	−0.48	−0.50	−0.47	−0.43	0.19	−0.05
$\ln(\psi_r)$	−0.41	−0.46	−0.46	−0.37	−0.31	0.39	0.30	0.21	−0.87	−0.84	−0.82	−0.74	−0.71	0.58	0.31
w_r	−0.17	−0.13	−0.10	−0.07	−0.05	0.71	0.34	0.32	−0.62	−0.50	−0.40	−0.32	−0.29	0.69	0.60
$\ln(\psi_t/\psi_a)$	−0.05	0.06	0.14	0.33	0.37	−0.33	0.37	0.38	−0.27	−0.11	0.04	0.17	0.22	0.62	0.46
$\ln(\psi_r/\psi_{a2})$	−0.06	−0.08	−0.09	−0.09	−0.09	0.35	−0.06	−0.05	−0.44	−0.40	−0.57	−0.50	−0.51	0.09	0.06

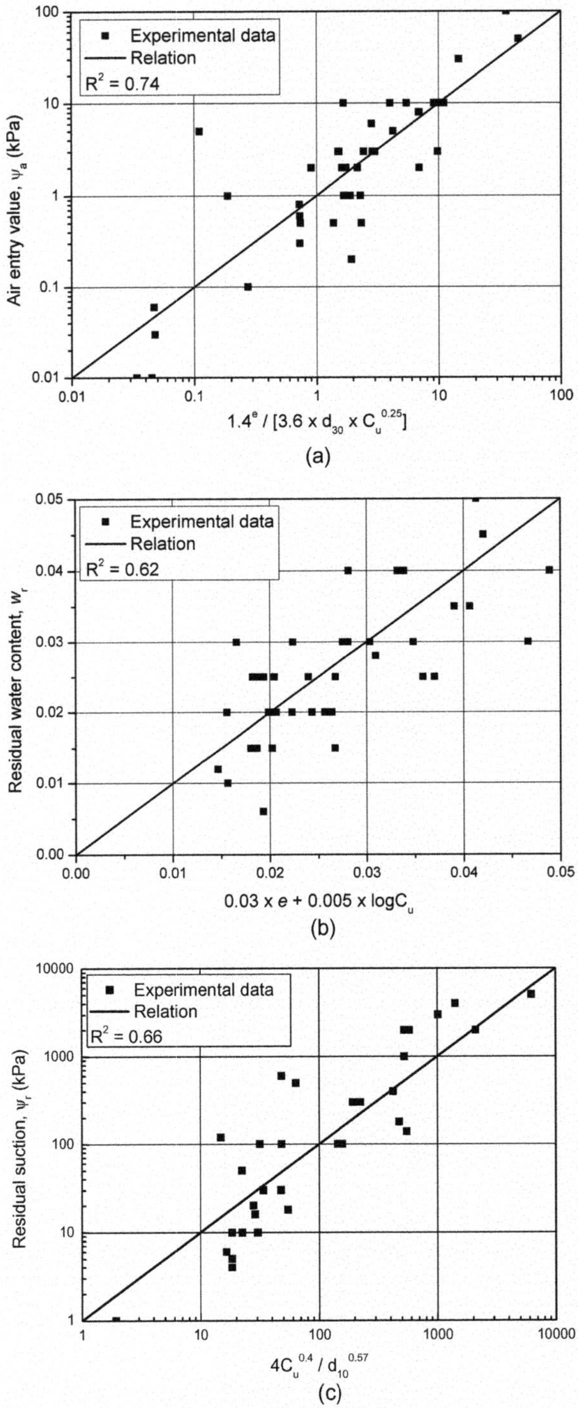

Fig. 8.14 Statistical relations between SWCC parameters and soil basic properties based on measured data in the literature: **a** 1st air entry value, **b** 2nd residual water content, **c** 2nd residual suction

After the 1st air entry value, ψ_a, the saturated water content, w_s, the 2nd residual water content, w_r, and the 2nd residual suction, ψ_r, have been determined, the remaining parameters, ψ_t and ψ_{a2} can be determined when the curve fitting to Eq. 8.11 is the best, i.e., when the summation of the squares of the differences between fitted data and measured data is the least. A statistical analysis is conducted based on the parameters obtained from the best-fit SWCCs and gives (Fig. 8.15)

$$\psi_t = 1.7c_u^{0.39}\psi_a \tag{8.16}$$

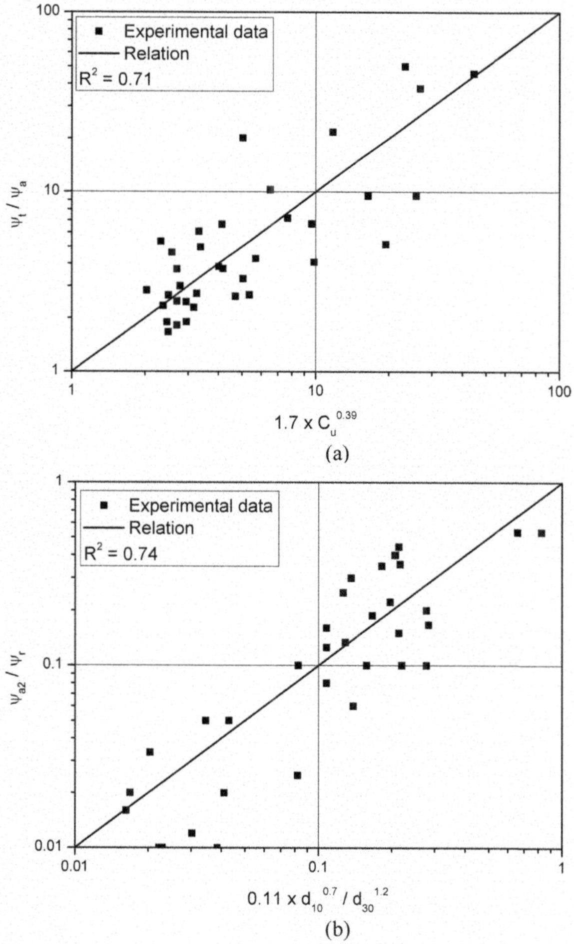

Fig. 8.15 Statistical relation between SWCC parameters and basic soil properties based on fitted data using the bimodal SWCC model: **a** 1st residual suction, **b** 2nd air entry value

$$\psi_{a2} = \frac{0.11 d_{10}^{0.7}}{d_{30}^{1.2}} \psi_r \tag{8.17}$$

8.3.4.2 Predicted SWCCs Based on GSD and Void Ratio

The bimodal SWCC model in Eq. 8.11 is used to predict SWCCs for the different soils based on their GSDs and void ratios. The parameters used in Eq. 8.11 are estimated by empirical equations, i.e., Eqs. 8.13–8.17. The detailed soil information and estimated parameters are listed in Table 8.1, which are extracted from SoilVision database (1997). The predicted SWCCs for the four different loam soils are shown in Fig. 8.16. The predicted SWCCs for four sandy or gravel soils are shown in Fig. 8.17. The predicted SWCCs in Figs. 8.16a, b, d, and 8.17a are close to the experimental data and show weak bimodal features. The predicted SWCCs in Figs. 8.16c and 8.17b, c and d show distinct bimodal features. In Figs. 8.16c and 8.17c, the predicted SWCCs match the measured SWCCs in the whole suction range. These results demonstrate that the prediction method based on grain-size distribution and void ratio is suitable

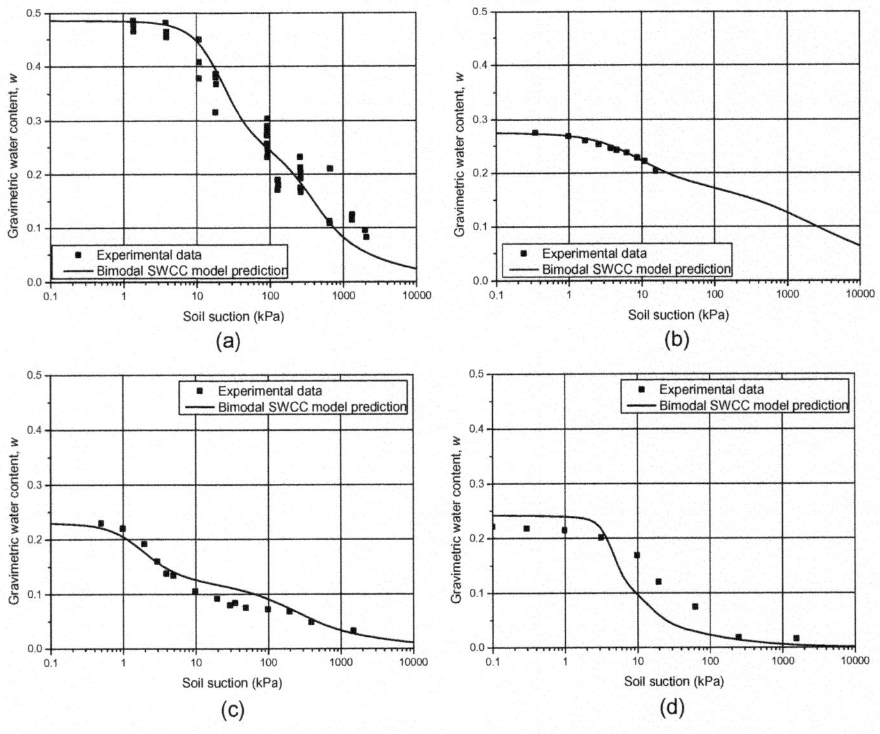

Fig. 8.16 Predicted SWCCs for loam soils using the bimodal SWCC model: **a** Silt loam, **b** loam, **c** sandy loam, **d** loamy sand

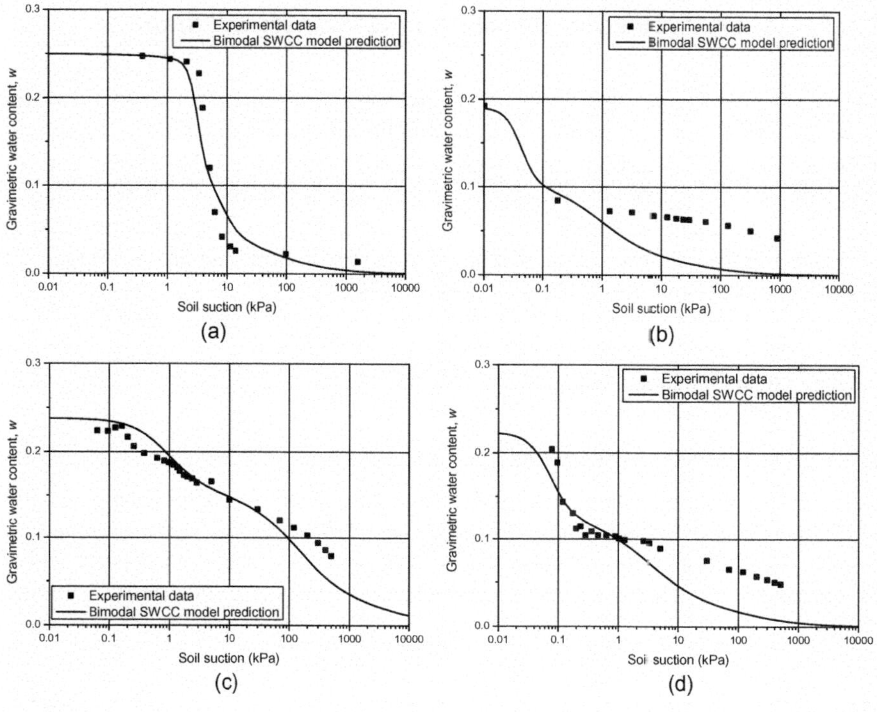

Fig. 8.17 Predicted SWCCs for sand and gravel soils using the bimodal SWCC model: **a** Sand, **b** poorly graded gravel, **c** SM with gravel, **d** GW-GM with sand

for granular materials such as loam soils, sandy soils, and gravel soils. In Fig. 8.17(b) and d, the predicted SWCCs match the measured SWCCs in the low suction range but deviate somewhat from the measured SWCCs in the high suction range. The influence of C_u on residual water content might have been underestimated for these two soils, as shown in Eq. 8.14. Under more realistic conditions, soil may exist in a state of high suction and the prediction is not good enough for these two soils in high suction ranges. As discussed, the bimodal or unimodal SWCCs shown in Figs. 8.16 and 8.17 can be distinguished by observing whether the ratio of ψ_{a2}/ψ_t is larger than 3 or not.

The bimodal SWCC predictions based on the soil property data rather than fitting are useful in the absence of measured SWCCs and will offer great convenience in seepage analysis and analysis of the diffusion of contamination. Clearly, the SWCC prediction is highly empirical and just a rough estimation to real SWCC. It cannot substitute the measured SWCC.

8.3.5 SWCC Prediction Based on Pore-Size Distribution

The direct SWCC measurement methods are normally limited in narrow measurement ranges and are time consuming. For example, the most popular SWCC measurement method, the axis-translation technique, is limited in 0–1500 kPa. One of byproduct of pore geometry measurement is to use the measured PSD to predict SWCC, which can cover a wider suction range (i.e., from several kPa to tens of MPa) comparing to the axis-translation technique. Such predictions can provide important additional information for directly measured SWCCs.

8.3.5.1 Basic Theory

Measurement range of pores in MIP

MIP is routinely and effectively used to measure the PSDs of powder and bulk materials with open and interconnected pore structures. In the MIP test, an assumption is made that cylindrical flow channels exist in the soil. The radius of the pores which are intruded by mercury under an applied pressure, p, can be calculated as (Diamond 1970):

$$r = \frac{2T_{\mathrm{s,m}} \cos \alpha}{p} \tag{8.18}$$

where $T_{\mathrm{s,m}}$ is the surface tension of the fluid (i.e., mercury); α is the contact angle between the mercury–air interface and the solid (i.e., soil particles). The surface tension of mercury is 472 μN/m^2 at 20 °C and α is taken as 135° in this study. The volume of the pores occupied by mercury as the mercury injection pressure is increased from p_1 to p_2 can be calculated as

$$\int_{r_1}^{r_2} f(r)\mathrm{d}r = \int_{p_2}^{p_1} f(p)\mathrm{d}p \tag{8.19}$$

where r_1 and r_2 are the pore radii at p_1 and p_2, as expressed in Eq. 8.18; $f(r)$ is the volume of pores with radii of r; and $f(p)$ is the volume of pores injected by mercury at pressure of p. Equation 8.19 demonstrates that the measurement range of pore radius is limited by the mercury injection pressure supplied by the MIP device. In this study, the intrusion pressure supplied by Micromeritics AutoPore IV 9500 V 1.04 MIP device is from 9.3 kPa to 210 MPa, and the measurement range of pore radius is from 0.003 to 73 μm.

SWCC prediction from PSD

The PSD measured in the MIP test can be effectively used for SWCC prediction. In a SWCC test, the relationship between the radius of the pores which are filled by water and the soil suction follows the following relation (Batchelor 1967),

$$r = \frac{2T_{s,w}\cos\beta}{\psi} \tag{8.20}$$

where the surface tension of water $T_{s,w}$ is 72.8 mN/m^2 at 20 °C and the contact angle between the air–water interface and the soil particle, β, is taken as 0° in this study. Therefore, SWCC is also a macro-performance indicator for PSD.

Linking Eqs. 8.18 and 8.20, the soil suction, ψ, can be related to the injected mercury pressure, p, through the pore radius, as

$$p = \psi\frac{T_{s,m}\cos\alpha}{T_{s,w}\cos\beta} \tag{8.21}$$

In the SWCC prediction, the gravimetric water content in a soil sample at suction ψ can be written as,

$$\begin{aligned} w(\psi) &= \int_0^{r(\psi)} f(r)\rho dr \\ &= \int_{r_{min}}^{r(\psi)} f(r)\rho dr + U_1\rho \end{aligned} \tag{8.22}$$

where ρ is water density; $w(\psi)$ is the gravimetric water content corresponding to soil suction, ψ; the minimum measurable pore radius, r_{min}, is equal to $2T_{s,m}$ cos α/p_{max}; p_{max} is the maximum mercury injection pressures supplied by the device; and $p(\psi)$ is the mercury intrusion pressure calculated from the soil suction ψ using Eq. 8.21; U_1 is the pore volume with radius smaller than r_{min}. U_1 must be estimated and corrected in the SWCC prediction. The gravimetric water content can also be calculated as,

$$\begin{aligned} w(\psi) &= \int_0^{\infty} f(r)\rho dr - \int_{r(\psi)}^{\infty} f(r)\rho dr \\ &= w_s - U_2\rho - \int_{r(\psi)}^{r_{max}} f(r)\rho dr \end{aligned} \tag{8.23}$$

where w_s is the saturated water content; the maximum measurable pore radius, r_{max}, is equal to $2T_{s,m}\cos\alpha/p_{min}$; p_{min} is the minimum mercury injection pressures supplied by the device; and U_2 is the pore volume with radius larger than r_{max}. U_2 must be estimated and corrected in the SWCC prediction.

8.3.5.2 Correction of Volume of Pores Out of the Measurement Range of MIP

For a fine-grained soil, Eq. 8.23 is more appreciated for SWCC prediction because the water stored in oversize pores (U_2 in Eq. 8.23) is negligible (as shown in Fig. 8.17). For a coarse-grained soil, U_2 could be large and is hard to be estimated accurately. Using another way, U_1 in Eq. 8.22 is estimated and used to correct the SWCC prediction. U_1 is the water stored pores smaller than 3 nm (i.e., the water content corresponds to a suction of 48.5 MPa). Because the soil residual suction is normally smaller than 48.5 MPa, U_1 is smaller than the residual water content. In SWCC, the water content changes very slowly with suction as reached the residual water content. Therefore, U_1, which should be a small and stable value, is more appreciated to be estimated and used to correct the SWCC prediction.

Normally, the atmospheric suction is about 1.4 (corresponding to a relative humidity of 99% at 20 °C) to 124 MPa (corresponding to a relative humidity of 40% at 20 °C). The soil exposed to the atmosphere over weeks will come to suction close to atmospheric suction. In our laboratory, the monitored average humidity is 62% (varied from 52 to 74%), which corresponds to an atmospheric suction of 72 MPa. The water contents of soils exposed to laboratory atmosphere over months (termed as the natural water content) are considered to be at suction around 72 MPa. Ignoring the difference between the natural water content and U_1, which is equal to the water content at a suction of 48.5 MPa, the natural water content is used as U_1 for correcting the SWCC prediction for coarse-grained soil.

In a MIP intrusion test, the larger pores are first filled by mercury, followed by the smaller pores. This process is similar to the drainage process in a SWCC test, where the larger pores are drained first, followed by the smaller pores. So the PSD measured in the MIP intrusion test is suitable for predicting the drying SWCC.

8.3.5.3 Correction of PSD Changes After Drying

The PSD may change during the drying process. A shrinkage model proposed by Simms and Yanful (2001) is used to consider the pore-size change during the soil drying process. In this model, the PSD of a saturated soil sample and the PSD of a dried soil sample were used. As an example, the simplified calculation procedure of SWCC prediction for CL with sand is shown in Table 8.5. As soil suction increases from ψ_1 to ψ_2, the pores are considered as two types: pores that do not shrink to smaller radii, which can be calculated from the PSD of the dried soil sample, and pores that shrink from r_1 to r_2, which can be calculated from the difference between the PSD of saturated soil sample and the dried soil sample. As soil suction increases from ψ_1 to ψ_2, the ratio between the drainage and the water stored in potential shrinkable pores is assumed to be calculated as

Table 8.5 Prediction of SWCC for CL with sand using the shrinkage model proposed by Simms and Yanful (2001)

Pore radius range r_1 to r_2 (μm)	Pore volume in drying sample $V_{d,n}$ (cm³/g)	Pore volume in saturated sample $V_{s,n}$ (cm³/g)	Potential shrinkable pore volume $V_{p,n}$ (cm³/g)	Water drained from potential shrinkable pores $V_{c,n}$ (cm³/g)	Water remained in pores shrunk $V_{r,n}$ (cm³/g)	Total drainage M (g/g)	Water content w_n (%)	Suction ψ_n (kPa)
			$V_{s,n} - V_{d,n} + V_{r,n-1}$	$V_{p,n} \times \xi$	$V_{p,n} \times (1-\xi)$	$\rho \times (V_{c,n} + V_{d,n})$	$w_{n-1} - M_n$	
> 59	0	0	0	0	0	0	0.38	2
59 ~ 21	0.002	0.029	0.027	0.017	0.01	0.019	0.361	7
21 ~ 3	0.003	0.087	0.094	0.080	0.014	0.083	0.278	49
3 ~ 0.16	0.141	0.174	0.047	0.044	0.003	0.185	0.093	909
0.16 ~ 0.08	0.034	0.031	0	0	0	0.034	0.059	1905
0.08 ~ 0.0035	0.04	0.035	–	–	–	0.035 [a]	0.024	41,453

ξ is calculated by Eq. 6.24; n is the step number [a] This value is taken as $V_{s,n}$

$$\xi = \frac{r_1^i - r_2^i}{r_1^i} \tag{8.24}$$

where r_1 and r_2 are the pore radius corresponding to ψ_1 and ψ_2, as shown in Eq. 6.20; i is a pore shape parameter. i is taken as 1 for laminar shape, 2 for cylindrical shape, and 3 for spherical shape. In this study, i is chosen as 2.

8.3.5.4 SWCC Prediction Results Based on PSD

The SWCC prediction results for five different colluvial soils are shown in Figs. 8.18, 8.19, 8.20, 8.21, and 8.22. The basic soil properties are shown in Chapter 2, Table 2.1 and Fig. 2.6. The sample information and measured PSD are shown in Chapter 3, Table 3.5 and Figs. 3.8 and 3.16–3.19. In this study, Eq. 8.23 was used for SWCC prediction. The water content of samples exposed to laboratory atmosphere over weeks was used as the second part in Eq. 8.23 for correction. The SWCC considering the PSD changes was further corrected using the PSD of the oven-dried soil sample with the help of a shrinkage model proposed by Simms and Yanful (2001). The predicted SWCCs were also compared with the SWCCs from the laboratory tests.

For CL with sand and sandy ML, the predicted SWCCs based on the saturated soil samples are quite similar to the respective measured SWCCs (Figs. 8.18 and 8.19). For SC with gravel (Fig. 8.20), SM with gravel (Fig. 8.21), and GW-GM with sand (Fig. 8.22), the SWCC predictions based on the saturated soil samples are close to the measured SWCCs at suctions smaller than 200 kPa but deviate from the

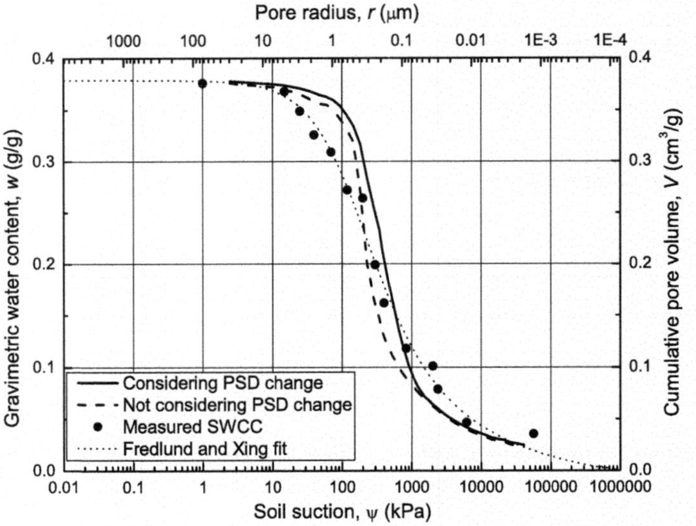

Fig. 8.18 SWCC prediction for CL with sand based on PSD

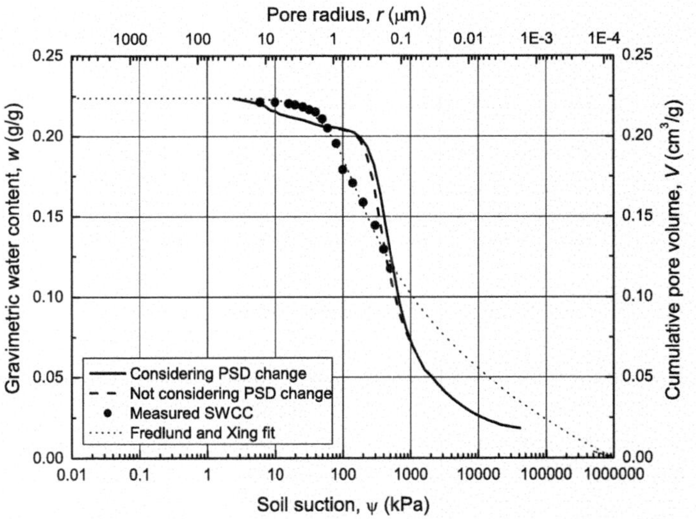

Fig. 8.19 SWCC prediction for sandy ML based on PSD

measured SWCCs at suctions larger than 200 kPa. One possible reason is relevant to the sample preparation method. In the SWCC test, the soil sample contained particles larger than 10 mm. But the soil sample used in the MIP did not contain particles larger than 10 mm. The PSDs of two samples may have some differences. However, the SWCC prediction method is effective for both fine soils and coarse soils. For the

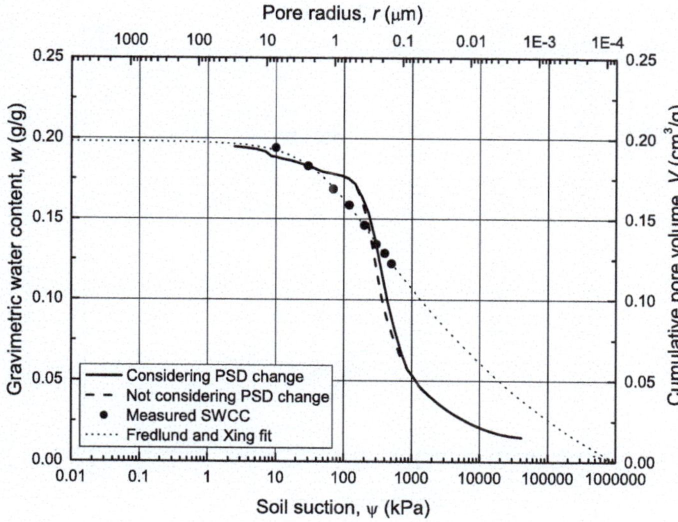

Fig. 8.20 SWCC prediction for SC with gravel based on PSD

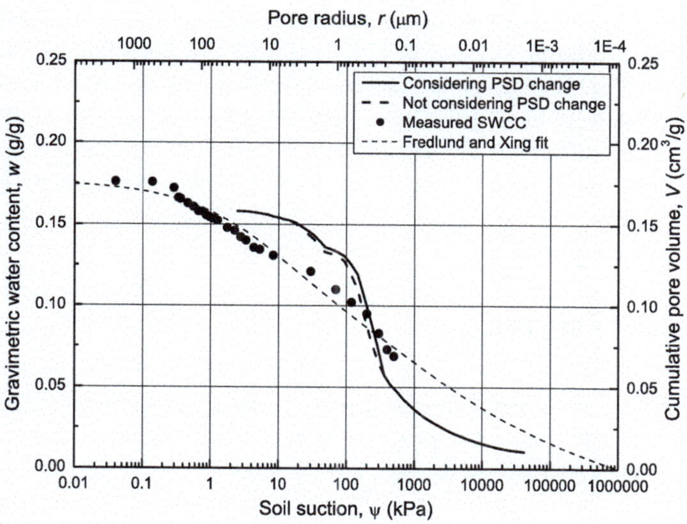

Fig. 8.21 SWCC prediction for SM with gravel based on PSD

coarse soils, the correction of pore volume beyond the measuring range of MIP is important and necessary (Figs. 8.19, 8.20, and 8.21).

For CL with sand, the correction considering the PSD change during the drying process is effective and results in good agreement with the measured data. But for other four coarser soils, the correction considering the PSD change during drying process is limited. Because of the high shrinkage ability of the clayey soil, the PSDs

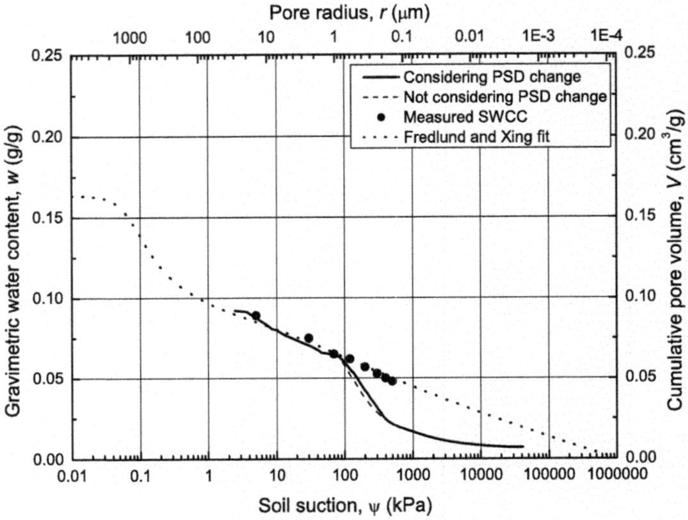

Fig. 8.22 SWCC prediction for GW-GM with sand based on PSD

of the saturated and oven-dried soil samples of CL with sand show large and consistent differences (referring to Chap. 3, Figs. 3.10 to 3.15). Therefore, the correction considering the PSD change is important for clayey soils. Due to the weak shrinkage ability of the sandy soil and gravel soil, the PSDs of the saturated and oven-dried sandy or gravel samples show limited differences (referring to Chap. 3, Figs. 3.16 to 3.19). Therefore, the correction considering the PSD change is less valued for sandy soils or gravel soils.

8.4 Bimodal SHCFs

As shown in Chap. 3, the SHCF for a dual-porosity soil may exhibit bimodal features. A new bimodal SHCF model is proposed in this section based on the concept of dual-porosity and the proposed bimodal SWCC model.

8.4.1 Link Between SHCF and SWCC

Considering the drainage process in a dual-porosity soil, all the pores are filled with water before the suction increases to the 1st air entry value, ψ_a. The flow-channel network is the same as that under saturated conditions. Therefore, the coefficient of hydraulic conductivity remains as the saturated hydraulic conductivity. Once the soil suction reaches the air entry value, ψ_a (seeing Fig. 8.11), the bulk water stored

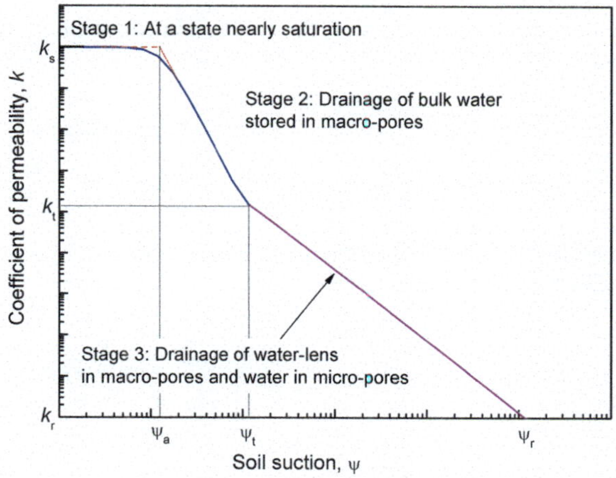

Fig. 8.23 Typical bimodal SHCF

in the macro-pores (inter-particles pores or inter-aggregates pores) starts to drain. Subsequently, the hydraulic conductivity will decrease. In this stage, the flow-channel network is considered to be a combination of a macro-pores network and a micropores network, with the former being more dominating. After the suction reaches the 1st residual suction, ψ_t (seeing Fig. 8.11), all of the bulk water stored in the macro-pores drains and the flow is controlled by the micropore network.

In the bimodal SWCC, the drainage of the bulk water in the macro-pores has a steeper slope than the drainage of the water lens in the macro-pores and the drainage of the bulk water in the micropores. Therefore, the SHCF controlled by the macro-pore network may also have a steeper slope comparing to the SHCF controlled by the micropores network. A turning point will be present on the SHCF at the 1st residual suction, ψ_t. Experimental evidence in the literature demonstrates that the SHCF has a log-linear relation with the soil suction in soils with a unimodal PSD. Figure 8.23 shows a typical bimodal SHCF.

8.4.2 *Bimodal SHCF*

An ideal bimodal SHCF is shown in Fig. 8.23. The function can be described by the following parameters: (1) the saturated hydraulic conductivity, k_s; (2) the hydraulic conductivity at the turning point, k_t; (3) the hydraulic conductivity at the residual suction, k_r; (4) the suction at which the hydraulic conductivity starts to decrease; (5) the suction at the turning point, ψ_t, which is considered to be equal to the 1st residual suction in a dual-porosity structure soil; and (6) the residual suction, ψ_r. After the soil suction increases to the residual suction, ψ_r, the water phase in the soil becomes discontinuous and water vapor diffusion becomes the major means of water exchange.

In the k-ψ coordinates, the linear relation from point (k_s, ψ_a) to point (k_t, ψ_t) can be described as

$$k(\psi) = \left(\frac{k_s}{k_t}\right)^{\frac{\psi_{c1}^h}{\psi^h + \psi_{c1}^h}} k_t \tag{8.25}$$

where h controls the slope; ψ_{c1} corresponds to a hydraulic conductivity of $(\psi_a\psi_t)^{0.5}$; h and ψ_c can be related to ψ_a and ψ_{c1} as follows:

$$\begin{cases} h = \frac{2}{\log\psi_t - \log\psi_a} \\ \psi_{c1} = \sqrt{\psi_a\psi_t} \end{cases} \tag{8.26}$$

Similarly, the linear relation from point (k_t, ψ_t) to point (k_r, ψ_r) can be simulated by

$$k(\psi) = \left(\frac{k_t}{k_r}\right)^{\frac{\psi_{c2}^j}{\psi^j + \psi_{c2}^j}} k_r \tag{8.27}$$

where

$$\begin{cases} j = \frac{2}{\log\psi_r - \log\psi_t} \\ \psi_{c2} = \sqrt{\psi_t\psi_r} \end{cases} \tag{8.28}$$

Thus, the typical bimodal SHCF shown in Fig. 8.23 can be expressed as,

$$k(\psi) = \left(\frac{k_s}{k_t}\right)^{\frac{\psi_{c1}^h}{\psi^h + \psi_{c1}^h}} \left(\frac{k_t}{k_r}\right)^{\frac{\psi_{c2}^j}{\psi^j + \psi_{c2}^j}} k_r \tag{8.29}$$

In Eq. 8.29, parameters h and j indicate the slopes of the SHCF controlled by the macro-pores network (the stage 2 in Fig. 8.23) and the micropores network (the stage 3 in Fig. 8.23), respectively. If the six meaningful parameters, k_s, k_t, k_r, ψ_a, ψ_t, and ψ_r, are determined, the bimodal SHCF will be fully defined.

In Eq. 8.29, the saturated hydraulic conductivity, k_s, can be measured in the laboratory or in situ. The air entry value, ψ_a, and the residual water content, ψ_r, can be used at that in the SWCC. ψ_t is the 1st residual suction in a bimodal SWCC. There are only two parameters, k_t and k_r, to be fitted to measured data or predicted empirically from soil properties, i.e., SWCC or grain-size distribution.

8.4.3 Fitting of Bimodal SHCFs

The bimodal SHCF model is used to fit the experimental data for GW-GM with sand and SM with gravel. The saturated hydraulic conductivity, k_s, is taken as the mean of the test data measured using constant water head hydraulic conductivity test (Li 2009). The air entry value, ψ_a, the 1st residual suction, ψ_t and the residual water content, ψ_r, are selected from the bimodal SWCC. The remaining two parameters, k_t and k_r, were determined so as to provide a good fit of the model to the experimental data.

The fitted bimodal SHCF for GW-GM with sand is shown in Fig. 8.24a. Referring to the bimodal SWCC shown in Fig. 8.13a, ψ_t is taken as 1.5 kPa; ψ_a is taken as 0.25 kPa; ψ_r is taken as 3000 kPa. The saturated hydraulic conductivity k_s is taken as 1×10^{-2} m/s. The fitted bimodal SHCF for GW-GM with sand shows a strong bimodal feature similar to the bimodal SWCC in Fig. 8.13a.

The fitted bimodal SHCF for SM with gravel is shown in Fig. 8.24b. Referring to the bimodal SWCC in Fig. 8.13b, ψ_t is taken as 9 kPa; ψ_a is taken as 0.2 kPa; and ψ_r is taken as 3000 kPa. The saturated hydraulic conductivity k_s is taken as 1×10^{-4} m/s. The fitted bimodal SHCF for SM with gravel shows a weak bimodal feature.

8.4.4 Estimation of Parameters Involved in Bimodal SHCFs

Empirical relations can be established for the estimation of parameters k_t and k_r involved in the bimodal SHCF. Dane and Klute(1977) express the SHCF as,

$$k_w = k_s \exp(b(\theta_w - \theta_r)) \tag{8.30}$$

where θ_w is the volumetric water content; θ_r is the residual water content; k_w is the hydraulic conductivity; k_s is the saturated hydraulic conductivity; and b is a coefficient related to the material. Denote the relative hydraulic conductivity k' as (k_w/k_s) and denote the relative volumetric water content θ' as $(\theta_w - \theta_r)$, which is the water content contributing to water flow, Eq. 8.30 can be modified as,

$$\ln k' = b\theta' \tag{8.31}$$

In the bimodal SHCF (PF), the stage 2 (Fig. 8.23) can be determined by two points (ψ_a, k_s) and (ψ_t, k_t). Point (ψ_a, k_s) corresponds to point (ψ_a, w_s) in bimodal SWCC and point (ψ_t, k_t) corresponds to point $(\psi_t, 0.25w_s + 3w_r)$. Thus, k_t is corresponding to $0.25w_s + 3w_r$ and k_s is corresponding to w_s. Hence, substituting these relations into Eq. 8.31, the relation for the PF in stage 2 can be obtained,

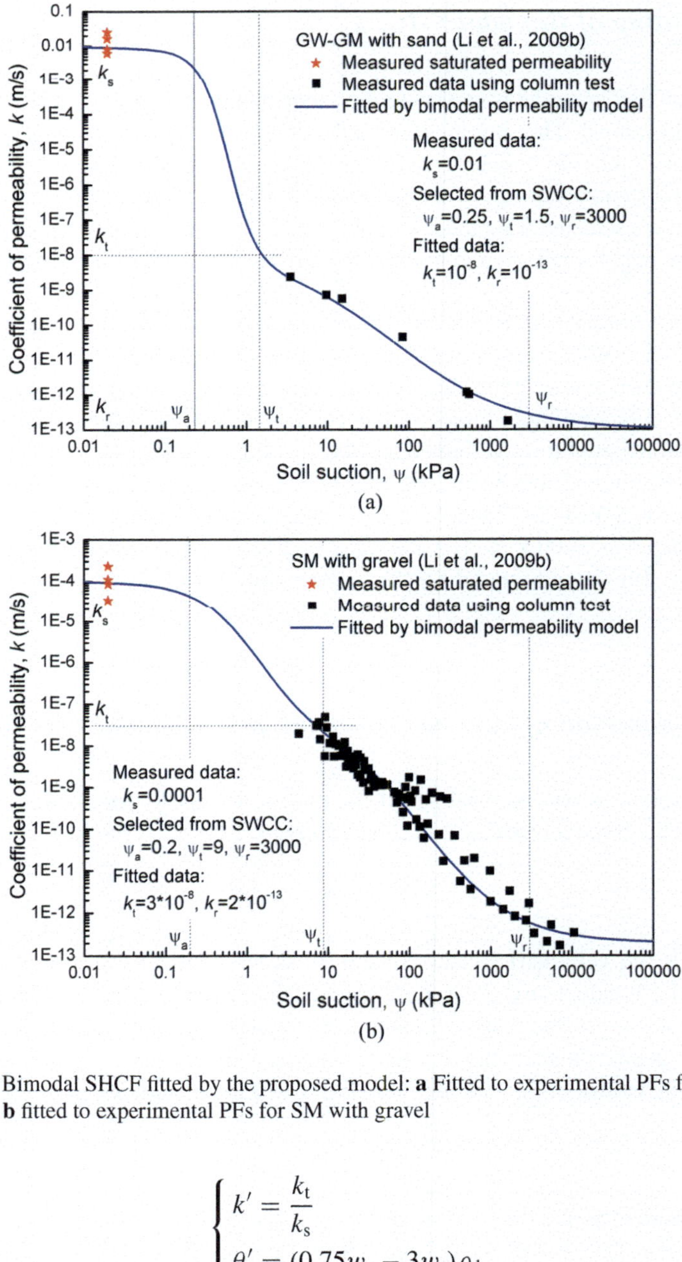

Fig. 8.24 Bimodal SHCF fitted by the proposed model: **a** Fitted to experimental PFs for GW-GM with sand, **b** fitted to experimental PFs for SM with gravel

$$\begin{cases} k' = \dfrac{k_t}{k_s} \\ \theta' = (0.75w_s - 3w_r)\rho_d \end{cases} \tag{8.32}$$

where ρ_d is the dry density; w_s is the saturated water content by weight; and w_r is the residual water content by weight. Similarly, the stage 3 (Fig. 8.23) can be determined by point (ψ_t, k_t) and (ψ_r, k_r), which are corresponding to (ψ_t, $0.25w_s + 3w_r$) and (ψ_r, w_r) in bimodal SWCC, respectively. Substituting these relations into Eq. 8.31, the relation for the PF in stage 3 can be determined as,

$$\begin{cases} k' = \dfrac{k_r}{k_t} \\ \theta' = [0.25w_s + 2w_r]\rho_d \end{cases} \tag{8.33}$$

Substituting Eqs. 8.32 and 8.33 into 8.31 yields

$$\begin{cases} \ln\left(\dfrac{k_t}{k_s}\right) = b\rho_d(0.75w_s - 3w_r) \\ \ln\left(\dfrac{k_r}{k_t}\right) = b\rho_d(0.25w_s + 2w_r) \end{cases} \tag{8.34}$$

Denote $(b\rho_d)$ as α, Eq. 8.34 is rewritten as

$$\begin{cases} k_t = k_s \exp[(0.75w_s - 3w_r)\alpha] \\ k_r = k_s \exp[(w_s - w_r)\alpha] \end{cases} \tag{8.35}$$

where α is a material parameter. The saturated water content, w_s, and the residual water content, w_r, can be found on the bimodal SWCC. Only the material parameter, α, which is an empirical coefficient, needs to be determined.

Using the fitting parameters for SM with gravel and GW-GM with sand (Fig. 8.24), α is found to be -154 for these two colluvial soils (Fig. 8.25). However, no regression analysis has been performed for other types of soil because of the lack of experimental data of bimodal SHCFs. The α value of -154 may be limited for coarse-controlled colluvial soils.

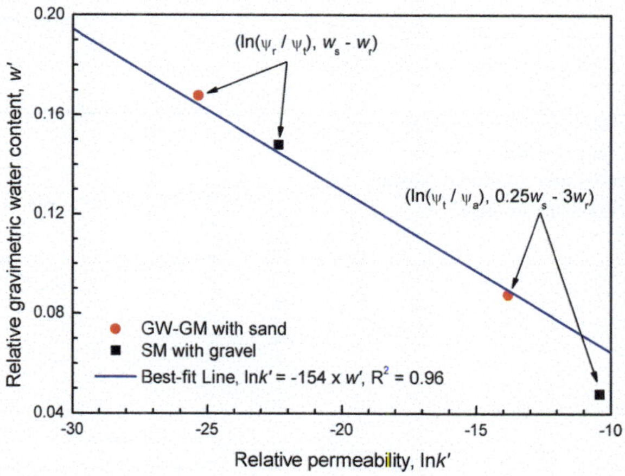

Fig. 8.25 Regression analysis for estimating parameter α in Eq. 8.35

8.5 Bimodal Shear Strength

8.5.1 *Variation of Compressibility with Suction*

The variation of specific volume v with net mean stress p' during the isotropic consolidation tests on the $f_c = 90\%$ specimen is shown in Fig. 8.26. The volume of the soil decreases as the net mean stress increases. Continuous increase in the mean net stress causes the soil to yield at some point. The compressibility index, $\lambda(s)$, is defined as the slope at large net mean stresses.

Figure 8.27 plots the variation of $\lambda(s)$ with suction for the $f_c = 90\%$ specimen and the kaolin clay. The parameter $\lambda(s)$ is matric suction-dependent. Matric suction is closely related to the amount of water in the voids of the soil. Hence, the compressibility of an unsaturated soil bears a close relation to the SWCC. The relation between $\lambda(s)$ of the $f_c = 90\%$ specimen and matric suction exhibits bimodal features. $\lambda(s)$

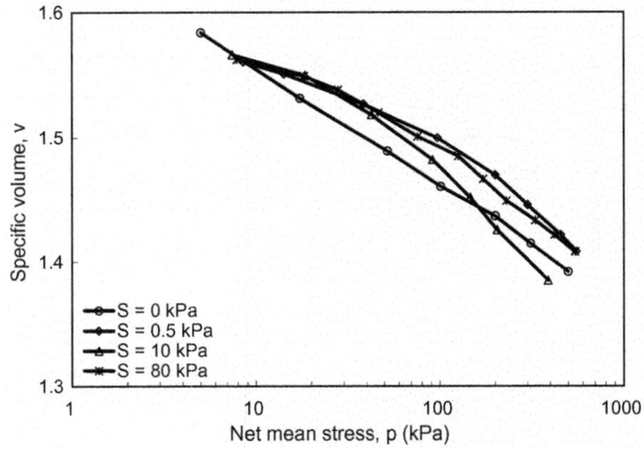

Fig. 8.26 Isotropic consolidation lines for the $f_c = 90\%$ specimen

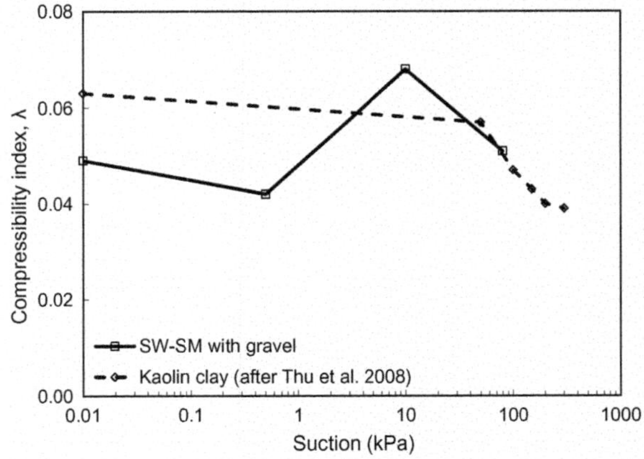

Fig. 8.27 Variations of compressibility of the $f_c = 90\%$ specimen and kaolin clay with suction

decreases when the suction increases from 0 to 0.5 kPa, as the soil state changes from stage *a* to stage *b*. The bulk water drains from the inter-aggregate pores, while the fines aggregates remain saturated and the water film forms a continuous phase. As drying proceeds, the capillary stress gradually increases, and the inter-particle contacts become more stable. The compressibility decreases with matric suction during stage *b*. The compressibility then increases in stage *c*. The meniscus water vanishes from the inter-aggregate pores, whereas the fines aggregates remain saturated. The water content in the $f_c = 90\%$ specimen at this condition is quite small and may not effectively transmit suction to the soil particles or aggregate contact points. This results in a decrease in compressibility. In stage *d*, the compressibility decreases with increasing matric suction due to the fact that the bulk water and meniscus water in the intra-aggregate pores start to drain and that the fines aggregates become unsaturated. The suction prevents the fines aggregates from breaking down, resulting in hardening of the packets and a decrease in compressibility.

However, the compressibility monotonously decreases with suction for the kaolin clay. During the drying process of the kaolin clay, water gradually drains from the intra-aggregate pores. The stiffness of the kaolin increases as the meniscus water forms during the drying process. Consequently, the compressibility decreases with suction. The kaolin clay is a highly plastic clay. Even at high values of suction, there could still be considerable water available to transmit suction between the soil particles. This behavior contributes to decreased compressibility. The bimodal compressibility behavior of the $f_c = 90\%$ specimen is related to the bimodal SWCC, and the compressibility of the clay is influenced by its unimodal SWCC.

8.5.2 Variation of Shear Strength with Suction

The results of drained triaxial tests carried out at selected matric suctions under constant net confining pressures of 5, 15 and 25 kPa are summarized in Fig. 8.28. The dilative behavior is more obvious with increasing suction. The critical state is reached at large strains. The shear strength at an axial strain of 25% is taken to be the critical state shear strength.

The shear strength of unsaturated soils is expressed as:

$$\tau = c' + (\sigma - u_a) \tan \phi' + c \tag{8.36}$$

where τ is the shear strength; c' is the effective cohesion; ϕ' is the effective friction angle with respect to net normal stress $(\sigma - u_a)$; and c is the apparent cohesion due to soil suction. The effect of soil suction on the shear strength is reflected by an increase in the apparent cohesion.

The apparent cohesion dependents on the soil suction, $(u_a - u_w)$, as well as the contact area of pore-water with soil particles (Han et al. 1995; Vanapalli et al. 1996). Fredlund et al. (1978) considered the apparent cohesion a linear function of soil suction:

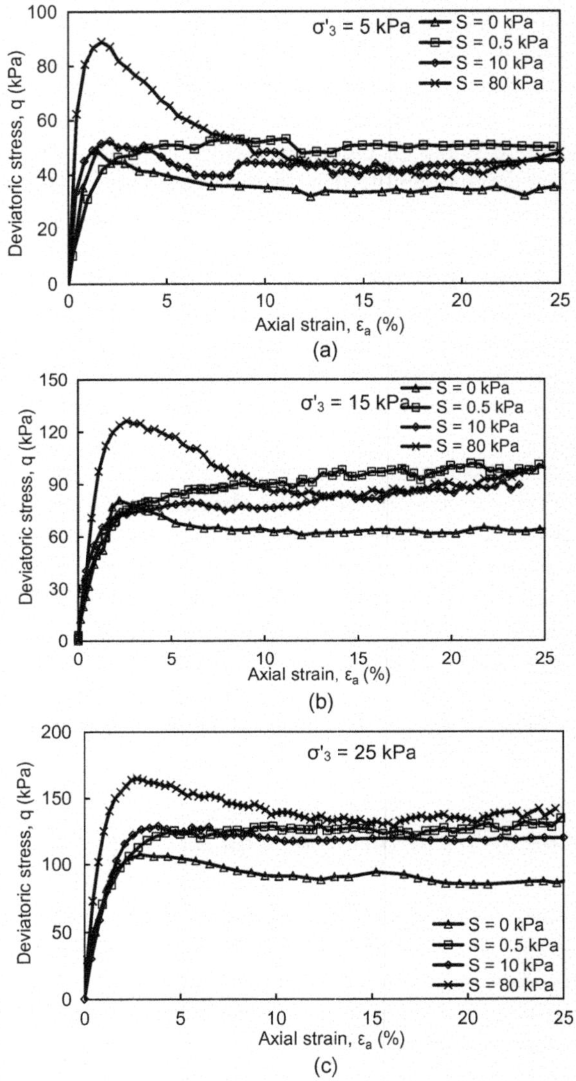

Fig. 8.28 Relationships between deviatoric stress and axial strain under different suctions: **a** 5 kPa; **b** 15 kPa; **c** 25 kPa

$$c = (u_a - u_w) \tan \phi^b \tag{8.37}$$

where ϕ^b is the friction angle that reflects the rate of change in the apparent cohesion with respect to soil suction. Vanapalli et al. (1996) pointed out that in the boundary stage, ϕ^b is equal or close to ϕ'. Therefore, the increment of shear strength follows a linear relationship at suctions lower than AEV. A nonlinear relationship between shear strength and matric suction was reported in the transition stage. When matric suction is larger than the residual matric suction, the shear strength of unsaturated soil may increase, decrease, or remain relatively constant depending on the soil type. The increase of shear strength occurs for clays in a large range of suctions. In contrast, in some cases, particularly for sands, the shear strength decreases with further desaturation beyond the residual state.

In this study, the effective cohesion of the soil is zero; the effective critical friction angle of the soil is $\phi_{cr}{}' = 41.4°$. Hence, the critical apparent cohesion values at various suction levels can be readily obtained from the test data in Fig. 8.28. Figure 8.29a shows the variation of apparent critical cohesion with soil suction under different net confining pressures (5, 15 and 25 kPa). To reflect the changes in the critical apparent cohesion under very low suctions, the log scale is used in this plot. A value of 0.01 kPa, which is smaller than the AEV (0.07 kPa), is used to represent the saturated state. The figure indicates that the apparent cohesion of the $f_c = 90\%$ specimen shows bimodal features. Figure 8.29b and c shows the variations of apparent cohesion of sand (Likos et al. 2012) and clay (Thu et al. 2007) with matric suction, respectively. The apparent cohesion of the sand and clay show a unimodal distribution. The apparent cohesion increases linearly up to the AEV. In the transition zone, the apparent cohesion increases nonlinearly as water drains from the soil. However, beyond the residual suction, the apparent cohesion decreases for the sand but continues to increase for the clay. Figure 8.29a shows that the critical apparent cohesion of the $f_c = 90\%$ specimen increases with increasing matric suction in the first drainage period and decreases at the first residual state. In the second drainage period, the shear strength increases monotonically with increasing suction. The shear strength behavior of the $f_c = 90\%$ specimen is similar to that of sand in the first drainage period, but similar to that of clay in the second drainage period.

The shear strength behavior of the $f_c = 90\%$ specimen is related to the bimodal SWCC. When the specimen changes from stage *a* to stage *b* as shown in Fig. 8.3, the inter-aggregate pores change from full saturation to an unsaturated condition. In stage *b*, the additional bonding force induced by suction increases as the suction increases monotonically. The shear strength increases as a consequence of bonding phenomena between soil particles. The soil then enters stage *c* which is the residual stage of the first drainage period. The water menisci vanish and may not transmit suction effectively to the soil particles or aggregate contact points. Thus, the shear strength decreases due to the de-bonding phenomena between soil particles. In the second drainage period, water starts to drain from the intra-aggregate pores. The fines aggregates become unsaturated and the stiffness increases. The fines aggregates behave more like individual particles. The possibility of slippage between the particles is partially reduced, which increases the shear strength.

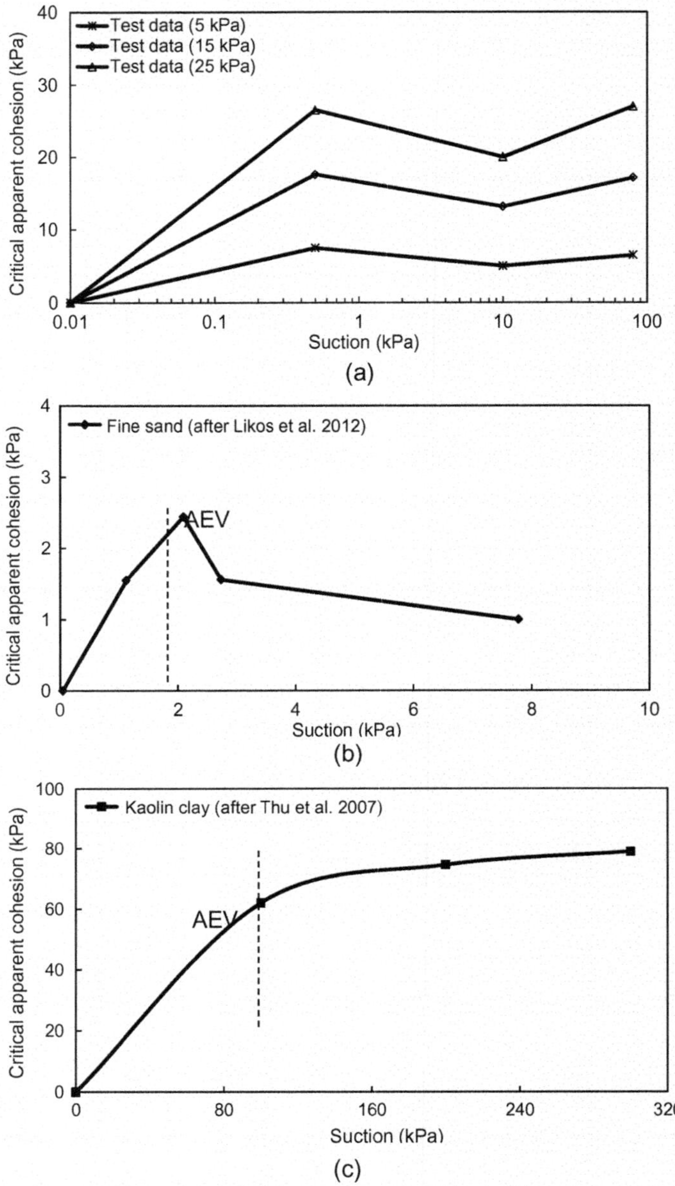

Fig. 8.29 Variation of critical apparent cohesion with suction: **a** the $f_c = 90\%$ specimen; **b** sand; **c** kaolin clay

Table 8.6 Published equations for predicting the shear strength of unsaturated soils

References	Apparent cohesion c
Fredlund et al. (1996)	$(u_a - u_w)(S_r)^k \tan\phi'$
Oberg and Sallfors (1997)	$(u_a - u_w)S_r \tan\phi'$
Vanapalli et al. (1996)	$(u_a - u_w)\left(\frac{\theta-\theta_r}{\theta_s-\theta_r}\right)\tan\phi'$
Khalili and Khabbaz (2003)	$(u_a - u_w)\left(\frac{s_e}{s}\right)^r \tan\phi'$

Notes $u_a - u_w$ is matric suction; S_r is the degree of saturation; θ is the volumetric water content; θ_r is the residual water content; θ_s is the saturated water content; and is ϕ' is the effect friction angle of saturated soil

8.5.3 *Evaluation of Various Shear Strength Equations for Unsaturated Soils*

Four published shear strength equations (Table 8.6) are evaluated using the test data of the $f_c = 90\%$ specimen, sand, and kaolin clay. The evaluation results are plotted in Fig. 8.30. Figure 8.30a shows the evaluation results for the $f_c = 90\%$ specimen. There is significant discrepancy between the experimental and predicted values under low suctions and between the observed and predicted trends in the first residual stage where the shear strength decreases with increasing suction. As shown in Fig. 8.30c, all the equations give reasonable estimates of the shear strength of the kaolin clay with the parameters in Table 8.7. The critical apparent cohesion increases as the suction increases. The changing rate decreases gradually as the residual suction conditions are approached. The predictions for the sand are significantly lower than the test data under low suction values in Fig. 8.30b.

Kim and Borden (2011) have discussed the limitations and difficulties associated with some of the shear strength equations. Most of the equations were developed to estimate the shear strength of fine-grained soils which have a unimodal SWCC and may not be applicable to granular soils that have a bimodal or multimodal SWCC. For example, the Khalili and Khabbaz (1998) method was developed based on the measured shear strengths of 14 fine-grained soils with AEVs ranging from 35 to 153 kPa. Therefore, this equation should be used with caution for coarse-grained soils and soils that have AEV values out of this range.

8.5.4 *Proposed Shear Strength Equations for Granular Soils with Bimodal SWCCs*

8.5.4.1 Microstructural Relative Degree of Saturation

As the drainage process on the bimodal SWCC demonstrates (Fig. 8.27), the pore water drains from different pore series in different stages. The pore water in the

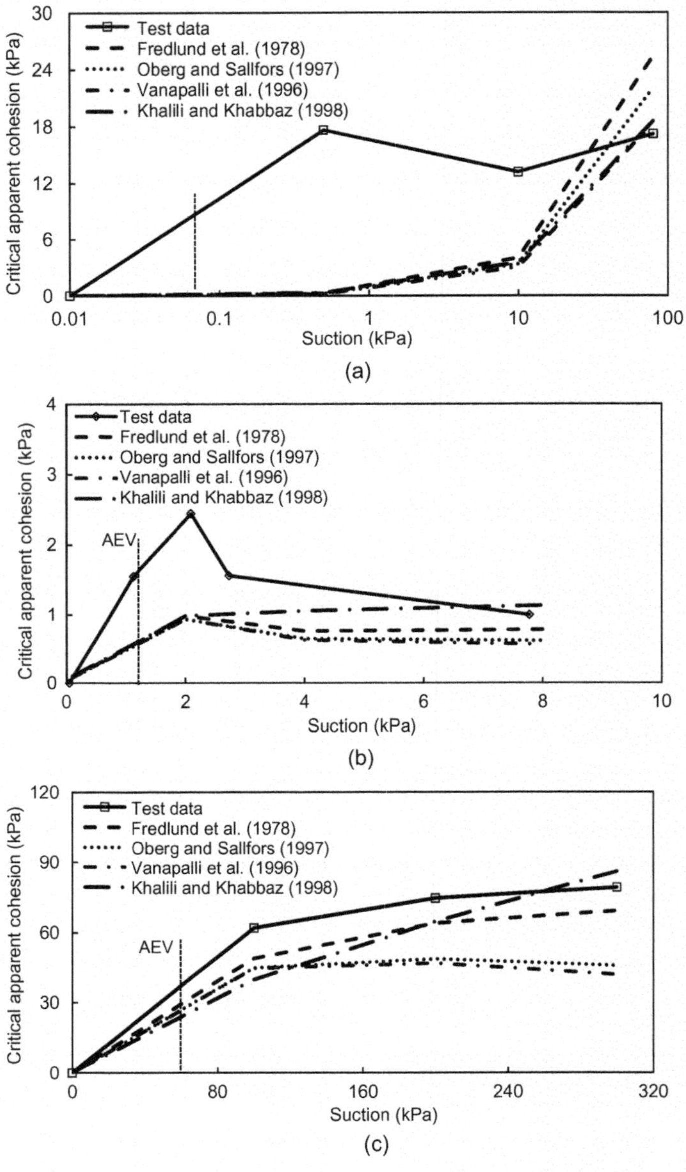

Fig. 8.30 Comparison between the shear strength envelopes from published equations and the experimental program: **a** the $f_c = 90\%$ specimen; **b** sand; and **c** kaolin clay

Table 8.7 Additional parameters for the shear strength equations for SW-SM with gravel, kaolin clay, and sand

References	Additional parameters		
	SW-SM with gravel	Kaolin clay	Sand
Fredlund et al. (1996)	$\phi_{cr}' = 41.4°$ $k = 1.2$, SWCC	$\phi_{cr}' = 31.1°$ $k = 0.7$, SWCC	$\phi_{cr}' = 38°$ $k = 0.9$, SWCC
Oberg and Sallfors (1997)	SWCC	SWCC	SWCC
Vanapalli et al. (1996)	$\theta_s = 0.91$, SWCC	$\theta_s = 0.99$, SWCC	$\theta_s = 0.93$, SWCC
Khalili and Khabbaz (2003)	$S_e = 0.1$ kPa, $r = 0.2$	$S_e = 25$ kPa, $r = 0.3$	$S_e = 1.2$ kPa $r = 0.1$

Notes ϕ_{cs}' is the effect friction angle; θ_s is the saturated water content; S_e is the air entry value; and k and r are two fit parameters

microstructure could be separated into three parts: (1) the bulk water in the inter-aggregate pores, (2) the meniscus water in the inter-aggregate pores, and (3) the water in the intra-aggregate pores. Normalized volumetric water content (degree of saturation) obtained from the SWCC is used as a controlling soil parameter for the secant slope between matric suction and unsaturated shear strength (Fredlund et al. 1996; Oberg and Sallfors 1997; Lee et al. 2005). In this study, the degree of saturation is used to quantify the pore water in the soil. As show in Fig. 8.31, parameter $S^1{}_{rs}$ represents the degree of saturation at the AEV of the first drainage period; $S^1{}_{rr}$ represents the first residual degree of saturation; $S^2{}_{rs}$ is the degree of saturation at the AEV of the second drainage period; and $S^2{}_{rr}$ is the degree of saturation in the second residual stage.

A microstructural relative degree of saturation, $S^{1e}{}_r$, is used to quantify the bulk water in the inter-aggregate pores in stages a and b and defined as:

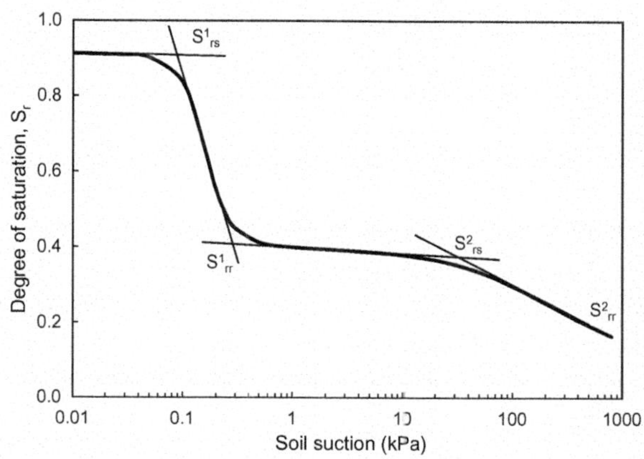

Fig. 8.31 Definition of four degrees of saturations of soil on a bimodal SWCC

$$S_{\mathrm{r}}^{1\mathrm{e}} = \frac{S_{\mathrm{r}} - S_{\mathrm{rr}}^{1}}{1 - S_{\mathrm{rr}}^{1}} \tag{8.38}$$

$S^{1e}{}_{r}$ is thus a measure of the freely available water filling the inter-aggregate pores, which ranges from 1 to 0 when the soil state changes from stage *a* to stage *c*.

The microstructural relative degree of saturation, $S^{2e}{}_{r}$, quantifies the meniscus water in the inter-aggregate pores at the first residual state (stage *c*) and can be defined as:

$$S_{\mathrm{r}}^{2e} = \frac{S_{\mathrm{r}} - S_{\mathrm{rs}}^{2}}{S_{\mathrm{rr}}^{1} - S_{\mathrm{rs}}^{2}} \tag{8.39}$$

$S^{2e}{}_{r}$ is therefore a measure of the meniscus water between the particle–particle contacts in stage *c*. Similarly, the microstructural relative degree of saturation in the second drainage period, $S^{3e}{}_{r}$, can be defined as:

$$S_{\mathrm{r}}^{3e} = \frac{S_{\mathrm{r}} - S_{\mathrm{rr}}^{2}}{S_{\mathrm{rs}}^{2} - S_{\mathrm{rr}}^{2}} \tag{8.40}$$

$S^{3e}{}_{r}$ describes the amount of water in the intra-aggregate pores using a scale that extends from 1 (when the intra-aggregate pores are fully saturated) to 0 (when the intra-aggregate pores are dry).

8.5.4.2 Apparent Cohesion for Granular Soils with Bimodal SWCCs

The shear strength of an unsaturated soil is strongly influenced by the amount of water in the void of the soil. Based on the drainage process of the bimodal SWCC and three microstructural relative degrees of saturation, shear strength equations under different drainage stages (*a*, *b*, *c* and *d*) are developed in this section.

In stage *a*, the soil is saturated; the negative pore water pressure increases the effective stress. The shear strength due to suction is defined as:

$$c = (u_{\mathrm{a}} - u_{\mathrm{w}}) \tan \varphi' \tag{8.41}$$

where $(u_{\mathrm{a}} - u_{\mathrm{w}})$ is matric suction; $\phi' = 41.1^{\circ}$ for SW-SM with gravel.

The suction-induced inter-particle bonding is a function of two factors: the number of water menisci per unit volume of the solid fraction and the intensity of the stabilizing normal force exerted at the inter-particle contact by a single water meniscus (Gallipoli et al. 2003). In stage *b*, the apparent cohesion, *c*, is defined as:

$$c = (u_{\mathrm{a}} - u_{\mathrm{w}})(1 - S_{\mathrm{r}}^{1\mathrm{e}}) \tan \varphi' \tag{8.42}$$

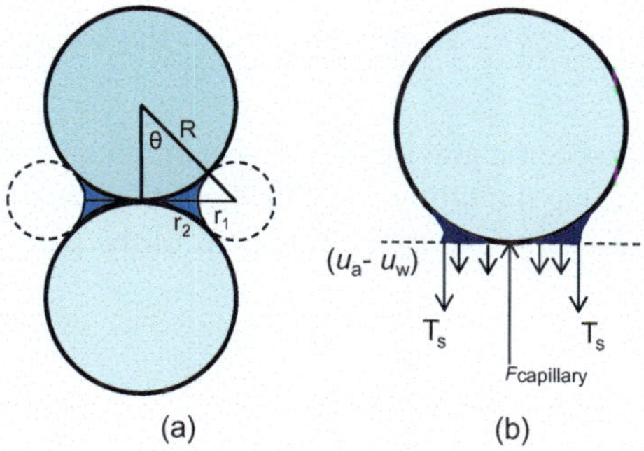

Fig. 8.32 Water–air menisci between two solid spheres: **a** contact model for the $f_c = 90\%$ specimen in stage *c*; **b** induced capillary force by suction

The factor $(1 - S^{1e}{}_r)$ accounts for the number of water menisci per unit volume of soil particles, which is equal to zero when the soil is saturated where the water menisci are absent.

At a particular level of saturation (stage *c*), the water phase becomes discontinuous. The pore-water pressure of isolated meniscus water causes local actions at the particle level. Numerous models have been developed to quantify the capillary stress developed between particles for a simplified two-particle system as shown in Fig. 8.32b (Fisher 1926; Likos and Lu 2004). The macroscopic consequence of capillary stress is a tendency for the particles to be pulled together. The capillary stress of granular unsaturated soil may be calculated based on the theory proposed by Likos and Lu (2004).

The relative compaction of the specimens of the $f_c = 90\%$ specimen is 83%, which means the soil is very loosely deposited. The soil structure is simplified by assuming that the grains are spherical and arranged in simple cubic packing. The coordination number of the particles is six. In two dimensions, any two contacting particles can be considered as an elemental volume. Referring to Fig. 8.32a, the radius of the particles is *R*, which can be determined by the mean particle radius (R_{50} = 1.75 mm) of the soil. As shown in Fig. 8.32b, the capillary forces along the center of the water lens between two particles are defined as:

$$F_{\text{capillary}} = (u_a - u_W)\pi r_2^2 + 2T_s\pi r_2 \tag{8.43}$$

where the first component refers to the difference in the pressures outside and inside the lens acting on the cross-section area, and the second component refers to the surface tension acting along the air–water contact.

The gravimetric water content (obtained from the SWCC) between two contacting particles can be written in terms of the filling angle, θ, and thus the water content of the soil–water system (Likos and Lu 2004):

$$w = \frac{9}{2G_s}\sin^2\theta - \frac{9}{2G_s}\sin^2\theta\cos\theta - \frac{3}{2G_s}(1-\cos\theta)^2(2+\cos\theta)$$

$$-\frac{9}{4G_s R^3}r^2\left[(\pi - 2\theta - \sin 2\theta)R\tan\theta - \frac{4}{3}R(1-\cos\theta)\cos^2\theta\right] \quad (8.44)$$

where $G_s = 2.64$ is the specific gravity of the soil particles.

Increasing or decreasing the filling angle changes the size of the water lens. The radii of the water meniscus r_1 and r_2 are functions of the particle radius R and the filling angle θ:

$$r_1 = R\frac{1-\cos\theta}{\cos\theta} \quad (8.45)$$

$$r_2 = R\tan\theta - r_1 \quad (8.46)$$

When the water menisci start to disappear from the particle contacts in the residual stage, the number of water menisci per unit volume of solid fraction can be expressed as a monotonically decreasing function of microstructural relative degree of saturation $S^{2e}{}_{r}$, so the magnitude of the capillary-induced shear strength due to the water menisci in the residual stage can be defined as follows:

$$c = \sigma_{\text{capillary}} S_r^{2e}\tan\phi' \quad (8.47)$$

where $\sigma_{\text{capillary}} = F_{\text{capillaty}}/\pi R^2$.

In stage d, the apparent cohesion, c, is determined by the clay aggregates. The apparent cohesion induced by the suction in this stage is a function of suction, microstructure relative degree of saturation $S^{3e}{}_{r}$, and the content of clay aggregates:

$$c = (1-f_c)(u_a - u_w)\left(1 - S_r^{3e}\right)\tan\phi' \quad (8.48)$$

where f_c is the coarse content.

In summary, the equations for the apparent cohesion of the soil with a bimodal SWCC can be defined as:

$$c = \begin{cases} (u_a - u_w)\tan\phi' & \text{stage } a \\ (u_a - u_w)(1 - S_r^{1e})\tan\phi' & \text{stage } b \\ \sigma_{\text{capillaty}} S_r^{2e}\tan\phi' & \text{stage } c \\ (1-f_c)(u_a-u_w)(1-S_r^{3e})\tan\phi' & \text{stage } d \end{cases} \quad (8.49)$$

8.5.5 Comparison Between Experimental and Predicted Results

The percentages of the measured pore volumes of the $f_c = 90\%$ specimen with radii above 73 mm and 2–73 mm are about 44% and 17%, respectively. The percentage of the volume of the intra-aggregate pores is 38%. The relative compaction of the specimen is 83%, which means the soil is loosely deposited. Li and Zhang (2009) pointed out that compaction changes the inter-aggregate structure of soil significantly. That is, the increase in net normal stress will cause the pore size of the unsaturated soil to decrease, and the soil will retain more water during desaturation. Ultimately, the contact area of the soil particles and water will be increased; thus, the shear strength of the soil matrix will increase. The apparent cohesion of the $f_c = 90\%$ specimen is highly stress-dependent as shown in Fig. 8.29.

The SWCC test performed at zero net normal stress is often used to predict the shear strength at different net normal stresses. The measurements of apparent cohesion under different stresses are normalized by the net confining pressure as shown in Fig. 8.33. The modeling results are plotted in terms of apparent cohesion and suction. The measured and predicted shear strengths show good agreement. The apparent cohesion almost does not change in stage *a* when the soil remains saturated and the contribution of suction is very little under the very low suction level. In stage *b*, the apparent cohesion increases sharply due to the formation of meniscus water. Afterward, the apparent cohesion decreases with increasing suction in stage *c*. In stage *d*, the apparent cohesion increases monotonically with suction.

The proposed equations are also used to predict the apparent cohesion of the sand and the kaolin clay. The proposed equations for stages *a*, *b* and *c* ($R_{50} = 0.125$ mm) are used to predict the apparent cohesion of the sand. The shear strength equation appears to provide a close prediction [Fig. 8.33b]. The predicted apparent cohesion increases linearly before the AEV, and reaches a peak at an unsaturated state. After the residual condition, the apparent cohesion decreases with increasing suction. The proposed equations in stages *a* and *b* can also be used to predict the apparent cohesion of the kaolin clay in the considered suction range. As shown in Fig. 8.33c, the shear strength of the kaolin clay is predicted satisfactorily. It appears that the proposed model can be used to effectively predict the shear strength of a variety of soils in the relevant stages.

8.6 Summary

In this chapter, a new bimodal SWCC model is presented, which uses five parameters with clear physical meanings to describe the drainage process in a dual-porosity soil structure, in which the water stored in the macro-pores (inter-aggregates pores or inter-particles pores) drains in the first drainage stage and the water stored in the micropores (intra-aggregate pores) drains in the second drainage stage. The new bimodal SWCC model can be used to fit different SWCCs very well, including both unimodal SWCCs and bimodal SWCCs.

A new bimodal SHCF model is also proposed and used to fit experimental data. The bimodal SHCF model uses six meaningful parameters. Among these parameters, three suction parameters can be found on the bimodal SWCC. The

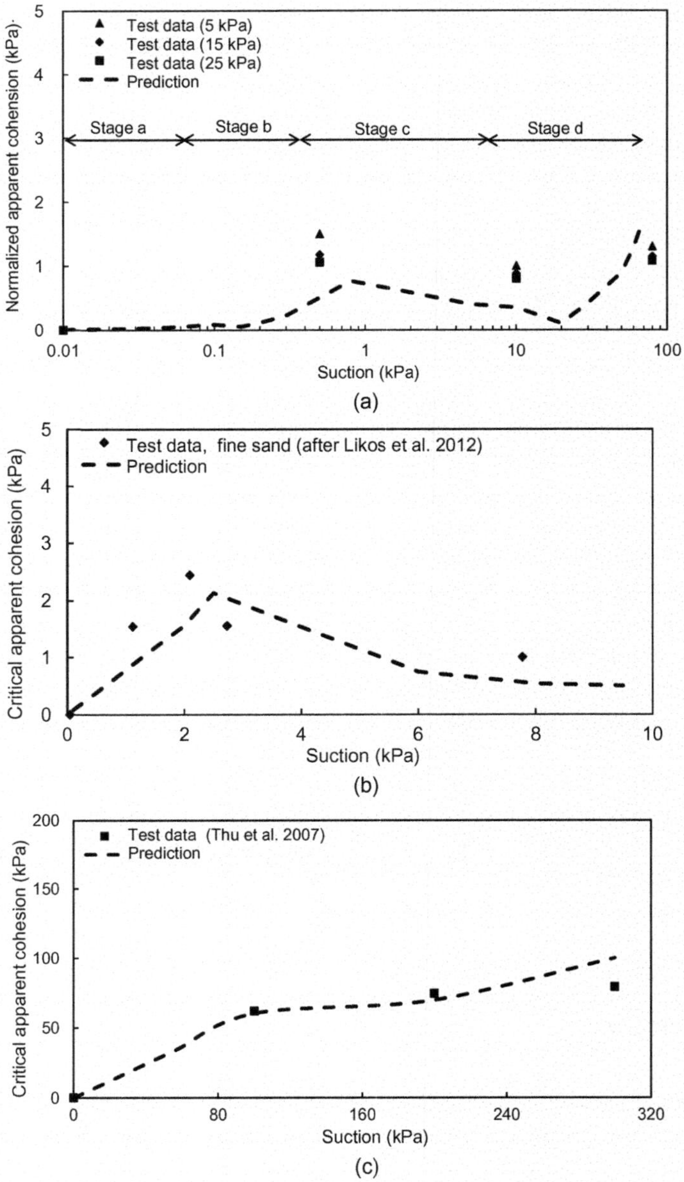

Fig. 8.33 Comparison between the shear strength envelopes from the proposed equations and the experimental program: **a** the $f_c = 90\%$ specimen; **b** sand; and **c** kaolin clay

other three hydraulic conductivity parameters can be calibrated from experimental measurements or estimated using empirical equations.

In response to the bimodal SWCC, the compressibility of the $f_c = 90\%$ soil also exhibits bimodal features. The parameter $\lambda(s)$ decreases as the soil state changes from stage a to stage b when the capillary stress increases. The compressibility increases in stage c because the reduced meniscus water area hinders effective transition of suction to the soil particle. In stage d, the compressibility decreases again with increasing suction as the clay aggregates become unsaturated. The shear strength also exhibits bimodal features. The apparent cohesion increases in stage b as the bonding force between soil particles increases due to the formation of meniscus water. In stage c, the water menisci are not continuous and start to disappear. The apparent cohesion decreases due to the de-bonding between soil particles. After that, the apparent cohesion increases again as the water starts to drain from the intra-aggregate pores. The overall shear strength behavior over the entire suction range shows two modes, which represent more general shear strength behavior not yet reported in the literature.

A shear strength model is proposed to calculate the shear strength of granular soils with a bimodal SWCC. The predicted results are in good agreement with the measured values. The proposed equations can also be degenerated to predict the shear strength of unimodal sand and kaolin clay.

References

Batchelor GK (1967) An introduction to fluid dynamics. Cambridge University Press

Brooks RH, Corey AT (1964) Hydraulic properties of porous media. Hydrology Paper, No. 3, Colorado State University, Fort Collins, Colorado

Brutsaert W (1966) Probability laws for pore size distributions. Soil Sci 101:85–92

Dane JH, Klute A (1977) Salt effects on the hydraulic properties of a swelling soil. Soil Sci Soc Am J 41(6):1043–1049

Diamond S (1970) Pore size distributions in clays. Clays Clay Miner 18:7–23

Domenico A, Schwartz FW (1997) Physical and chemical hydrogeology, 2nd edn. Wiley, New York

Fisher RA (1926) On the capillary forces in an ideal soil: correction of formula given by W. B Haines. J Agric Sci 16:492–505

Fredlund DG, Morgenstern NR, Widger RA (1978) Shear strength of unsaturated soils. Can Geotech J 15(3):313–321

Fredlund DG, Xing AQ (1994) Equations for the soil-water characteristic curve. Can Geotech J 31:521–532

Fredlund MD, Fredlund DG, Wilson GW (2002) Use of the grain-size distribution for the estimation of the soil-water characteristic curve. Can Geotech J 39(5):1103–1117

Gallipoli D, Gens A, Sharma R, Vaunat J (2003) An elasto-plastic model for unsaturated soil incorporating the effects of suction and degree of saturation on mechanical behavior. Geotechnique 53(1):123–135

Guan GS, Rahardjo H, Choon LE (2010) Shear strength equations for unsaturated soil under drying and wetting. J Geotech Geoenviron Eng 136(4):594–606

Gvirtzman H, Roberts PV (1991) Pore scale spatial-analysis of 2 immiscible fluids in porous-media. Water Resour Res 27(6):1165–1176

Han KK, Rahardjo H, Broms BB (1995) Effect of hysteresis on the shear strength of a residual soil. In: Proceedings of the first international conference on unsaturated soils/UNSAT, Balkema, Rotterdam, The Netherlands, vol 2, pp. 499–504
Houston SL, Perea-Garcia N, Houston WN (2008) Shear strength and shear-induced volume change behavior of unsaturated soils from a triaxial test program. J Geotech Geoenviron Eng 134(11):1619–1632
Juang CH, Holtz RD (1986) A probabilistic permeability model and the pore size density function. Int J Numer Anal Meth Geomech 10:543–553
Khalili N, Khabbaz MH (1998) A unique relationship for the determination of the shear strength of unsaturated soils. Geotechnique 48(5):681–687
Kim WS, Borden RH (2011) Influence of soil type and stress state on predicting shear strength of unsaturated soils using the soil-water characteristic curve. Can Geotech J 48:1886–1990
Lee IM, Sung SG, Cho GC (2005) Effect of stress state on the unsaturated shear strength of a weathered granite. Can Geotech J 42:624–631
Li X, Zhang LM (2009) Characterization of dual-structure pore-size distribution of soil. Can Geotech J 46(2):129–141
Li X, Zhang LM, Li JH (2009) Development of a modified axis translation technique for measuring SWCCs for gravel soils at very low suctions. Geotech Test J 32(6):478–488
Li XS (2005) Modelling of hysteresis response for arbitrary wetting/drying paths. Comput Geotech 32:133–137
Likos WJ, Lu N (2004) Hysteresis of capillary stress in unsaturated granular soil. J Eng Mech 130(6):646–655
Likos WJ, Bowders JJ, Gates WP (2010) Mineralogy and engineering properties of bentonite. In: Bouazza A, Bowders JJ (eds) Geosynthetic clay liners for waste containment facilities. CRC Press, Taylor and Francis, Boca Raton, FL, pp 31–53
Likos WJ, Wayllace AW, Godt JG, Lu N (2012) Modified direct shear apparatus for unsaturated sands at low suction and stress. Geotech Test J 33(4):1–13
Likos WJ, Wayllace A (2010) Porosity evolution of free and confined bentonites during interlayer hydration. Clays Clay Miner 58:399–414
Oberg A, Sallfors G (1997) Determination of shear strength parameters of unsaturated silts and sands based on the water retention curve. Geotech Test J 20(1):40–48
Pedroso DM, Williams DJ (2010) A novel approach for modelling soil–water characteristic curves with hysteresis. Comput Geotech 37(3):374–380
Redlund DG, Xing A, Fredlund MD (1996) The relationship of the unsaturated soil shear strength to the soil-water characteristic curve. Can Geotech J 33(3):440–448
Sheng DC, Zhou AN, Fredlund DG (2011) Shear strength criteria for unsaturated soils. Geotech Geol Eng 29:145–159
Simms H, Yanful EK (2001) Measurement and estimation of pore shrinkage and pore distribution in a clayey till during soil-water characteristic curve tests. Can Geotech J 38(4):741–754
SoilVision database (1997) SS1996 dataset (dataset ID), Scott Sillers (dataset lead), Scott Sillers Dataset (dataset name), US2000 dataset (dataset ID), Alex Kozlow (dataset lead), Soils from University of Saskatchewan (dataset name), SP1022 dataset (dataset ID), Ward Wilson (dataset lead), Mining tailings (dataset name), RS2000 dataset (dataset ID), Murray Fredlund (dataset lead), Soils obtained from research papers (dataset name), SP1020 dataset (dataset ID), Feike Leij (dataset lead), UNSODA (dataset name), SoilVision Systems Ltd., Saskatoon, SK
Thu TM, Rahardjo H, Leong EC (2007) Critical state behavior of a compacted silt specimen. Soils Found 47(4):749–755
Tsiampousi A, Zdravković L, Potts DM (2013) A new Hvorslev surface for critical state type unsaturated and saturated constitutive models. Comput Geotech 48:156–166
Vanapalli SK, Fredlund DG, Pufahl DE (1996) Model for the prediction of shear strength with respect to soil suction. Can Geotech J 33(3):379–392
Wang YH, Siu WK (2006) Structure characteristics and mechanical properties of kaolinite soils. I. Surface charges and structural characterizations. Can Geotech J 43(6):587–600

Zhao HF, Zhang LM, Chang DS (2013) Behavior of coarse widely graded soils under low confining pressures. J Geotech Geoenviron Eng 139(1):35–48

Chapter 9
Convergence Criterion for Highly Nonlinear Unsaturated Seepage Analysis

9.1 Introduction

Seepage in unsaturated media is a common phenomenon in geotechnical and geoenvironmental engineering, such as rainfall infiltration into slopes (Fredlund et al. 2006; Ng and Shi 1998; Wu et al. 2020a), water flow through capillary barrier covers (Chen et al. 2020; Zhan et al. 2020), and contamination migration in unsaturated soils (Bai et al.2020; Wei et al.2021). The governing equation of this process, namely Richards' equation (Richards 1931), is an elliptic parabolic partial differential equation with high nonlinearity (List and Radu 2016; Vereecken et al. 2016). The analytical solution to Richards' equation is extremely difficult to derive except under very few special and simple conditions (Miller et al. 1998). Therefore, at present, the solution to most unsaturated seepage problems depends on numerical methods (Miller et al. 2013; Farthing and Ogden 2017) such as the finite element method (FEM). In FEM analysis of unsaturated seepage, the mesh size, time step, and solver settings critically influence the calculation accuracy (Tan et al. 2004; Phoon et al. 2007; Chen et al. 2008; Zhu et al. 2020). An improper setting will reduce the simulation accuracy and may even cause a numerical problem (Pan et al. 1996; Ju and Kung 1997; Karthikeyan et al. 2001; Vogel and Ippisch 2008; Delfs et al. 2009; Carrera-Hernández et al. 2012; Miller et al. 2013; Cui et al. 2015; Vereecken et al. 2016; Wu et al. 2020b).

Studies on the appropriate selection of mesh size and time step in the FEM analysis of unsaturated seepage have been performed. With respect to mesh size, Ju and Kung (1997) indicated that to achieve convergence, mesh refinement should be prioritized over an arbitrarily decreasing time step. Several later scholars illustrated that coarsening the mesh size increases the numerical solution error in unsaturated seepage (Delfs et al. 2009; Carrera-Hernández et al. 2012; Vereecken et al. 2016), but the determination of the proper mesh size was not mentioned in their research. Vogel and Ippisch (2008) presented the concept and simple estimation of "critical spatial discretization" for solving Richards' equation, which is expected to be the maximum mesh size for avoiding a biased solution. Zhang et al. (1999) discussed the relationship between mesh size and accuracy of the differencing scheme in the

X. Li et al., *Fundamental Behavior of Unsaturated Widely-Graded Soil*,
https://doi.org/10.1007/978-981-19-3402-5_9

solution of finite-difference method with respect to the nonlinear Navier–Stokes equation, which is expected to provide a reference for the estimation of mesh size in the numerical solution of Richards' equation. Overall, in practical unsaturated seepage analysis, determining a reasonable mesh size remains a challenge (Fredlund 2006). In terms of time step, Thomas and Zhou (1997) provided a time-step setting criterion to avoid numerical oscillations in diffusion problems, and Karthikeyan et al. (2001) further confirmed that this criterion is also applicable to unsaturated seepage simulations. Cui et al. (2015) derived the non-oscillation criterion of the time step for the coupled seepage-heat conduction problem. Although these criteria can avoid numerical oscillations induced by extremely small time steps, they cannot eliminate or minimize the numerical error induced by an inappropriately large time step.

In summary, the existing studies provide valuable information on the effect of mesh size and time step on the FEM solution accuracy of unsaturated seepage. Yes, the upper limits of mesh size and time step for ensuring convergence in the numerical results are yet to be defined. This chapter focuses on such a critical point in the FEM analysis of unsaturated seepage and aims to establish selection criteria for determining the critical values of mesh size and time step that ensures a convergent result.

In this chapter, through theoretical analysis and numerical tests, selection criteria are proposed to predict the critical mesh size and critical time step leading to a convergent solution in different degrees of nonlinearity. Additionally, the criteria are verified on multiple scales and dimensions based on the analytical solution, laboratory model test, and quasi-engineering problem. Finally, suggestions for selecting mesh size and time step in the practice of unsaturated seepage simulation are put forward.

9.2 Numerical Problems Caused by Mesh Size and Time Step in Unsaturated Seepage Analysis

9.2.1 Highly Nonlinear Properties of Unsaturated Soils

A wetting front is often encountered in the numerical simulation of unsaturated seepage. The wetting front is defined as the region with a rapid decrease in water content and is often accompanied by changes of several magnitudes in the soil–water characteristic curve (SWCC) and soil hydraulic function curve (SHFC, Li et al. 2009; Wu et al. 2017). Such variations often occur over a very short time and distance, as is shown in Fig. 9.1, resulting in much greater nonlinearity than that encountered in other common geotechnical problems, as shown in Table 9.1 (Gibson et al. 1967; Li et al. 2018; Tai et al. 2021).

Thus, convergence difficulty and accuracy problems arise from the wetting front in the numerical simulation of unsaturated seepage (Karthikeyan et al. 2001; Tan et al. 2004; Cui et al. 2015). During the simulation process, if the mesh density or time step is inappropriate, mesh size-dependent or time step-dependent problems

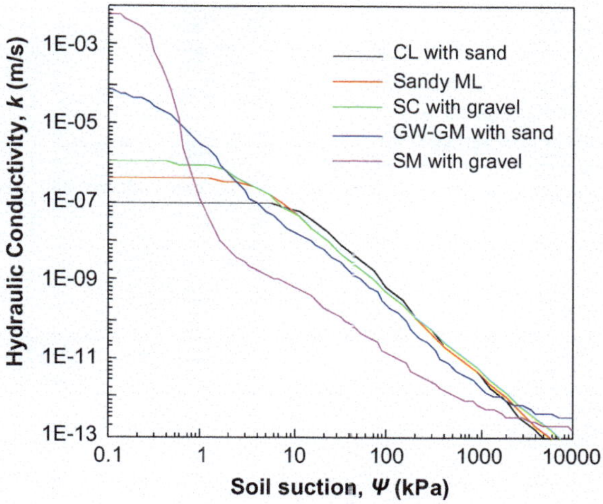

Fig. 9.1 Highly nonlinear hydraulic conductivity function of unsaturated soils (after Li et al. 2009)

Table 9.1 Comparison of the nonlinear degree of common geotechnical problems

Common problems	Governing equation	Key parameter	Change magnitude	Nonlinear degree
Saturated seepage	$\frac{\partial}{\partial z}\left[k\frac{\partial h}{\partial z}\right]=0$	Saturated hydraulic conductivity, k	0	None
Heat conduction with phase change	$\frac{\partial T}{\partial t}=\frac{\partial}{\partial x}\left[k\frac{\partial T}{\partial x}\right]+Q$	Thermal conductivity, $k(T)$	< 1	Weak
Consolidation of saturated clays	$\frac{\partial e}{\partial t}=\frac{\partial}{\partial a}\left[c\frac{\partial e}{\partial a}\right]$	Consolidation coefficient, $c(e)$	< 2	Middle
Unsaturated seepage	$\frac{\partial \theta}{\partial t}=\frac{\partial}{\partial z}\left[D(\theta)\frac{\partial \theta}{\partial z}\right]+\frac{\partial k(\theta)}{\partial z}$	Unsaturated hydraulic conductivity, $k(\theta)$	5–10	Strong

may be encountered in such a highly nonlinear problem, as discussed in the following section.

9.2.2 *Mesh Size-Dependent Problem in the Simulation of Unsaturated Seepage*

9.2.2.1 Numerical Model for Simulating the Capillary Process

Soil column test is often conducted to study the water infiltration or capillary process in single-layer or multilayer soils (Li et al. 2009, 2021; Ng and Leung 2012).

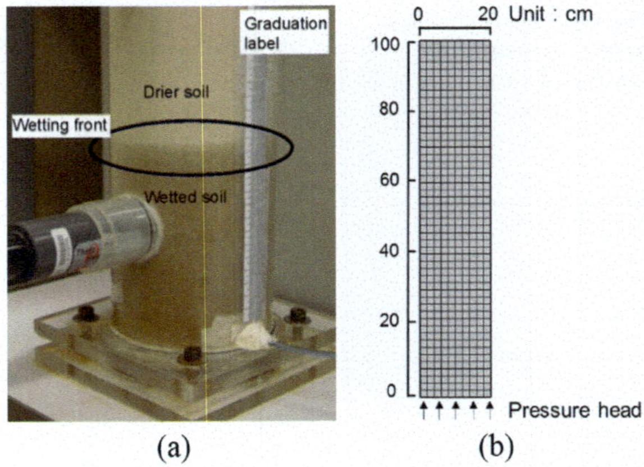

Fig. 9.2 Capillary test of unsaturated soil column and corresponding FEM model: **a** capillary test in an unsaturated soil column; **b** numerical model of capillary rising test

Figure 9.2a shows a capillary column with one layer of compacted lean clay (CL), in which a clear boundary appeared between the wetted soil and the initially dry soil. This demarcation is called the "wetting front" and can be identified and measured during the wetting process.

Numerical simulation is a valuable method to study the infiltration or capillary process. Figure 9.2b illustrates a numerical model simulating the capillary test, which is shown in Fig. 9.2a. The essential settings include the mesh size L, time step Δt, and solving method. If these factors are reasonable, the numerical simulation can obtain a converged solution, that is, consistent results from the settings of various factors. However, because of the extremely high nonlinearity, numerical simulation of the wetting front may encounter problems with accuracy, as suggested by Ju and Kung (1997), Karthikeyan et al. (2001) and Zhu et al. (2019).

9.2.2.2 Mesh Size-Dependent Problem in the Simulation of Wetting Front Advancing Process

An example of mesh size-dependent problem in the simulation of wetting front advancing process is shown in Fig. 9.3. Figure 9.3a shows the suction profile at time $t = 5e6$ s through different mesh sizes, where the suction decreases from 10,000 to 100 kPa within a thickness of no more than 5 cm. Such a zone can be recognized as the wetting front, where a rapid increase in volumetric water content accompanied by an increase in hydraulic conductivity by several orders of magnitude occurs, as shown in Fig. 9.1.

Evidently, at $t = 5e6$ s, the wetting front height changes significantly with different mesh size. If the mesh size L is set as 0.1 cm, the wetting front height is 75 cm. However, when L is set as 5 cm, the wetting front height is only 20 cm. In other words, the advancing speed of the wetting front may be several times slower if an oversized element is used. This phenomenon is evident from the moisture content distribution in Fig. 9.3b. An oversized element contributes to both inaccurate wetting

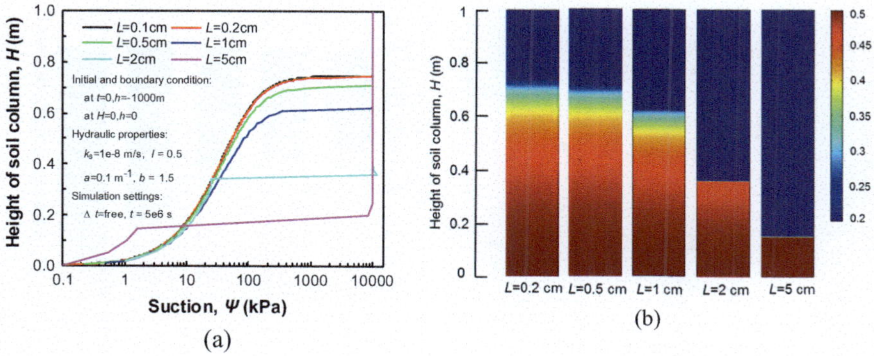

Fig. 9.3 Mesh size-dependent problem in unsaturated seepage simulation: **a** suction profile; **b** volumetric moisture profile

front height and unconventional moisture distribution in the wetting front zone. Zhu et al. (2019) also reported the mesh size-dependent error of unsaturated seepage problems induced by high nonlinearity as shown in Fig. 9.4, where the absolute error represents the absolute difference between the numerical and analytical solutions, and Dz is the spatial interval in a 1D and 10 m-high numerical model of unsaturated infiltration. As shown in the figure, the absolute error significantly decreases with decreasing mesh size.

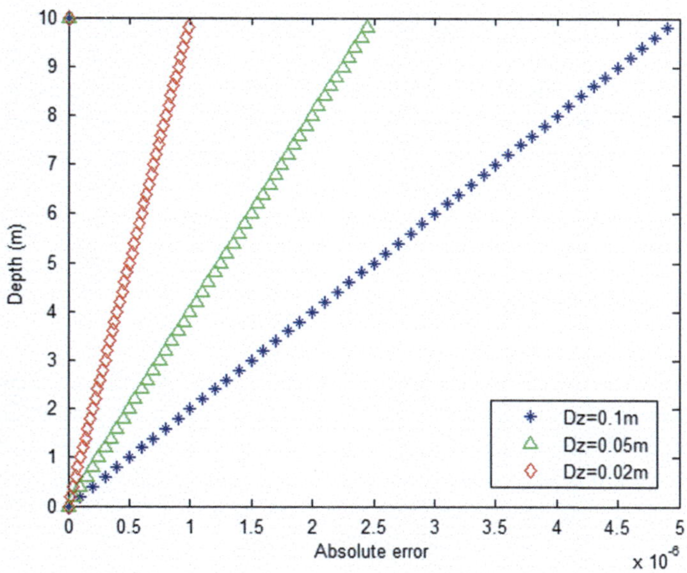

Fig. 9.4 Numerical error depended on mesh size (after Zhu et al. 2019)

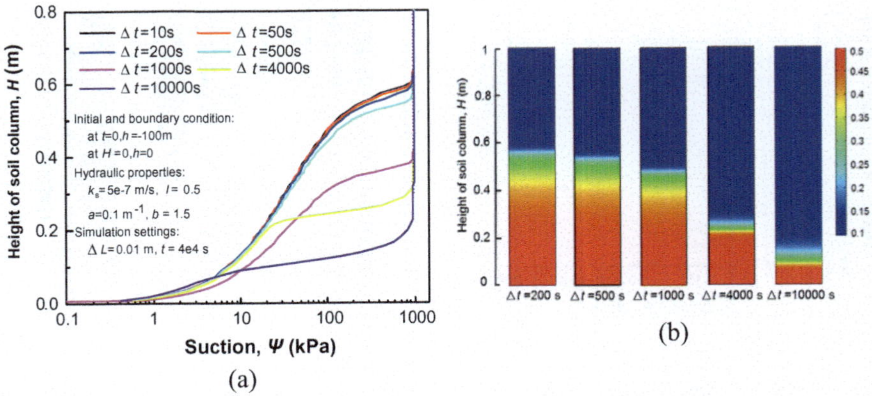

Fig. 9.5 Time step-dependent problem in unsaturated seepage simulation: **a** suction profile; **b** volumetric moisture profile

9.2.3 *Time Step-Dependent Problem in the Simulation of Unsaturated Seepage Analysis*

An example of time step-dependent problem in unsaturated seepage simulation is shown in Fig. 9.5. Figure 9.5 indicates the suction profile, wetting front height, and volumetric water content distribution, respectively, at a time when $t = 4e4$ s under different time steps Δt.

Evidently, when $t = 4e4$ s, the wetting front height changes with the time step. If Δt is set as 10 s, the wetting front height is approximately 60 cm. While if Δt is set to 10,000 s, the wetting front height is only 20 cm. Similar to the mesh size, the advancing speed of wetting front may be several times lower if an excessive time step is applied. The moisture distribution in Fig. 9.5b verifies this time step-dependent problem.

9.2.4 *Effect of Solving Method*

Recently, the under-relaxation method and its improvement were explored to yield acceptable results through relatively loose requirements on mesh size and time step (Phoon et al. 2007; Cheng et al. 2008). Besides, the improved Jacobi iterative method with two-side equilibration was proved to be a good choice for solving the problem of the highly nonlinear unsaturated flow (Zhu et al. 2019).

All of the aforementioned methods need a secondary development of the existing software or a self-programming platform. However, the unsaturated seepage process is usually simulated by commercial software and what most users or engineers can do is adjusting the mesh and time step with no access to the iteration methods embedded in software. Therefore, a proper criterion for mesh size and time step is essential to obtain the accurate solution to the wetting front height and moisture content distribution.

9.2.5 *Summary*

As shown in Figs. 9.3, 9.4, and 9.5, the mesh size causes significant errors at $L =$ 5 cm and negligible errors at $L = 0.2$ cm, and a similar case is true with the time step turning from 10 into 10,000 s. The mesh size and time step show decisive influence on the capillary rising or infiltration process in numerical simulation. It seems that the error of wetting front height can be eliminated through a sufficiently small L and Δt. So far, a key issue is to quantitatively determine a "small enough" mesh size and time step, that is, a criterion for the upper limit value of mesh size and time step. A theoretical relation between L, Δt, hydraulic properties, and boundary and initial conditions should be derived.

9.3 Theoretical Analysis on the Accuracy of Numerical Simulation of Unsaturated Seepage

9.3.1 *Definition of Critical Mesh Size and Time Step*

In solving the problem of unsaturated seepage, the first concern is the position of the wetting front. Also, the pressure head distribution should not oscillate too much. Therefore, the correct wetting front height is taken as the primary control criterion for convergence. Tan et al. (2004) and Cheng et al. (2008) have well demonstrated that the numerical oscillation of unsaturated seepage is the most serious at the wetting front, so the acceptable oscillation at the wetting front is taken as the secondary control criterion.

Based on the above factors, the convergence of the numerical simulation of unsaturated seepage is defined in Eqs. 9.1 and 9.2.

$$\frac{H_{\mathrm{s}} - H_x}{H_{\mathrm{s}}} < 0.05 \tag{9.1}$$

$$\frac{h_s - h_x}{h_s} < 0.05 \tag{9.2}$$

where H_{s} is the standard solution for the wetting front height, and h_{s} is the standard solution for the pressure head at the wetting front. When the analytical solution exists, the two can be determined by the analytical solution. Otherwise, they can be obtained through the numerical solution with a sufficiently dense mesh or small time step. x indicates the parameters for performing refinement, i.e., the mesh size L and the time step Δt. It should be pointed out that in the convergence criterion, Eqs. 9.1 and 9.2 must be satisfied simultaneously.

9.3.2 Critical Mesh Size in the Finite-Difference Method

Richards' equation of one-dimensional unsaturated seepage flow in the volume water content form is shown as:

$$\frac{\partial \theta}{\partial t} = \frac{\partial}{\partial z}\left[D(\theta)\frac{\partial \theta}{\partial z}\right] + \frac{\partial k(\theta)}{\partial z} \tag{9.3}$$

where θ is the volume water content; $D(\theta)$ is the diffusion coefficient, which is the ratio of the hydraulic conductivity $k(\theta)$ to the specific water capacity $c(\theta)$; $k(\theta)$ is the hydraulic conductivity; z is the seepage path; t is the time.

Equation 9.3 indicates the water storage of soil element under the influence of pressure head gradient and gravity gradient. In unsaturated seepage, when the pressure head gradient is much higher than the gravity gradient, the major driving potential of water flow will be the former. Thus, the rightmost item representing the gravity potential in Eq. 9.3 is ignored, and write Eq. 9.3, which is equivalent to the differential format of m-order precision (Zhang et al. 1999, 2016), is expressed as:

$$\frac{\partial \theta}{\partial t} = \frac{\partial}{\partial z}\left[D(\theta)\frac{\partial \theta}{\partial z}\right] + O(L^{\alpha}) \tag{9.4}$$

where L is the length of the discretized mesh; $O(L^{\alpha})$ represents the truncation error, which is a small amount of L with the order no less than α; α is the first integer greater than mesh order, which represents the order of the field variable within the element after the Taylor expansion of pressure head along the seepage path. For example, when $\alpha = 2$, the mesh is first-order, and there are only two nodes on one edge of the element. When these two nodes' pressure heads are determined, the heads inside the mesh are linearly distributed.

Equation 9.4 points out that when the soil's initial water content is low, $D(\theta)$ is relatively small. To correctly calculate the contribution of the item of the $D(\theta)$ to the pressure head change, it is necessary to require the truncation error to be smaller than the item of the $D(\theta)$. Then Eq. 9.4 can be written accordingly as follows:

$$\frac{\partial}{\partial z}\left[D(\theta)\frac{\partial \theta}{\partial z}\right] \gg O(L^{\alpha}) \tag{9.5}$$

Due to the strong nonlinearity on the left side of the inequality, its analytical solution cannot be obtained. However, a correlation between reasonable mesh size and $D(\theta)$ can be inferred, and this correlation can be expressed as follows:

$$\Delta L < f(D(\theta)^{1/\alpha}) \tag{9.6}$$

In general seepage analysis, the mesh is linear, i.e., $\alpha = 2$, and this is also the case in the subsequent studies in this article. For unsaturated soils, $D(\theta)$ is not a constant. It increases as the suction decreases. That is, drier soil has a smaller diffusion coefficient. The minimum diffusion coefficient in the seepage process is defined as D_0, namely:

$$D_0 = \min(D(\theta)) \tag{9.7}$$

Unfortunately, in this study, we have not obtained an analytical expression of ΔL through mathematical derivation. Since Eq. 9.6 is not an explicit expression, this formula has only reference significance for the actual mesh size selection. The critical mesh size that can ensure the convergence of the seepage FEM analysis can be determined through numerical tests and a semi-theoretical and semi-empirical ΔL criterion can be drawn.

9.3.3 Critical Time Step in the Finite-Difference Method

For unsaturated seepage, the most challenging region in need of a solution, the most radical change in soil properties, and the most significant error are in the interface of dry and wet soil (also known as the wetting front). If the calculation error at the dry–wet interface can be controlled even within a small range, the entire unsaturated soil seepage field can be accurately solved. Therefore, the capillary rise shown in Fig. 9.2 can be regarded as an extreme case in unsaturated soil seepage analysis.

The wetting front advancing problem illustrated in Fig. 9.2 is explored here. In the one-dimension (1D) seepage analysis utilizing the finite-difference method, h_n^m notes the pressure head at n-th node in m step calculation. Before the water advances, the pressure head in the dry zone will not change. Let's consider the most critical position is in the interface between the wetted soil and dry soil. At this position, the relationship of pressure head between adjacent nodes and time steps are shown in Eqs. 9.8 and 9.9, as is shown below,

$$h_{n+1}^m = h_n^m \tag{9.8}$$

$$\frac{h_{n-1}^m + h_{n+1}^m - 2h_n^m}{2L} k_n^m t = C_n^m L \Delta h_n^m \tag{9.9}$$

where C_n^m is the specific water capacity at m moment, n node.

At moment m, the soil node at n and node $n + 1$ is dry, and that at $n - 1$ node are wet; at moment $m + 1$, with the advancing wetting front, the soils at nodes n and

$n - 1$ are accordingly wet. Equation 9.9 indicates that the flux of soil at node n is conserved during the humidification process.

The following can be obtained based on Eq. 9.9:

$$\frac{h_{n-1}^m - h_n^m}{2L} k_n^m t = C_n^m L(h_n^{m+1} - h_n^m) \tag{9.10}$$

$$t = \frac{2C_n^m L^2 (h_n^{m+1} - h_n^m)}{k_n^m (h_{n-1}^m - h_n^m)} = 2L^2/D_n^m \frac{(h_n^{m+1} - h_n^m)}{(h_{n-1}^m - h_n^m)} \tag{9.11}$$

In a smooth wetting front advancing process, the pressure head profile should fulfill Eqs. 9.12 and 9.13 (Li et al. 2009, 2021), as are shown below.

$$h_n^{m+1} \geq h_n^m \tag{9.12}$$

$$h_{n-1}^m \geq h_n^{m+1} \tag{9.13}$$

Based on Eqs. 9.12 and 9.13, the necessary condition for convergent time step is:

$$t < t_c = 2L^2/D_n^m \tag{9.14}$$

According to Eq. 9.14, the critical time step is related to the mesh size, which can be determined after a reasonable mesh size is set. Besides, Eq. 9.14 also shows that the critical time step is related to the diffusion coefficient D of the node at the wetting front. Since D of unsaturated soil is not a constant, if $D_{\max}$ in Eq. 9.12 is taken as the diffusion coefficient under saturated state, the calculation result is conservative. It is the lower limit of the time step that can ensure the convergence of the problem and accurate calculation, but it may lead to extremely small time step with high calculation costs.

Therefore, this study aims to perform numerical tests to find the maximum time step ensuring the convergence of the seepage simulation. Subsequently, a semi-theoretical and semi-empirical formula can be obtained.

Notably, Eq. 9.14 is an original contribution of this study for determining the maximum time step for ensuring the numerical convergence of unsaturated seepage simulation. In previous studies, some classic criteria resembling Eq. 9.14 have been proposed, but they are essentially a lower bound of the time step to avoid numerical oscillation in transient finite element analysis (Thomas and Zhou 1997; Karthikeyan et al. 2001; Cui et al. 2015).

9.4 Numerical Test Plan and Test Procedure

9.4.1 Numerical Model of Nonlinear Unsaturated Seepage Process

A commercial FEM software, COMSOL 4.3, is used to study the capillary process in a 1D column. Transient seepage analysis is conducted to a homogenous soil column with an initial state of soils (h_0, θ_0, k_0). Considering a 1 m soil column, constant boundary conditions are used. Specifically, (1) constant suction is applied to the bottom of the soil column, i.e., $h(z = 0) = h_{\text{bot}}$; (2) the flux at the side edge of the column is 0; (3) free boundary is applied at the top surface of the soil column.

In this wetting process, the maximum suction and minimum diffusivity are within the dry zone and depend on the initial condition. The maximum hydraulic conductivity depends on the saturated hydraulic conductivity and may be present in the wetted zone.

The van Genuchten model (van Genuchten 1980) is used to describe the soil–water characteristic curve (SWCC), which is expressed as the following equation:

$$S_e = \frac{1}{[1 + |ah|^b]^c} \tag{9.15}$$

where S_e stands for the effective degree of saturation, h is the value of the pressure head, and α, b, and c are the fitting parameters of SWCC. Usually, parameters b and c are related as is shown in this equation: $c = 1 - 1/b$.

The van Genuchten and Mualem model (Mualem 1976) is used to describe the soil hydraulic conductivity function,

$$k_r = S_e^l[1 - (1 - S_e^{1/c})^c]^2 \tag{9.16}$$

where k_r stands for the relative hydraulic conductivity and l stands for the fitting parameter.

Generally, quadrilateral elements are widely used in unsaturated seepage analysis due to their stability. Therefore, the element uses a quadrilateral with two orders in this study, which is also the default mesh order of most seepage simulation software. In the solver settings, the relative tolerance is 0.001, the time step difference method is the backward difference method, and the nonlinear equations are solved by the Newton–Raphson method.

9.4.2 Numerical Test Plan

A series of numerical tests that simulate different soil properties are conducted with the aim of obtaining L_c and Δt_c under different nonlinear conditions. Then based

Table 9.2 SWCC and hydraulic conductivity parameters of test soils

Soil type	a (m^{-1})	b	l	k_s (m/s)	θ_s	θ_r
Clay	0.1	1.5	0.5	1e–9	0.496	0.159
Silty clay	0.1	1.5		1e–8		
Silt 1	0.5	2		1e–6	0.4	0.032
Silt 2	0.5	2		1e–5		
Silty sand	2	4		1e–4	0.367	0.057

on the theoretical framework of Eqs. 9.6 and 9.14, the semi-empirical and semi-theoretical criteria for L_c and Δt_c can be derived.

9.4.2.1 Soil Characteristics

Three typical types of soils are selected: clay, silt, and sand. Accordingly, three different SWCCs are used. In order to broaden the range of the diffusion coefficient D, five types of SHCFs are used. In total, the numerical test uses five soil test groups with different hydraulic characteristics: clay, silty clay, silt 1, silt 2, and sand. According to Lu and Likos (2004), the soil parameters are determined and summarized in Table 9.2, and the SWCC and SHCF of the five soils are shown in Fig. 9.6.

9.4.2.2 Initial and Boundary Conditions

To obtain L_c and Δt_c under different initial diffusion coefficients, D_0, five different initial degrees of saturation were selected for each test group. The initial effective saturation degree of the silt and sand groups is set as 0.02, 0.1, 0.2, 0.4, and 0.6, respectively, and that of the clay group is 0.1, 0.2, 0.3, 0.4, and 0.6, respectively. Through the preliminary simulation, when the S_{e0} is set to 0.02, it will lead to extreme nonlinearity and rugged solution, e.g., suction changes of hundreds of MPa at dry–wet interface, resulting in failure to obtain convergent L and Δt. On the other hand, clay close to the residual saturation state is rarely encountered in practice. Therefore, the minimum S_{e0} of the clay group is taken as 0.1. The initial suction of each test group is shown as the black mark in Fig. 9.6a.

Generally, the boundaries of seepage problems are divided into two categories: the first is the Dirichlet boundary (i.e., constant pressure head), and the second is the Neumann boundary (i.e., constant infiltration rate). The nonlinearity of a reasonable Neumann boundary is less than that of the Dirichlet boundary with a zero-pressure head. This study aims to obtain the L_c and Δt_c under the most extreme conditions. Therefore, to simulate this condition, the numerical experiment's boundary is taken as the zero-pressure head ($h_{bot} = 0$ m).

Fig. 9.6 SWCC and SHCF of test soils: **a** SWCC curve; **b** SHFC curve

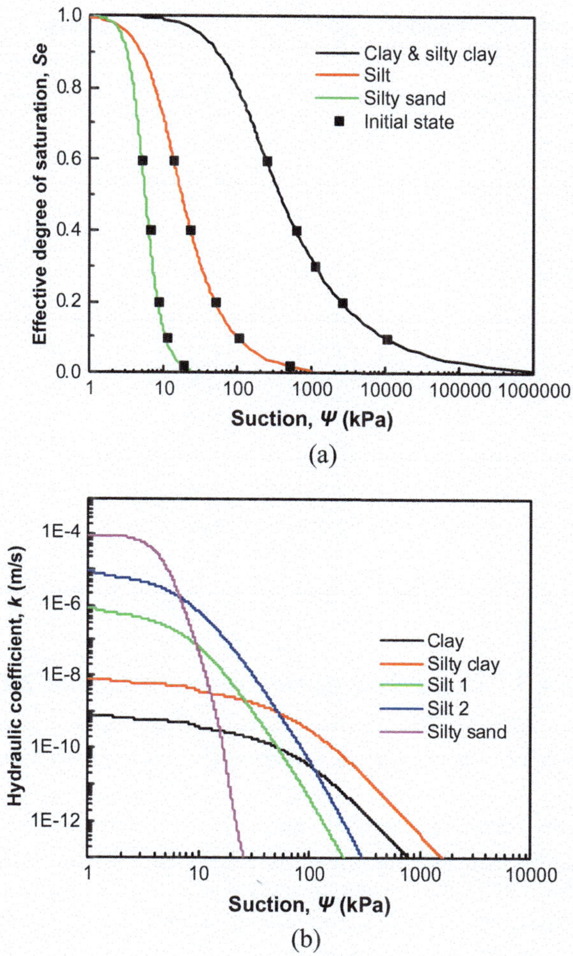

9.4.2.3 Numerical Test Plan

The groupings, boundary conditions, and initial conditions of the numerical test are summarized in Table 9.3. The test is carried out for 25 times in 5 soil groups in total. In each test, the critical mesh size and critical time step in the corresponding initial state and boundary are obtained. The detailed process will be illustrated in the following section.

Table 9.3 Numerical test plan

Soil type	S_{e0}	h_0 (m)	h_{bot} (m)	Top boundary
Clay and silty clay	0.1	1000	Zero pressure	Free boundary
	0.2	250		
	0.3	110		
	0.4	60		
	0.6	24		
Silt	0.02	100		
	0.1	20		
	0.2	10		
	0.4	4.6		
	0.6	2.7		
Silty sand	0.02	1.8		
	0.1	1.1		
	0.2	0.8		
	0.4	0.6		
	0.6	0.5		

9.4.3 *Procedure of Determining L_c and Δt_c*

9.4.3.1 *Approximate Solution of L_c and Δt_c in Numerical Tests*

Obviously, the criteria defined in Eqs. 9.1 and 9.2 are suitable for judging the convergence of a specific L and Δt, while it is difficult to find out values for L_c and Δt_c that satisfy the convergence criterion under various initial conditions. Therefore, an increment of Δx (i.e., increment of mesh size L and time step Δt) is introduced to approximately determine the critical value. "Critical" means that at x, the calculation result is convergent, whereas at $x + \Delta x$, the calculation result exceeds the convergent criterion in Eqs. 9.1 and 9.2.

To narrow the difference between the searched values of L_c and Δt_c and their "true" values, for the operability of experiments, the value range of Δx is introduced as follows:

$$\Delta x \leq \begin{cases} 0.2, L \leq 1 \\ 2, 1 \leq L < 10 \\ 4, 10 \leq L < 20 \end{cases} \tag{9.17}$$

$$\Delta x \leq \begin{cases} 2, \Delta t \leq 10 \\ 20, 10 < \Delta t \leq 100 \\ 200, 100 < \Delta t \leq 1000 \end{cases} \tag{9.18}$$

In the numerical simulation of unsaturated seepage, it is usually recommended that the geometric size is not less than 5–7 times of the mesh size (Geo-Slope International Ltd. 2007). Therefore, for the 1 m soil column in this experiment, the mesh size should not be less than 20 cm. To avoid the violent shock caused by the instantaneous application of boundary conditions, the test results within the height range of 0.3–0.7 m in the soil column are selected to search for values of L_c and Δt_c.

9.4.3.2 *Procedure of Determining L_c*

The search process of L_c adopts the idea of dichotomy. The initial L is 2 cm, and the search direction is determined through the calculation results of $L = 1$ cm and $L = 4$ cm. If $L = 2$ cm converges, the convergence of $L = 4$ cm will continue to be assessed. Otherwise, the convergence of $L = 1$ cm will be determined until a L_c is found satisfying Eqs. 9.1, 9.2, and 9.17. Specific search cases are presented in Sect. 9.5.

In the search process of L_c, the adaptive time step (Kavetski et al. 2001; Söderlind 2006) was applied, which can automatically determine the time interval for each calculation step based on the nonlinearity degree of the problem and relative tolerance setting. This method demonstrates the following two advantages: (a) The adaptive time step can automatically control the time step to a reasonable degree when determining a reasonable time step is difficult. In other words, when the tolerance setting of each calculation is constant, it can be considered that Δt is reasonable in each test of different L. (b) The adaptive time step can significantly reduce the calculation time in the L_c search process, especially for the low initial water content soils.

9.4.3.3 *Procedure of Determining Δt_c*

By using L_c determined in Sect. 9.4.3.2, a process similar to searching for L_c will be adopted to determine Δt_c. During the searching process, the mesh size and time step of each numerical experiment are manually set as determined critical value and assessed values, respectively. As a result, in this stage, it is confirmed that an inappropriate Δt rather than L induces the non-convergence problem.

With the wide range of reasonable time steps, a trial calculation is firstly conducted to obtain an initial time step, Δt_0. Subsequently, the time steps of $2\Delta t_0$ and $0.5\Delta t_0$ are applied to determine the search direction. If the result of Δt_0 converges, the performance of $2\Delta t_0$ will be tested; otherwise, the performance of $0.5\Delta t_0$ will continue to be assessed until it determines Δt_c that satisfies Eqs. 9.1, 9.2, and 9.18. Specific search cases will be illustrated in Sect. 9.5.

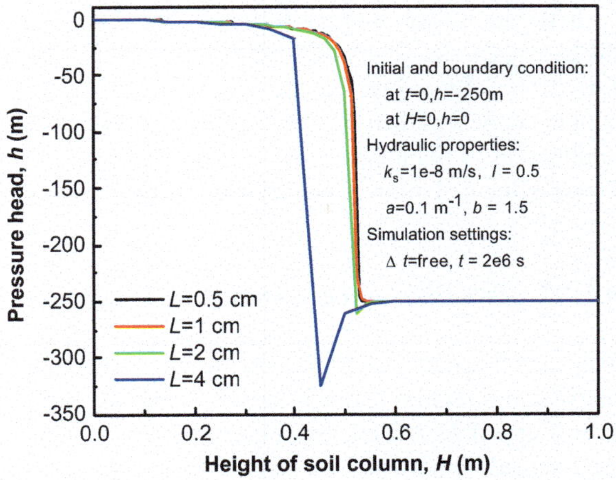

Fig. 9.7 Searching process of L_c

9.5 Numerical Study on the Convergence Criterion of Critical Mesh Size L_c

9.5.1 Numerical Test Results of L_c Values

9.5.1.1 *An Example for Searching for L_c*

Taking the $S_{e0} = 0.2$ test in the clay group as an example, the search process of the critical mesh size L_c is depicted here, and the pressure head distribution with different mesh sizes at $t = 2e6$ s is summarized in Fig. 9.7. First, taking $L = 2$ cm for calculation, the pressure head distribution of soil column can be obtained, followed by the solution of models with $L = 1$ cm and $L = 4$ cm. Obviously, compared with $L = 2$ cm, when $L = 4$ cm, the position of the wetting front has a significant deviation, i.e., non-convergence. When $L = 2$ cm, compared with the results of $L = 1$ cm, the wetting front position meets the convergence criterion in Eq. 9.1. The improved numerical oscillation at the wetting front, however, still exceeds the relative error of 5% in Eq. 9.2. Therefore, it is necessary to confirm whether the result of $L = 1$ cm is the standard solution. Furthermore, taking $L = 0.5$ cm for calculation, it is found that the results of $L = 0.5$ cm and $L = 1$ cm are consistent, so it can be inferred that both $L = 0.5$ cm and $L = 1$ cm lead to the standard solutions. Finally, L_c for this test is thus determined as 1 cm.

9.5.1.2 *Results for L_c*

According to the numerical test plan, the L_c of different test groups is obtained and summarized in Fig. 9.8. From Fig. 9.8, with the increase of D_0 (or the increase of initial water content), the critical mesh sizes of soils increase accordingly. For

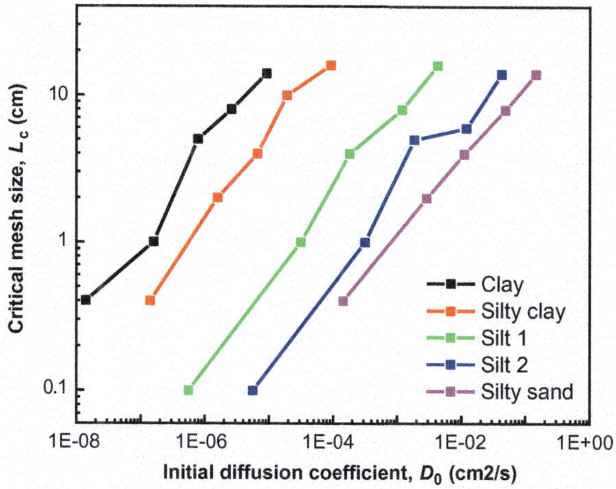

Fig. 9.8 Test results of L_c for various soils with corresponding D_0

each kind of soil, the L_c and D_0 have an approximately linear trend in the double logarithmic coordinate system. That is, L_c can be expressed in the form of a power function of D_0.

For different soils, the "slope" of L_c versus D_0 is basically the same, which means that the powers of L_c and D_0 are approximately equal. However, for different types of soils, the L_c to be obtained is different even if the initial D_0 is the same. This phenomenon shows that D_0 is not the only factor that determines L_c.

Previous studies have shown that in the simulation of unsaturated seepage, the influence of SHFC is more significant than that of SWCC (Li et al. 2009). Based on Fig. 9.8, when D_0 is the same, the smaller the saturated hydraulic conductivity is, the larger the allowable L_c is. The saturated hydraulic conductivity k_s is expected to be another coefficient determining L_c. This can be explained as follows: For different soils, when the initial D_0 is the same, the soil with a small hydraulic conductivity has a smaller seepage velocity in each element. Correspondingly, the wetting front advances more slowly, resulting in a lower nonlinearity of the governing equation and larger allowable mesh size.

9.5.2 A Semi-empirical and Semi-theoretical Criterion for L_c

Based on the experimental data in Sect. 9.5.1, the programming solver in Excel is used to determine the relationship between L_c, D_0, and k_s. The prediction formula of L_c under the minimum error condition is obtained as Eq. 9.19.

$$L_c = 4.05 \times \frac{D_0^{0.48}}{k_s^{0.41}} \tag{9.19}$$

The fitted power of D_0 is 0.48, and α in the theoretical analysis should be 0.5 since a second-order mesh is used. The order of D_0 obtained from the numerical test is highly consistent with the theoretical value, which confirms the reliability of the theoretical analysis. Furthermore, comparing the calculated value through Eq. 9.19 with the test result, the correlation coefficient can reach 0.95, as is shown in Fig. 9.9a. From the perspective of prediction accuracy, the formula seems to be good.

To formally unify the prediction formula, taking D_0 and k_s to the same power and performing programming solver again, another prediction formula for L_c can be obtained as Eq. 9.20.

Fig. 9.9 Prediction results of formulas for L_c: **a** Prediction results of L_c by Eq. 9.19; **b** prediction results of L_c by Eq. 9.20

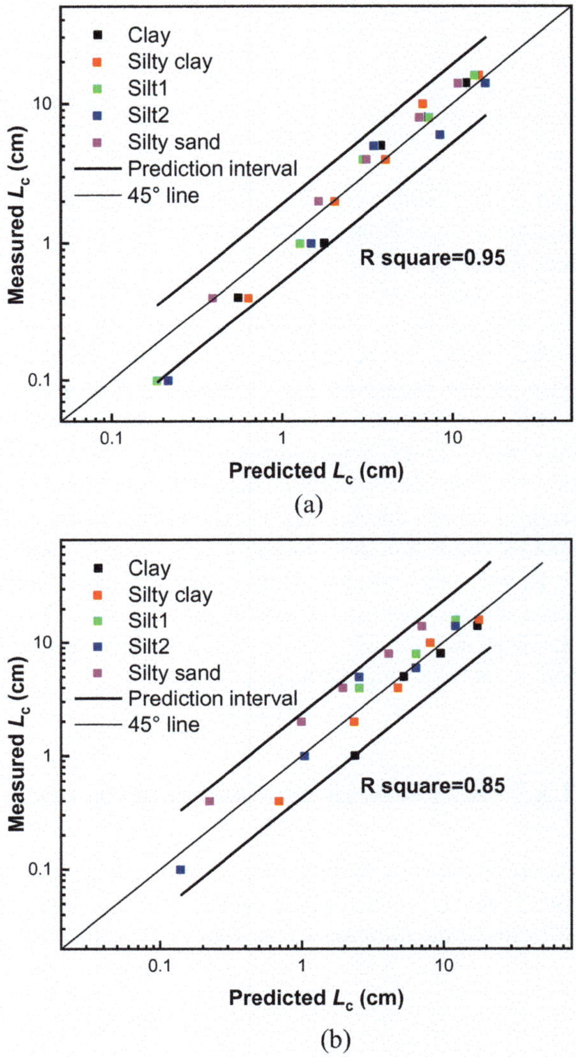

$$L_c = 1.87 \times \left(\frac{D_0}{k_s}\right)^{0.5} \tag{9.20}$$

Similarly, comparing the predicted value of Eq. 9.20 with the measured data, the correlation coefficient between the two reaches 0.85, as is shown in Fig. 9.9b, which is also an acceptable result.

The prediction formula Eq. 9.20 realizes the relative unity of the three dimensions of L_c, D_0, and k_s. That is, the critical mesh size is related to the highest order of mesh discretization. In addition, this formula reflects the positive correlation of L_c with D_0, and its negative correlation with k_s, which is also in line with the general law of seepage simulation: the larger the D_0, the higher the initial water content; the smaller the k_s, the lower the flow rate. Correspondingly, both factors lead to a lower nonlinearity degree and a larger mesh size.

Furthermore, considering the discreteness among the experimental critical mesh sizes, the prediction interval is introduced to better estimate L_c. Assuming that the probability function of L_c obeys a normal distribution and taking the measured L_c as the mean value with a confidence interval of 95%, the prediction interval of L_c can be obtained, as is shown in Fig. 9.9. The prediction interval of Eqs. 9.19 and 9.20 is expressed as Eqs. 9.21 and 9.22, respectively. Among them, the prediction interval is 0.52–1.91L_c in Eq. 9.21 and 0.42–2.36L_c in Eq. 9.22.

$$L_c = 4.05 \times \frac{D_0^{0.48}}{k_s^{0.41}} \times 10^{\pm 0.27} \tag{9.21}$$

$$L_c = 1.87 \times \left(\frac{D_0}{k_s}\right)^{0.5} \times 10^{\pm 0.37} \tag{9.22}$$

9.5.3 Verification of L_c Criterion

In addition to the five types of soils in the numerical test, two more soils, new kinds of clay and silt, which are commonly used in the one-dimension soil column infiltration test, were selected as the verification group to check the performance of the proposed criterion. The physical parameters of the soil are shown in Table 9.4. The FEM model settings (e.g., boundary conditions, solver settings) are the same as is described in the numerical test in Sect. 9.4.1.

Table 9.4 SWCC and SHCF parameters of verification soils

Soil type	a (m^{-1})	b	l	k_s (m/s)	θ_s	θ_r	S_{e0}	h_0 (m)
Clay	0.2	2.2	0.5	1e–7	0.463	0.139	0.1	– 38.8
Silt	1	2		1e–5	0.322	0.03	0.1	– 10

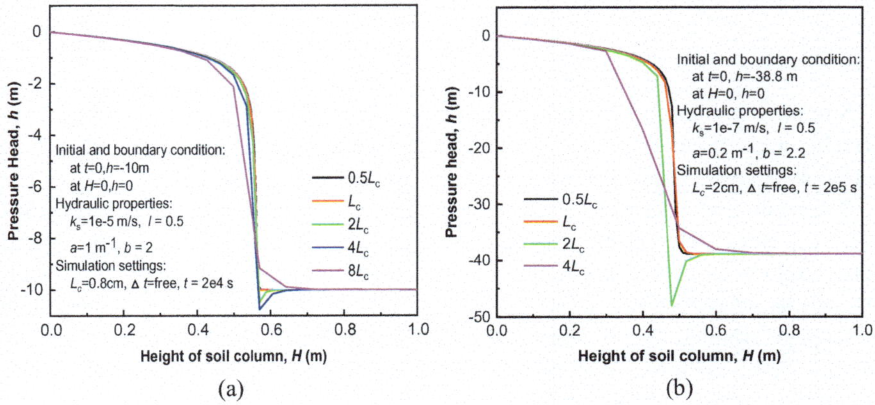

Fig. 9.10 Verification of proposed criterion for L_c: **a** L_c verification case (silt); **b** L_c verification case (clay)

According to the proposed criteria in Eq. 9.19, for two verified soils, the L_c is 0.88 cm and 2 cm, respectively, and the L_c calculated by Eq. 9.20 is 1.14 cm and 1.8 cm, respectively. It can be seen that in practice, the calculation results of Eqs. 9.19 and 9.20 are not much different. Therefore, in the verification model, the critical mesh sizes of the two soils are taken as 0.8 cm and 2.0 cm, respectively. The pressure head distributions are shown in Fig. 9.10.

It can be illustrated from Fig. 9.10a that: (a) the tests using L_c and $0.5L_c$ get the same result, which can signify a standard solution; (b) the result of $2L_c$ has a negligible error in the wetting front position, while the pressure head has an acceptable numerical oscillation (less than 5%); (c) when the mesh size is set as $4L_c$, the position of the wetting front is acceptable, but there exist non-convergent numerical oscillations at wetting front (over 5%). As L increases to $8L_c$, both the position of the wetting front and the pressure head produce significant errors.

The results in Fig. 9.10b are similar to those in Fig. 9.10a. That is, the standard solution results can be ensured when $L = L_c$ and $L = 0.5L_c$. As L increases, the pressure head has severe numerical oscillations when the mesh size is set as $2L_c$, followed by huge errors because of the mesh size of $4L_c$.

From the two verification cases, it can be concluded that: (a) in the L_c prediction interval, from the lower limit to the mean value, the convergence of numerical solution can be guaranteed; (b) from the mean value to the upper limit, the numerical solution may be acceptable, though errors may occur. Its applicability depends on the accuracy requirements of the problem. (c) when the mesh size exceeds the upper limit of prediction interval, the convergent solution cannot be guaranteed. In the actual simulation, it is necessary to avoid selecting a mesh size that exceeds the upper limit of L_c prediction interval.

Fig. 9.11 Searching process of Δt_c

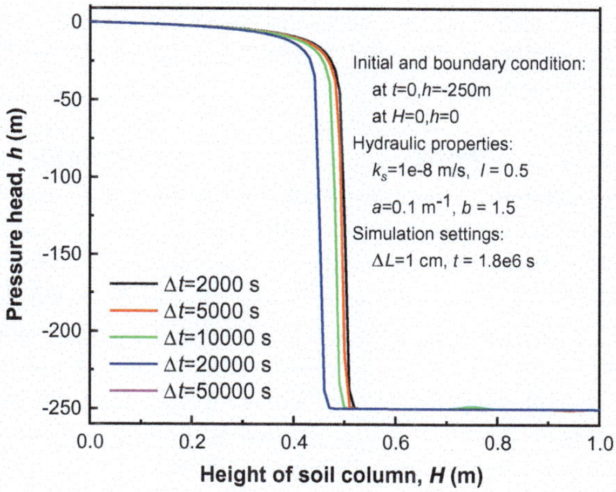

9.6 Numerical Study on the Convergence Criterion of Time Step Δt_c

9.6.1 *Numerical Test Results of Δt_c Values*

9.6.1.1 *An Example for Searching Δt_c*

Consistent with the L_c search case, the test of $S_{e0} = 0.2$ in the clay group is selected as an example to explain the searching process of the critical time step Δt_c. The results at 1.8e6 s are summarized in Fig. 9.11. Based on the former test results of critical mesh sizes, the L is set as L_c, i.e., 1 cm, and it could be concluded that $\Delta t =$ 10,000 s is the critical time step in this test.

9.6.1.2 *Results for Searching Δt_c*

According to previous theoretical derivation, Δt_c is inversely proportional to D_m^n. Therefore, for the most extreme conditions when one side of the mesh is dry and the other is wet, D_m^n Dnm should take the maximum possible value, and then a Δt_c that strictly obeys the convergence law can be obtained. However, for the diffusion coefficient of unsaturated soil, its smooth mathematical expression is a function that increases monotonically with the degree of saturation. When $S_e = 1$, the hydraulic conductivity is k_s and the specific water capacity $c(h)$ approaches 0, resulting in infinite D and unsolvable Δt_c. To overcome this limitation, an effective diffusion coefficient at the air entry value is taken as D_{max}, which is also in line with the physical characteristics of unsaturated soil: In the natural state, due to the dissolution of air bubbles in water, the soil's degree of saturation is almost impossible to reach

100%. Moreover, it is generally accepted that the soil has already exhibited saturated soil properties when the degree of saturation reaches the air entry value.

The relationships between the numerical measurement of Δt_c and results through different types of D are summarized in Fig. 9.12a and b. It can be found that: (a) When Dnm D_m^n is taken as D_{max}, Δt increases with rising L^2/D_{max}, but the result is not generally clear for different soils. When the same L^2/D_{max} is present, compared with granular soils, the fine-grained soil has a greater upper limit of Δt. (b) When D_m^n Dnm is taken as D_0, the results obtained are more discrete. Notably, D_0 is more suitable for determining the lower limit of Δt rather than the upper limit.

Considering that when D_m^n is singly selected as D_0 or D_{max}, the general expression of Δt_c under different soils cannot be obtained, D_{mean} is introduced as the representative of the mesh $D_m^n D_m^n D_m^n$, and is expressed as is shown in Eq. 9.23. The relationship between Δt_c and L^2/D_{mean} in each test group is plotted in Fig. 9.12c, indicating that Δt_c has a good correspondence with L^2/D_{mean} in different soils.

$$D_{mean} = \sqrt{D_{max} \times D_0} \tag{9.23}$$

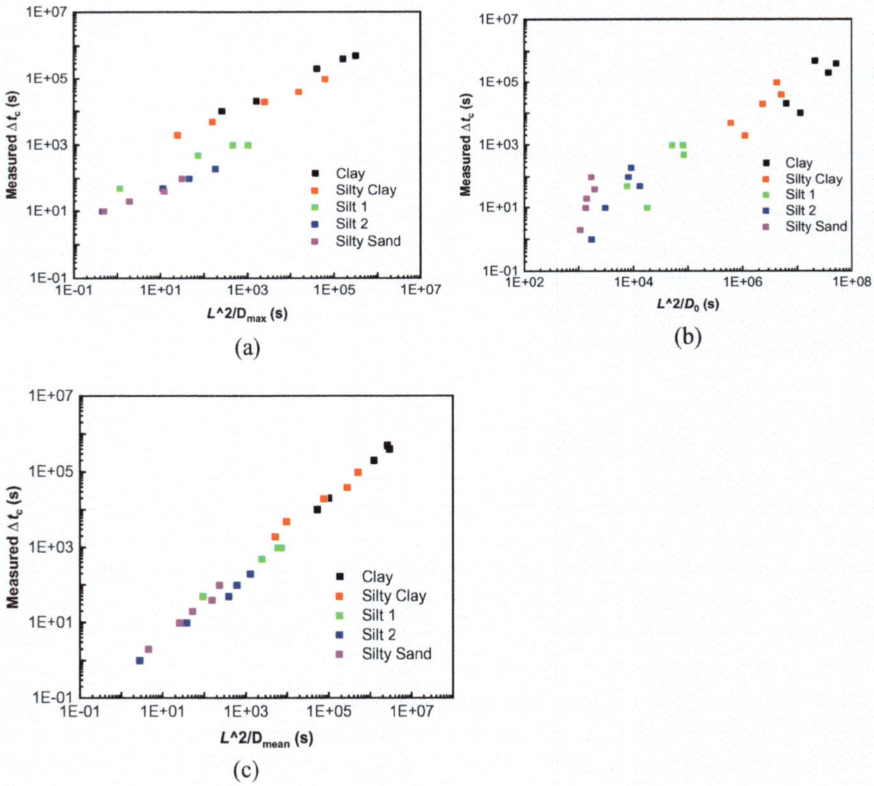

Fig. 9.12 Test results of Δt_c for various soils with different hydraulic diffusivity: **a** relation between Δt_c and D_{max}; **b** relation between Δt_c and D_0; **c** relation between Δt_c and D_{mean}

9.6.2 *A Semi-empirical and Semi-theoretical Criterion for* Δt_c

In Fig. 9.12c, the test results of Δt_c are linearly fitted and summarized in Fig. 9.13a. The fitted R-square reaches 0.97, indicating that the relationship between Δt_c and D_{mean} has a good consistency for various soils. The empirical formula of Δt_c, L, and D_{mean} is obtained from the fitting result as shown in Eq. 9.24. This formula shows that the reasonable time step is positively correlated with the mesh size L, which is consistent with previous studies (Kavetski et al. 2001; Cui et al. 2015). In addition, considering the dimension consistency, the units of Eq. 9.15 on both sides are seconds.

Fig. 9.13 Estimation of Δt_c for various soils: **a** fitting result of Δt_c; **b** prediction interval of Δt_c

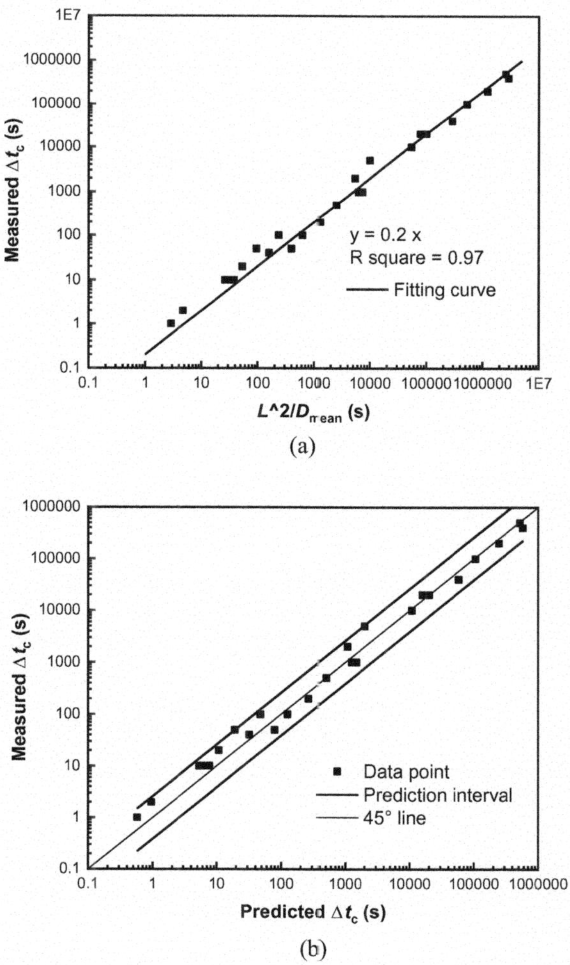

$$\Delta t_c = 0.2 \times \frac{L^2}{(D_{max} \times D_0)^{0.5}} \tag{9.24}$$

Furthermore, considering L in Eq. 9.24 as L_c, and bringing Eqs. 9.21 into 9.24, another formula for Δt_c can be obtained as Eq. 9.25. Equation 9.25 well explains the changing trend of Δt_c. (a) The advancing speed of the wetting front increases with a rising k_s, requiring a smaller time step. (b) When the initial water content increases or the boundary pressure head decreases (i.e., higher D_0 and lower D_{max}, respectively), the difference of pressure head between the dry–wet interface falls, resulting in a looser time step.

$$\Delta t_c = 0.7 \times k_s^{-1} \sqrt{\frac{D_0}{D_{max}}} \tag{9.25}$$

From the Δt_c test results, assuming that the probability density function of Δt_c obeys a normal distribution, and the measured Δt_c is taken as the mean value with the confidence interval of 95%, the prediction interval of Δt_c can be obtained as depicted in Eq. 9.26. The prediction interval of Δt_c expressed in Eq. 9.26 is shown in Fig. 9.13b.

$$\Delta t_c = 0.2 \times \frac{L^2}{(D_0 \times D_{max})^{0.5}} \times 10^{\pm 0.41} \tag{9.26}$$

9.6.3 Verification of Δt_c Criterion

The verification of Δt_c criterion is carried out through a wetting front rising test using the verification soils in Table 9.4. The soil physical parameters, initial and boundary conditions, and solution settings are the same as those in the L_c verification case. Corresponding with the measured L_c, the Δt_c calculated according to Eq. 9.24 is about 10 s and 1000 s, respectively. The verification results of the two soils are shown in Fig. 9.14.

From Fig. 9.14a, it can be demonstrated that: (a) Δt_c and $0.5\Delta t_c$ share the same result, and thus the two solutions can be regarded as the standard solution. (b) As the time step increases, the wetting front position, when $5\Delta t_c$ is applied, has an acceptable error (less than 5%). It can be considered that the solution at this point is still convergent, while at the $10\Delta t_c$ time step, the wetting front position does not converge. Similar to Fig. 9.14a, the results in Fig. 9.14b obtained by the time step of Δt_c and $0.5\Delta t_c$ can be regarded as accurate solutions, and the results after time step exceeding $2\Delta t_c$ do not converge.

With the two verification cases in mind, the following conclusions can be drawn. (a) The convergence of the numerical solution can be strictly guaranteed within the

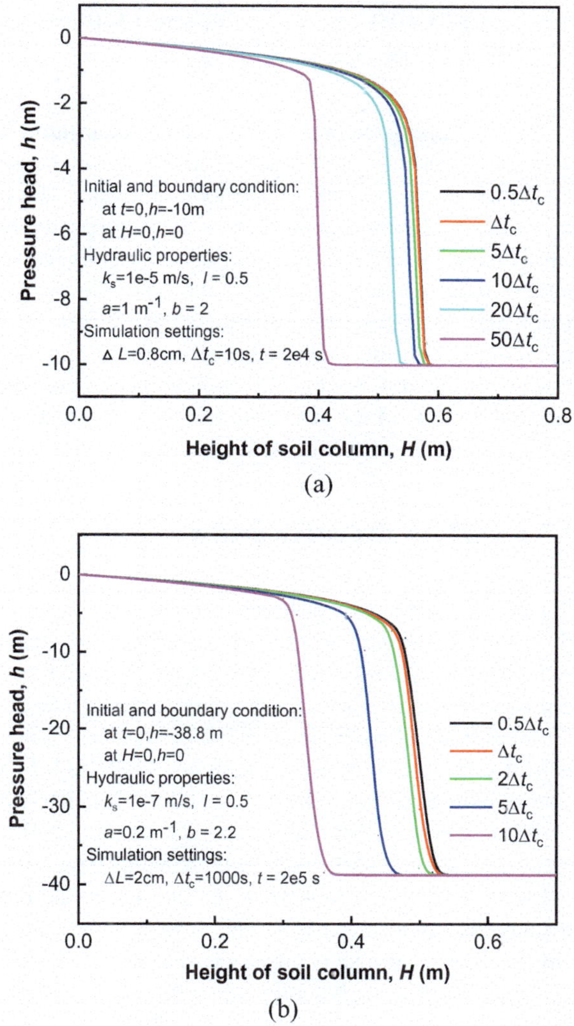

Fig. 9.14 Verification of proposed criterion for Δt_c: **a** Δt_c verification case (silt); **b** Δt_c verification case (clay)

prediction interval from the lower limit to the mean value. (b) The numerical solution may have acceptable errors within the range from the mean value to the upper limit. (c) When Δt exceeds the upper limit, the obtained solution may have a significant error, which should be avoided in practice. (d) Compared with the mesh size, the selection range of the time step is broader, and the accuracy still could be maintained within the interval from the mean value to the upper limit.

9.7 Application of the Proposed Convergence Criterion of L_c and Δt_c

To check the performance of the L_c and Δt_c criteria in multiple scales and dimensions, the proposed criteria were applied to the numerical simulations of a 1D analytical solution (Philip 1957), 2D model test data (Wu et al. 2017), and a 2D quasi-engineering problem (Lu and Likos 2004), respectively. Usually, an analytical result can be regarded as a precise solution of unsaturated seepage in specific conditions; the test data is a reliable source for verifying the accuracy of numerical results, which extends the present study to a field engineering scale. As such, by comparing the numerical results with and without the proposed criteria in various problems, the correctness, reliability, and applicability of the convergence criteria can be well clarified. Details of the verification cases are summarized in Table 9.5.

9.7.1 1D Analytical Solution

Philip (1957) proposed a classic analytical solution for the moisture distribution of 1D infinite soil under transient vertical infiltration, and used the Yolo light clay, which hydraulic properties of Yolo light clay were tested by Moore (1939), as an example to provide a solution under a zero-pressure boundary.

According to Eqs. 9.20 and 9.24, L_c and Δt_c of the model are 0.85 cm and 736 s, respectively. For convenience in actual modeling, L_c was chosen as 1 cm, and the corresponding Δt_c was adjusted to 1000 s. The detailed parameters of the numerical model are presented in Table 9.5. The results obtained by applying the proposed criteria are shown in Fig. 9.15a and b, respectively.

Figure 9.15 shows that when the mesh size and time step are small enough (for instance less than or equal to the critical value), the numerical simulation results and the analytical results are in good agreement, which confirms the reliability of the proposed criteria. Notably, when the mesh size and time step are slightly larger than the critical value, such as $2L_c$ and $2\Delta t_c$, the numerical result may have a slight but acceptable error. However, this process cannot be controlled. Specifically, when the mesh size and time step are smaller than the critical value, convergence can be guaranteed; when the mesh size and time step are greater than the critical value, the calculation result may converge within a specific range but it tends to be divergent. In other words, the criteria proposed in this study are sufficient but not necessary conditions for the convergence of unsaturated seepage calculations.

Table 9.5 Summary of verification cases

Cases	Reference	Properties of soils								Suggested set	
		Soil	A (m^{-1})	b	l	k_s (m/s)	θ_s	θ_r	Initial state	L_c (cm)	Δt_c (s)
Analytical solution	Philip (1957)	Yolo light clay	4	1.25	0.5	$5e^{-7}$	0.49	0.14	$\theta_0 = 0.237$	1	1000
Laboratory test	Wu et al. (2017)	Loess	1	1.56		$7.2e^{-6}$	0.43	0.04	$\psi_0 = 264$ kPa	1	50
Quasi-slope	Lu and Likos (2004)	Clay	0.1	1.5		$1e^{-9}$	0.49	0.16	Steady flow	10	4000

Note the SWCC and SHCF are fitted by VG and Mualem model, the boundary condition is zero water pressure

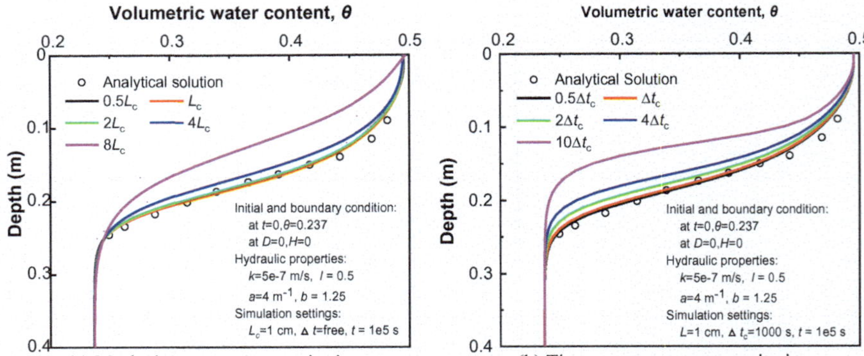

(a) Mesh size convergence criterion (b) Time step convergence criterion

Fig. 9.15 Verification of proposed convergence criteria based on the analytical solution

9.7.2 2D Model Test

Wu et al. (2017) conducted a model test of slope failure under heavy rainfall infiltration, which provided valuable data on the suction change over time at a specific test point inside the slope. As done by Wu et al. (2017), the data from test B in their study was selected to conduct numerical inversion, and T3 point data were chosen as the typical solution to compare the numerical results from different mesh sizes and time steps.

The establishment of the numerical model in this section is consistent with Wu et al. (2017), and the detailed parameters of the numerical model are listed in Table 9.5. According to Eqs. 9.20 and 9.24, L_c and Δt_c of the model are calculated to be 0.66 cm and 18.13 s, respectively. For convenience in actual modeling, L_c was set to 1 cm, and the corresponding Δt_c was adjusted to 50 s. The results obtained by applying the proposed selection criteria are shown in Fig. 9.16a and b, respectively.

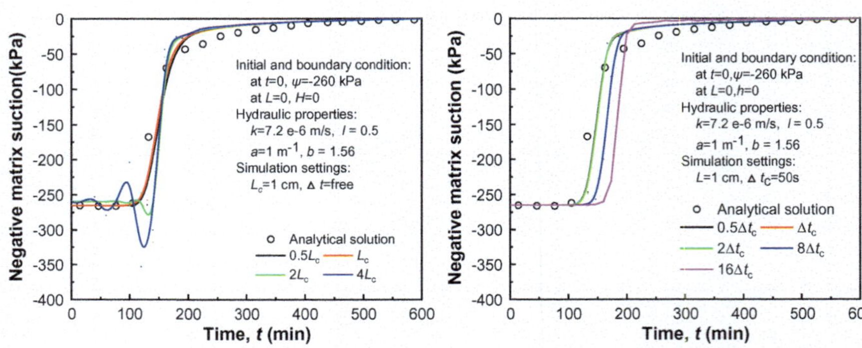

(a) Mesh size convergence criterion (b) Time step convergence criterion

Fig. 9.16 Verification of proposed convergence criteria based on 2D model test

Figure 9.16 shows that when the mesh size and time step are less than or equal to the critical value, the numerical and test results are in good agreement, which further confirms the reliability of the proposed criteria from the laboratory test scale and dimension. Figure 9.16a shows that when the mesh size is taken as $2L_c$, the numerical result starts to fluctuate, but the arrival time of the wetting front is the same. The numerical result will have an unacceptable error when the mesh size continues to increase. Combining Fig. 9.16a and b, it can be seen that the requirement of the time step is looser than the mesh size, and it is not until $8\Delta t_c$ that the calculated moisture change curve begins to show apparent errors. The above results provide a reasonable interpretation of the sufficient but not necessary characteristics of the proposed convergence criterion.

9.7.3 2D Engineering Problem

To verify the applicability of the L_c and Δt_c criteria at the engineering scale, the clay in the verification group is selected to conduct a two-dimensional (2D) slope infiltration analysis.

A 2D model with 25 m in width and 15 m in height is built, and the size of the slope is 15 m in width and 7.5 m in height. The soil is set as homogeneous, and its constitutive parameters are listed in the clay group in Table 9.3. The two sides of the model are set as symmetrical boundaries, and the bottom is an impervious boundary simulating the bedrock. Firstly, the calculation of the steady-state groundwater flow is performed, then a transient rainfall infiltration is conducted based on the steady-state results. In the steady-state stage, the total water head in the left boundary is 4 m, and that of the right boundary is 2 m, simulating the groundwater flow field. In the transient stage, the top and the surface of the slope are set as zero-pressure head, simulating heavy rain ponding conditions.

According to the proposed selection criteria, the mesh size and time step are respectively taken as $L_c = 0.1$ m and $\Delta t_c = 4000$ s, and the result at 3.2e5 s is obtained. The contour diagrams of the volumetric water content of the slope under various mesh sizes and time steps are shown in Fig. 9.17.

The following conclusions can be drawn. (a) In light of Fig. 9.17a and d, improper selection of mesh size could lead to severe problems, e.g., the inconsistent position of the wetting front, the singular point of moisture distribution. (b) In Fig. 9.17a and b, when the mesh size is selected as L_c, an oversized time step will decrease the wetting front position but not bring about a significant error in the water content distribution. (c) In Fig. 9.17a and c, when L and Δt increase simultaneously, faster infiltration will occur, while the local water content distribution will not be singular. (d) Comparing Fig. 9.17b and d, L problem is much more severe than Δt problem under the same degree of magnification. Overall, the "accuracy" order of impact from various L and Δt on the seepage solution is "convergent L with non-convergent Δt", "non-convergent but still matching L and Δt", and "Δt_c with non-convergent L". That is to say, when L and Δt cannot meet the criterion, L should be ensured

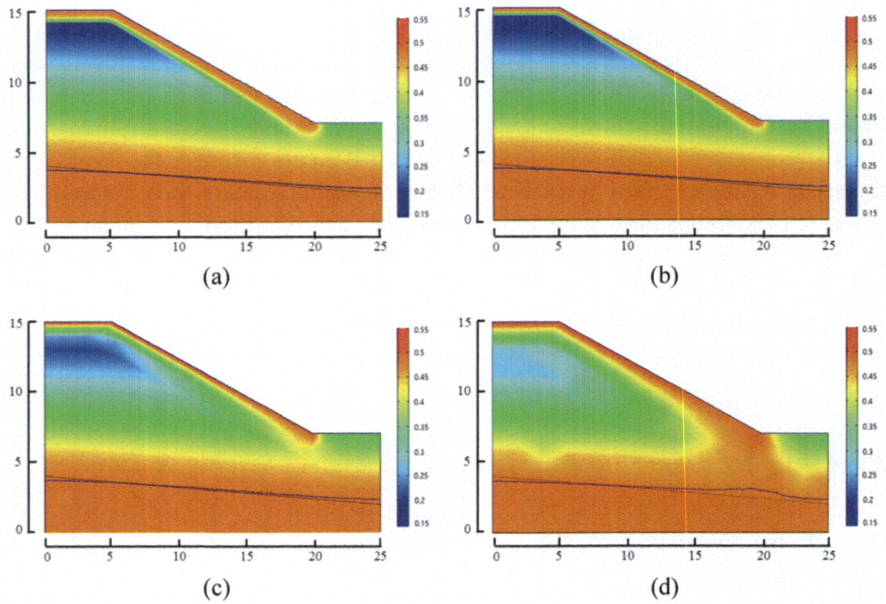

Fig. 9.17 Volumetric moisture distribution of 2D slope inflation ($t = 3.2e^5$ s): **a** $L = L_c$, $\Delta t = \Delta t_c$; **b** $L = L_c$, $\Delta t = 20\Delta t_c$; **c** $L = 20L_c$, $\Delta t = 20\Delta t_c$; **d** $L = 20L_c$, $\Delta t = \Delta t_c$

first. When L cannot be satisfied, it should be ensured that the two are matched with each other to control the error, e.g., they are enlarged by similar times.

9.8 Selection of Mesh Size and Time Step in Unsaturated Seepage Simulation

9.8.1 Recommended Values in Various Seepage Conditions

Based on the former research, a general recommended table can be given for the selection of mesh size and time step in practical FEM simulation of unsaturated seepage. To cover as many situations as possible, three typical soils with seven saturated hydraulic conductivities were summarized (Lu and Likos 2004). The SWCC and SHCF parameters are shown in Table 9.6. The van Genuchten model and the Mualem model are applied to SWCC and SHCF, respectively, and the residual water content of the soil is ignored. Based on Eqs. 9.20 and 9.25, the critical mesh sizes and the critical time steps under different initial conditions can be obtained.

The practical recommendation table for mesh size and time step is shown in Fig. 9.18. For the first type of boundary condition (zero-pressure boundary in the extreme state), L_c and Δt_c can be determined directly according to the table. For the

Table 9.6 Typical SHCF and SWCC parameters for common soils

Soil type	a (m^{-1})	b	θ_s	k_s (m/s)
Clay1	0.1	1.5	0.496	1e^{-9}
Clay2	0.15			1e^{-8}
Silty clay1	0.2		0.45	1e^{-7}
Silty clay2	0.3			1e^{-6}
Silt 1	0.5	2	0.4	1e^{-5}
Silt 2	0.7			1e^{-4}
Sand	2	4	0.367	1e^{-3}

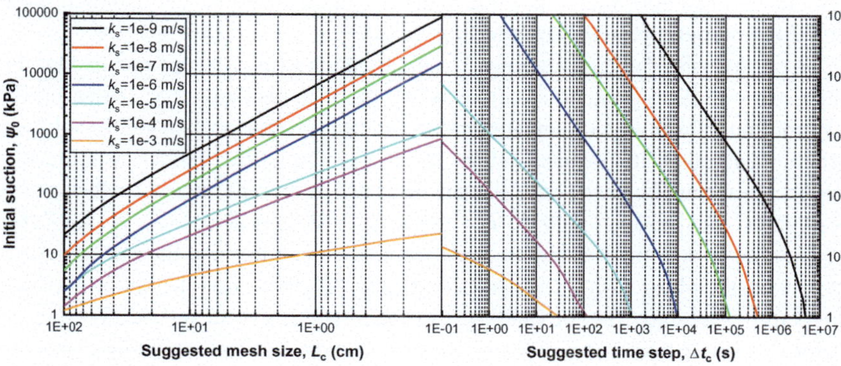

Fig. 9.18 Selection table for mesh size and time step in practical simulation. *Note* for zero-pressure boundary, L_c and Δt_c can be determined directly according to the table; to avoid the singularity of D_{max} at the saturated condition, D_{max} can take air entry value (e.g., a degree of saturation around 0.95). For an infiltration boundary with $q_{in} \leq k_s$, k_s should be used as q_{in}, and D_{max} should also be determined according to the value of q_{in}

second type of boundary condition (rainfall infiltration boundary where $q_{in} \leq k_s$), the mesh size and time step can be selected by having k_s and D_{max} replaced according to the infiltration rate, q_{in}. The detailed calculation and verification processes are shown in the appendix.

For realistic simulations, the scale of the model also needs to be considered. In the laboratory test scale, L_c and Δt_c are recommended to be select from the lower limit to the mean value of the prediction interval. In the engineering scale, the upper limit of the prediction interval can be taken to reduce the calculation cost.

9.8.2 *Discussion*

A one-dimensional model was used to conduct theoretical analysis and numerical tests for convenience. Then, 2D model tests and quasi-field engineering were used in the verification process of the proposed criteria, which demonstrates the applicability of the criteria to 2D problems. In the local area with the strongest nonlinearity (wetting front), the nonlinearity degree under one-dimensional and two-dimensional conditions is similar, indicating that the criteria obtained under 1D conditions are also suitable for 2D conditions. In other words, for strongly nonlinear unsaturated seepage, the influence of dimensionality on the convergence criterion is likely to be minor.

In the practical application of the selection criteria, it should be mentioned that, as is shown in Fig. 9.17, if the mesh size is not good as in the case of failure of Eqs. 9.21 or 9.22, the adjustment of time step alone cannot lead to a convergent result. By contrast, when the mesh size obeys Eq. 9.21, the range of time step may be broadened in light of the magnitude. That is to say, the mesh size is much more essential in comparison with the time step.

In FEM analysis, the "convergence" in the mathematical concept means that the iterative error in solving linear equations is less than the residual error. However, for highly nonlinear unsaturated seepage, mathematical convergence can yield a solution but cannot guarantee its physical reliability. For example, the results in numerical problems shown in Figs. 9.3 and 9.5 all meet the requirements of mathematical convergence, but they evidently do not conform to the physical laws. As such, for highly nonlinear and unsaturated seepage problems, in addition to the mathematical convergence, the physical convergence should also be given enough attention, which is the focus of this study. The advancing position of wetting front and the distribution of water content do not change with the mesh size and time step, which indicates the satisfaction of physical convergence. In fact, the "convergence" in this study is the "physical convergence", and the proposed criterion ensures the physical convergence (i.e., a reliable solution) of unsaturated seepage.

9.9 Conclusions

This chapter proposes a theoretical framework of the critical mesh size L_c and time step Δt_c, thus introducing a convergent solution in the FEM analysis of unsaturated seepage. Under this theoretical framework, the empirical formulas for predicting those two key parameters are proposed and verified through a series of numerical tests. Additionally, the application of these criteria to practical problems is explored and discussed through numerical models in laboratory and engineering scales. Finally, the appropriate mesh size and time step in different initial and boundary conditions are given as a recommendation table. The results establish the selection criteria of the mesh size and time step that ensure a convergent result in FEM analysis of highly nonlinear unsaturated seepage. The main conclusions are as follows.

(1) The proposed criteria can facilitate the calculation of L_c and Δt_c. Among the range of L_c and Δt_c, when the mesh size and time step are taken from the

lower limit to the mean value, the accuracy of the wetting front position can be ensured, and the numerical oscillations can be effectively controlled.

(2) The proposed criteria of L_c and Δt_c can be applied to both the laboratory test scale and engineering scale, but the two have their application scopes. When the mesh size and time step are taken from the mean value to the upper limit, acceptable results can also be obtained for the engineering scale.

(3) The mesh size and the time step are interdependent. Reasonable mesh size is more important than the time step. The mesh size criterion is suggested to be ensured first, followed by the setting of an appropriate time step.

Appendix

Application of Proposed Criteria in Infiltration Boundaries

Due to different hydraulic gradients, the wetting front advancing speed of saturated infiltration boundary is less than that of zero-pressure boundary during the initial seepage stage. This law can also be illustrated in Fig. 9.19, where the wetting front height under the $q_{\text{in}} = k_s$ boundary is lower than that under the zero-pressure boundary. As a result, compared to the zero-pressure boundary, the infiltration boundary has a lower nonlinearity degree. That is to say, the proposed convergence criterion obtained by zero-pressure condition brings a stricter value than the infiltration boundary.

For a given initial state, the soil's D_0 has been determined. When the boundary condition is set as the infiltration rate, q_{in}, the critical mesh size L_c can be calculated by having k_s in Eq. 9.20 replaced with q_{in}, as is shown in Eq. 9.27.

$$L_c = 1.87 * \left(\frac{D_0}{q_{\text{in}}}\right)^{0.5} \tag{9.27}$$

According to Eq. 9.24, in order to determine the critical time step, the value of $D_{\max}$ needs to be determined. As discussed in the main text, since $q_{\text{in}} < k_s$, there will be no water ponding at the boundary (i.e., pressure head > 0). Therefore, q_{in} can be used as the actual hydraulic conductivity to obtain the maximum saturation degree of boundary soil, and thus $D_{\max}$ can be determined. This process can be depicted as in Eq. 9.28.

$$\Delta t_c = 0.2 * \frac{L^2}{[D(S_e|k = q_{in}) * D_0]^{0.5}} \tag{9.28}$$

The verification soils in Table 9.3 are used, and the infiltration boundary is set to verify the performance of the proposed criterion of mesh size and time step. For each verification test, q_{in} is taken as k_s and $0.1k_s$, respectively, and the critical mesh

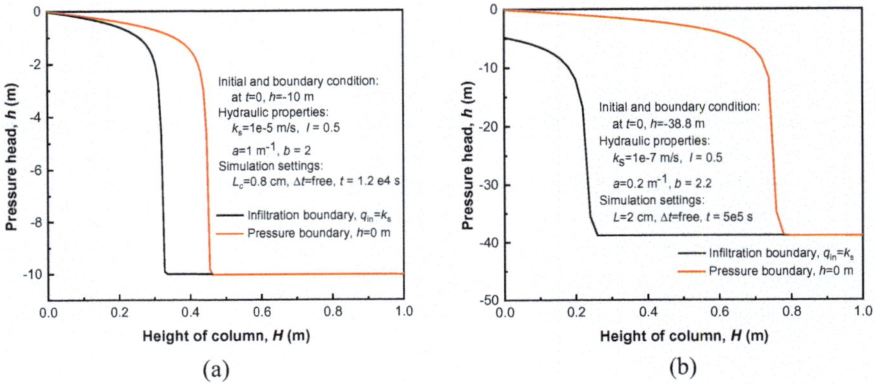

Fig. 9.19 Comparison of wetting front height under zero-pressure and saturated infiltration boundary: **a** verification silt; **b** verification clay

size and critical time step are calculated based on Eqs. 9.27 and 9.28, respectively. The results are shown in Figs. 9.20, 9.21, 9.22, and 9.23.

According to Figs. 9.20 and 9.21, under the infiltration boundary, the critical mesh size calculated with Eq. 9.27 can ensure the convergent solution of the wetting front. When the mesh size is set at $2L_c$, the numerical error is also acceptable. Figures 9.22 and 9.23 demonstrate that under the infiltration boundary, the critical time step based on Eq. 9.28 can also lead to the convergent solution. Overall, in the infiltration boundary condition, the critical mesh size and critical time step can be determined through Eqs. 9.27 and 9.28, respectively.

List of Symbols

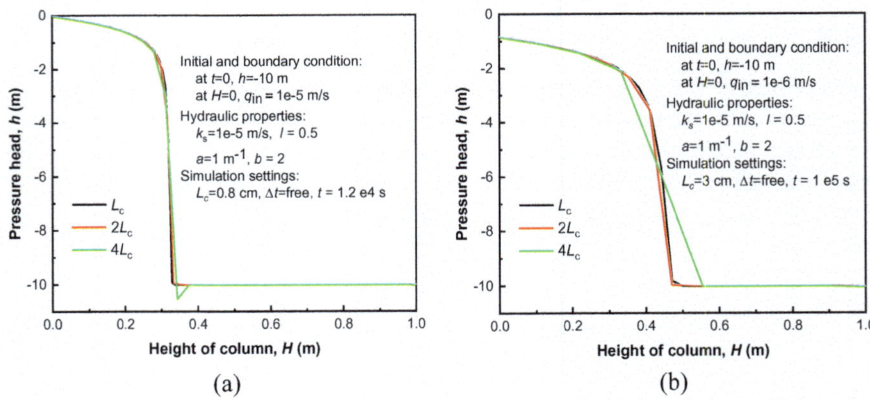

Fig. 9.20 Performance of L_c criterion under infiltration boundary (verification silt): **a** $q_{in} = k_s$; **b** $q_{in} = 0.1k_s$

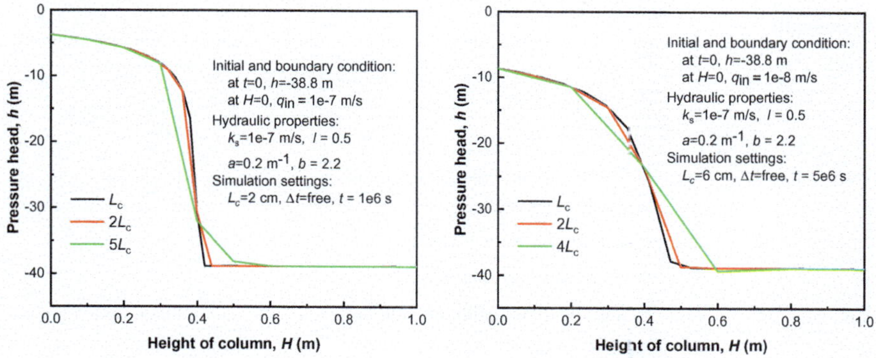

(a) $q_{\text{in}}=k_s$ (b) $q_{\text{in}}=0.1k_s$

Fig. 9.21 Performance of L_c criterion under infiltration boundary (verification clay): **a** $q_{\text{in}} = k_s$; **b** $q_{\text{in}} = 0.1k_s$

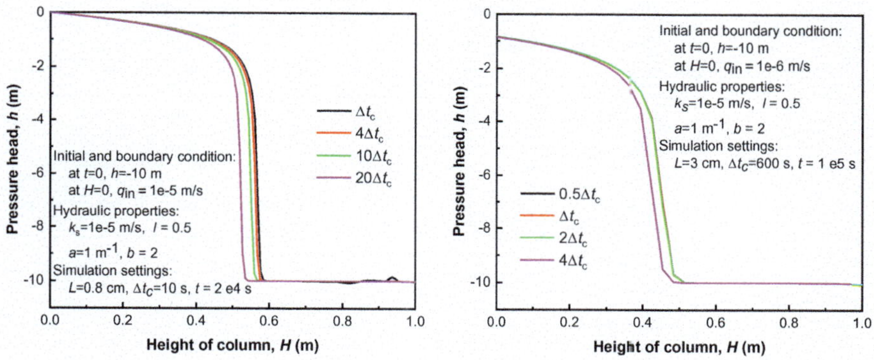

(a) (b)

Fig. 9.22 Performance of Δt_c criterion under infiltration boundary (verification silt): **a** $q_{\text{in}} = k_s$; **b** $q_{\text{in}} = 0.1k_s$

Symbol	Meaning	Unit
t	Infiltration time	s
Δt	Time step	s
Δt_c	Critical time step	s
L	Mesh size	cm
L_c	Critical mesh size	cm
H	Height of soil column	m
h	Pressure head	
h_0	Initial pressure head	

(continued)

(continued)

Symbol	Meaning	Unit
h_{bot}	Bottom pressure head	
ψ	Matric suction	kPa
ψ_0	Initial matric suction	
θ	Volume water content	
θ_0	Initial volume water content	
S_e	Effective degree of saturation	
S_{e0}	Initial effective degree of saturation	m/s
k_s	Saturated hydraulic conductivity	
k_r	Relative hydraulic conductivity	
k_0	Initial hydraulic conductivity	
q_{in}	Infiltration flow rate	1/m
c	Specific water capacity	1/m
D	Diffusion coefficient	m^2/s
D_0	Minimum diffusion coefficient	m^2/s
D_{max}	Maximum diffusion coefficient	m^2/s
D_{mean}	Mean diffusion coefficient	m^2/s
z	Seepage path	m
α	Order of mesh	
a, b, c, l	Parameters of SWCC and SHCF	
x	L and Δt	

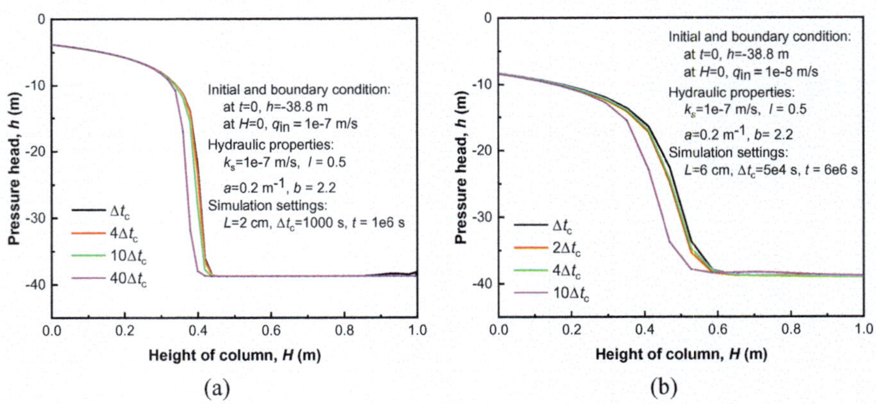

Fig. 9.23 Performance of Δt_c criterion under infiltration boundary (verification clay): **a** $q_{in} = k_s$; **b** $q_{in} = 0.1k_s$

References

Bai B, Xu T, Nie Q, Li P (2020) Temperature-driven migration of heavy metal Pb2+ along with moisture movement in unsaturated soils. Int J Heat Mass Transf 153:119573
Carrera-Hernández JJ, Smerdon BD, Mendoza CA (2012) Estimating groundwater recharge through unsaturated flow modelling: sensitivity to boundary conditions and vertical discretization. J Hydrol 452:90–101
Chen Y, Gao Y, Ng CWW (2020) A new capillary barrier system for retaining wall backfilled with fine-grained soil. Comput Geotech 118:103324
Chen Y, Zhou C, Zheng H (2008) A numerical solution to seepage problems with complex drainage systems. Comput Geotech 35(3):383–393
Cheng YG, Phoon KK, Tan TS (2008) Unsaturated soil seepage analysis using a rational transformation method with under-relaxation. Int J Geomech 8(3):207–212
Cui W, Gawecka KA, Taborda DMG, Potts DM, Zdravković L (2015) Time-step constraints in transient coupled finite element analysis. Int J Numer Meth Eng 106(12):953–971
Delfs JO, Park CH, Kolditz O (2009) A sensitivity analysis of Hortonian flow. Adv Water Resour 32(9):1386–1395
Farthing MW, Ogden FL (2017) Numerical solution of Richards' equation: A review of advances and challenges. Soil Sci Soc Am J 81(6):1257–1269
Fredlund DG (2006) Unsaturated soil mechanics in engineering practice. J Geotech Geoenviron Eng 132(3):286–321
Geo-Slope International Ltd. (2007) Seepage Modeling with SEEP/W 2007, Fourth Edition. Geo-Slope International Ltd. Calgary, Alta, Canada. ISBN: 978-7-5024-5693-1
Gibson RE, England GL, Hussey MJL (1967) The theory of one-dimensional consolidation of saturated clays: 1. finite non-linear consildation of thin homogeneous layers. Geotechnique 17(3):261–273
Ju SH, Kung KJS (1997) Mass types, element orders and solution schemes for the Richards equation. Comput Geosci 23(2):175–187
Karthikeyan M, Tan T-S, Phoon K-K (2001) Numerical oscillation in seepage analysis of unsaturated soils. Can Geotech J 38(3):639–651
Kavetski D, Binning P, Sloan SW (2001) Adaptive time stepping and error control in a mass conservative numerical solution of the mixed form of Richards equation. Adv Water Resour 24(6):595–605
Li B, Fang YG, Ou ZF (2018) Asymptotic solution for the one-dimensional nonlinear consolidation equation including the pore evolution effect. Int J Geomech 18(10):04018125
Li X, Zhang LM, Fredlund DG (2009) Wetting front advancing column test for measuring unsaturated hydraulic conductivity. Can Geotech J 46(12):1431–1445
Li X, Zhang Z, Zhang L, Zhang L, Wu L (2021) Combining two methods for the measurement of hydraulic conductivity over a wide suction range. Comput Geotech 135:104178
List F, Radu FA (2016) A study on iterative methods for solving Richards' equation. Comput Geosci 20(2):341–353
Lu N, Likos WJ (2004) Unsaturated soil mechanics. Wiley
Miller CT, Christakos G, Imhoff PT, McBride JF, Pedit JA, Trangenstein JA (1998) Multiphase flow and transport modeling in heterogeneous porous media: challenges and approaches. Adv Water Resour 21(2):77–120
Miller CT, Dawson CN, Farthing MW, Hou TY, Huang J, Kees CE, Langtangen HP (2013) Numerical simulation of water resources problems: models, methods, and trends. Adv Water Resour 51:405–437
Moore R (1939) Water conduction from shallow water tables. Hilgardia 12(6):383–426
Mualem Y (1976) A new model for predicting the hydraulic conductivity of unsaturated porous media. Water Resour Res 12(3):513–522
Ng CWW, Leung AK (2012) Measurements of drying and wetting permeability functions using a new stress-controllable soil column. J Geotech Geoenviron Eng 138(1):58–68

Ng CWW, Shi Q (1998) A numerical investigation of the stability of unsaturated soil slopes subjected to transient seepage. Comput Geotech 22(1):1–28
Pan L, Warrick AW, Wierenga PJ (1996) Finite element methods for modeling water flow in variably saturated porous media: numerical oscillation and mass-distributed schemes. Water Resour Res 32(6):1883–1889
Philip JR (1957) The theory of infiltration: 1. The infiltration equation and its solution. Soil Sci 83(5):345–358
Phoon KK, Tan TS, Chong PC (2007) Numerical simulation of Richards equation in partially saturated porous media: under-relaxation and mass balance. Geotech Geol Eng 25(5):525–541
Richards LA (1931) Capillary conduction of liquids through porous mediums. Physics 1(5):318–333
Söderlind G (2006) Time-step selection algorithms: adaptivity, control, and signal processing. Appl Numer Math 56(3–4):488–502
Tai B, Yue Z, Qi S, Wang P (2021) Experimental and numerical investigation on thermal-moisture-mechanical behaviors on a new anti-frost cutting bed of high-speed railway in deep seasonally frozen ground regions under extreme climate. Comput Geotech 136:104251
Tan TS, Phoon KK, Chong PC (2004) Numerical study of finite element method based solutions for propagation of wetting fronts in unsaturated soil. J Geotech Geoenviron Eng 130(3):254–263
Thomas HR, Zhou Z (1997) Minimum time-step size for diffusion problem in FEM analysis. Int J Numer Meth Eng 40(20):3865–3880
van Genuchten MT (1980) A closed-form equation for predicting the hydraulic conductivity of unsaturated soils. Soil Sci Soc Am J 44(5):892–898
Vereecken H, Schnepf A, Hopmans JW, Javaux M, Or D, Roose T, Young IM (2016) Modeling soil processes: review, key challenges, and new perspectives. Vadose Zone J 15(5)
Vogel HJ, Ippisch O (2008) Estimation of a critical spatial discretization limit for solving Richards' equation at large scales. Vadose Zone J 7(1):112–114
Wei Y, Xu X, Zhao L, Chen X, Qiu H, Gao B, Cao X (2021) Migration and transformation of chromium in unsaturated soil during groundwater table fluctuations induced by rainfall. J Hazard Mater 126229
Wu LZ, Huang J, Fan W, Li X (2020a) Hydro-mechanical coupling in unsaturated soils covering a non-deformable structure. Comput Geotech 117:103287
Wu LZ, Zhou Y, Sun P, Shi JS, Liu GG, Bai LY (2017) Laboratory characterization of rainfall-induced loess slope failure. CATENA 150:1–8
Wu LZ, Zhu SR, Peng J (2020b) Application of the Chebyshev spectral method to the simulation of groundwater flow and rainfall-induced landslides. Appl Math Model 80:408–425
Zhan LT, Li GY, Jiao WG, Lan JW, Chen YM, Shi W (2020) Performance of a compacted loess/gravel cover as a capillary barrier and landfill gas emissions controller in Northwest China. Sci Total Environ 718:137195
Zhang HX, Guo C, Zong WG (1999) Problems about grid and high order schemes. Acta Mech Sin 31(4):398–405 (in Chinese)
Zhang SH, Li Q, Zhang LP, Zhang HS (2016) The history of CFD in China. Acta Aerodynamica Sinica 34(2):157–174
Zhu SR, Wu LZ, Peng JB (2020) An improved Chebyshev semi-iterative method for simulating rainfall infiltration in unsaturated soils and its application to shallow landslides. J Hydrol 590:125157
Zhu SR, Wu LZ, Shen ZH, Huang RQ (2019) An improved iteration method for the numerical solution of groundwater flow in unsaturated soils. Comput Geotech 114:103113

Chapter 10
Assessing the Stability of Widely-Graded Soils

10.1 Introduction

Failure of soil slopes is often associated with instability of soils. Instability refers to a behavior in which large plastic strains are generated rapidly when a soil element sustains a given load or stress. Currently, the research related to instability of soils is primarily conducted at saturated conditions through undrained triaxial tests on loose saturated soils (e.g., Lade 1992; Leong et al. 2000; Yang 2002) and drained constant shear tests on saturated medium and dense sands (Chu et al. 2003). However, many natural soil deposits encountered in engineering practice are often unsaturated. During rainfall infiltration, a reduction in soil suction causes a decrease in the shear strength, which leads to the development of plastic strains and ultimately to the instability of the soil. This process can be idealized as a wetting path along which the shear stress and net mean stress keep constant, but the suction decreases over time. However, the instability behavior of unsaturated granular soils along the wetting path has seldom been investigated.

The experimental study shown in Chap. 5 demonstrates that the hydraulic properties of colluvial soils are highly related to their coarse contents. The soils with gravel proportions above 44% (e.g., GW-GM with sand and SM with gravel) show typical dual-porosity structures, which manifest bimodal SWCCs and bimodal SHCFs. The soils with gravel proportions smaller than 28% (e.g., CL with sand and sandy ML) show substantial intra-aggregate pore structures, which manifest unimodal SWCCs and unimodal SHCFs. A question to ask is how stable are slopes composed of colluvial soils with different gravel proportions under a rainfall event.

In this chapter, first, the instability of unsaturated granular soils along the wetting path is investigated experimentally. A method for evaluating the instability of saturated and unsaturated granular soils is proposed. Then, the infiltration process and the stability of colluvial slopes are investigated by numerical simulation. Finally, a case study is performed on a natural landslide along Province Road 303 (PR303) in Sichuan Province.

X. Li et al., *Fundamental Behavior of Unsaturated Widely-Graded Soil*,
https://doi.org/10.1007/978-981-19-3402-5_10

10.2 Instability of Saturated and Unsaturated Coarse Granular Soils

10.2.1 Experimental Program

In this study, three series of tests were conducted on granular soils with $f_c = 50$, 68, and 90%. The first series (i.e., isotropic consolidation tests and mercury intrusion porosimetry tests) was conducted to investigate the instability in the microstructures of three types of soil with different coarse contents. The second series (i.e., drained and undrained triaxial tests) was performed to study the instability of these granular soils under saturated undrained shearing condition. The third series (i.e., wetting tests) was carried out to investigate the instability of unsaturated granular soils under water infiltration (Fig. 7.3).

In the wetting tests, the shear stress and the net mean stress were kept constant, but the suction was decreased over time. Such a stress path (Fig. 10.1) simulates the stress path of a slope element subject to rainfall infiltration (Brand 1981). The test plans are summarized in Table 9.3. Four specimens were wetted under a constant confining pressure of 15 kPa with different stress ratios ($\eta = 0$, 1.0, 1.5, and 1.87) to investigate the effect of stress ratio on the instability behavior of unsaturated soils. Three specimens were wetted under the same stress ratio of $\eta = 1.5$ but different confining pressures (5, 15, and 25 kPa) to investigate the effect of confining pressure on the instability behavior of unsaturated soil.

A GDS triaxial apparatus with a three-bar air entry disk was used to conduct the wetting tests on unsaturated specimens with $f_c = 68\%$. The axis-translation technique was used to control the matric suction. A precision differential pressure transducer with a double cell was used to measure the total volume change of the unsaturated specimen. A flushing system was used to remove the diffused air that might accumulate below the ceramic disk during the test. Figure 10.1 summarizes the stress path for wetting tests. Before conducting the wetting tests, the compacted specimens were saturated first. Then, the specimens were equilibrated to a high suction value (280 kPa) under specified constant net confining pressures (e.g., 5, 15, and 25 kPa). After that, deviator stress was increased to selected stress ratios, η (e.g., 0, 1.0, 1.5, 1.85) at constant net confining pressures. The deviator stress was increased at a rate of 0.5 kPa per hour to provide a uniform suction distribution throughout these triaxial specimens. After reaching the required deviator stress, the specimen was subjected to a reduction of suction in several steps at constant net mean stress and deviator stress. The cell pressure and the pore-air pressure were reduced at the same rate, while the net mean stress and the backwater pressure were kept constant. Each suction decrement was maintained for a sufficiently long time to allow suction equalization in the specimens. Suction equilibration was considered achieved when the drainage of water in 8 h was smaller than 0.01% of the total dry soil weight. The tests were terminated when the travel limit of the piston of the apparatus was reached (25%).

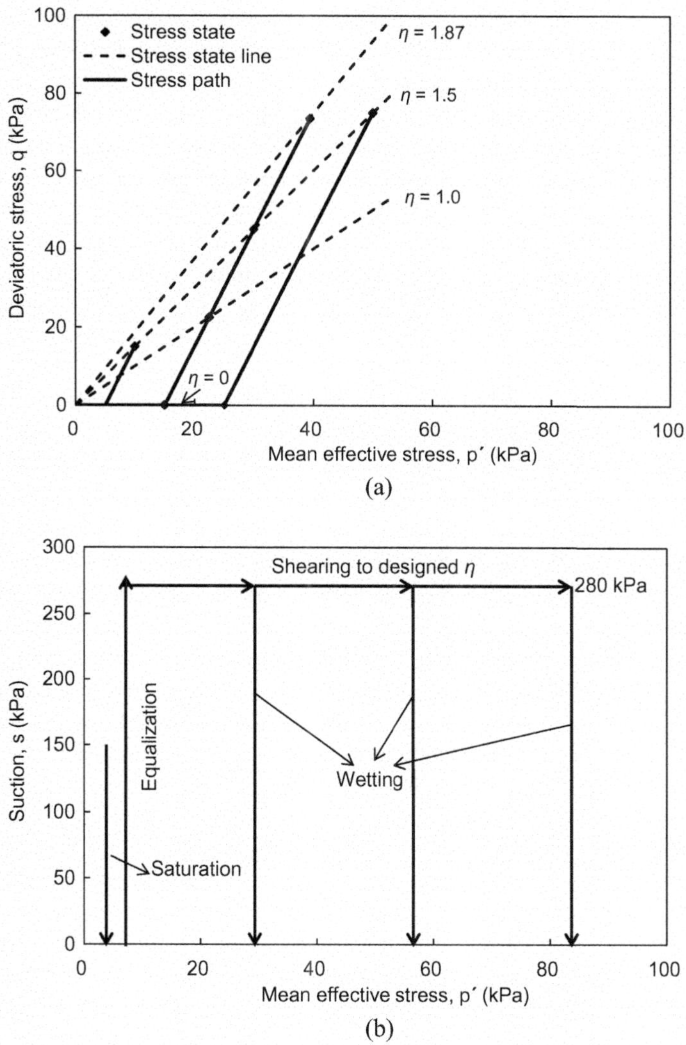

Fig. 10.1 Wetting stress paths: **a** in the *p–q* space; **b** in the *s–q* space

10.2.2 Instability in the Microstructures of the Granular Soils

The microstructure of a soil varies with the coarse content of the soil (Zhang and Li 2010) and affects the mechanical behavior of the soil. Yamamuro et al. (2008) remarked that a stable undrained stress–strain response is due to stable contacts in the soil skeleton. Specimens exhibiting a more contractive undrained response may contain a greater number of unstable contacts. In this chapter, the instability in microstructure was studied by analyzing the compressibility and pore-size distributions (PSDs) of the granular soils.

Figure 10.2 shows the isotropic compression curves for the $f_c = 50\%$, $f_c = 68\%$, and $f_c = 90\%$ specimens with different initial void ratios. Isotropic normal consolidation lines (NCLs) are identified within the range of the applied mean effective stress. The compression index, λ, and the swell index, k, which are the slopes of the NCL and the unloading curve, respectively, are used to describe the compressibility of the soils. As shown in Table 10.1, at the same initial void ratio, the $f_c = 68\%$ specimen has the highest values of λ and λ/k, whereas the $f_c = 90\%$ specimen has the lowest values of λ and λ/k. The skeleton of the $f_c = 68\%$ specimen is the most compressible.

Soils with different coarse contents may change from a fines-controlled structure to a coarse-controlled structure. As discussed in Chap. 3, it is observed that the soils with $f_c = 50\%$, 68%, and 90% are fines-controlled soil, transition soil, and coarse-controlled soil, respectively. Figure 10.3a shows the corresponding PSDs of the three types of soil at a void ratio 0.62. The PSD of the $f_c = 50\%$ specimen exhibits a unimodal pore-size distribution as the skeleton is formed by clay aggregates. The low compressibility of the $f_c = 50\%$ specimen is attributed to the dominance of the intra-aggregate pores which are difficult to compress. The structure of the transition soil with $f_c = 68\%$ is controlled by both clay aggregates and coarse particles. The PSD shows a distinct dual-porosity structure with a large number of inter-aggregate pores 2–75 μm in size between the clay aggregates and between the clay aggregates and the coarse particles. During compression, the clay aggregates break and collapse into the large inter-aggregate pores, leading to the highest compressibility of the $f_c = 68\%$ specimen. The $f_c = 90\%$ specimen forms a skeleton with grain-grain contacts, which has the lowest compressibility. The instability potential in the microstructures is in accordance with the compressibility, which initially increases with the coarse content, reaches a maximum at a transition state, and decreases afterward.

A soil is an assemblage of particles and aggregates. Although the particle-size distribution is controlled in this study, the arrangement of particles of different sizes and the distribution of pores within different samples are different depending on the compaction void ratio. The pore-size distributions of the $f_c = 68\%$ specimens with different initial void ratios are shown in Fig. 10.3b. The two loose specimens with $e = 0.87$ and 0.62 exhibit bimodal distributions due to the presence of a large number of inter-aggregate pores between the clay aggregates or between the clay aggregates and the coarse particles. The skeleton is unstable because the clay-aggregate contacts are very sensitive to stresses. However, as the void ratio decreases to 0.52, the pore-size distribution of the soil changes to unimodal as the inter-aggregate pores are compressed significantly. With decreasing void ratio, the clay aggregates break down and collapse into the large inter-aggregate pores, more stable grain-grain contacts develop, and the microstructure, hence, becomes more stable. This observation is in agreement with the fact that λ and λ/k decrease with decreasing void ratio.

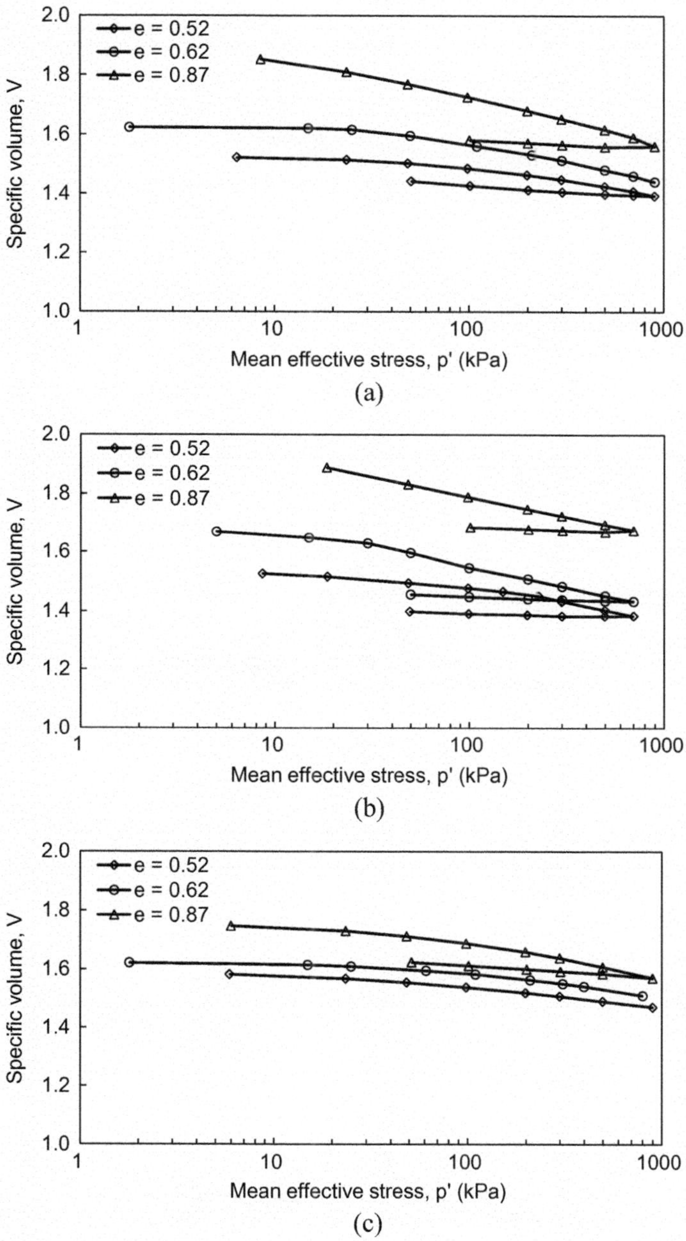

Fig. 10.2 Isotropic consolidation curves for soils of different void ratios: **a** $f_c = 50\%$; **b** $f_c = 68\%$; **c** $f_c = 90\%$

Table 10.1 Parameters of the isotropic normal consolidation lines for the three types of soil

Coarse content (%)	Swelling index, k	$e = 0.87$		$e = 0.62$		$e = 0.52$	
		compression index, λ	λ/k	compression index, λ	λ/k	compression index, λ	λ/k
$f_c = 50$	0.019	0.059	3.11	0.053	2.79	0.043	2.26
$f_c = 68$	0.01	0.059	5.90	0.059	5.90	0.059	5.90
$f_c = 90$	0.017	0.053	3.12	0.036	2.12	0.031	1.82

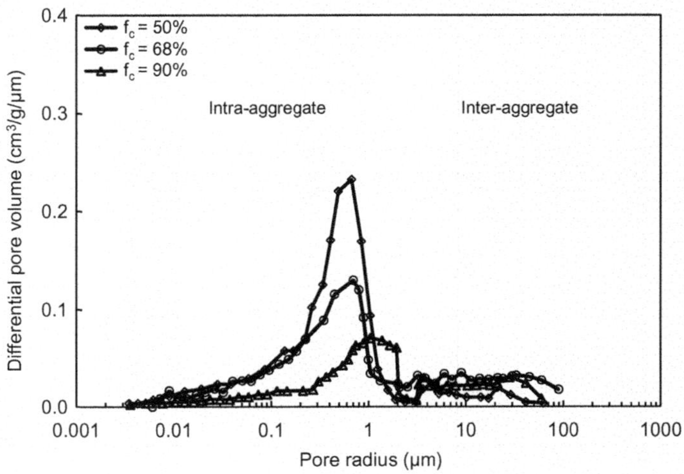

(a)

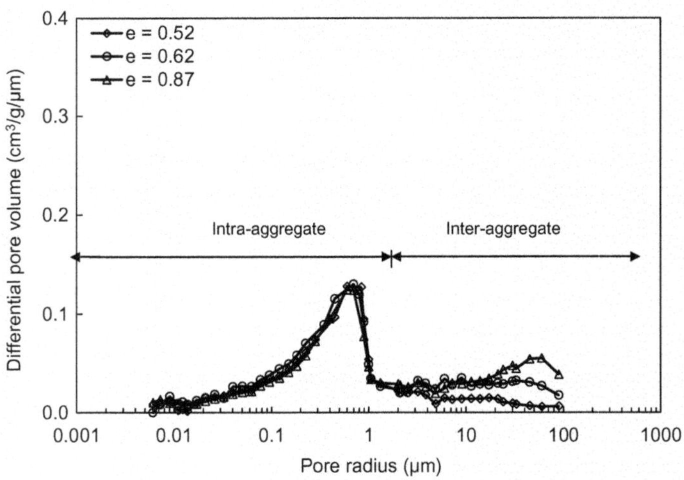

(b)

Fig. 10.3 Pore-size distributions: **a** three types of soil under the same void ratio 0.62; **b** $f_c = 68\%$ specimens with different void ratios

10.2.3 Instability of Saturated Granular Soils During Shearing

Instability is defined as behavior in which large plastic strains develop rapidly when a soil element sustains a given load (Chu and Leong 2002). For large plastic strains to develop, the soil must yield. Therefore, yielding is considered as the necessary condition for instability.

One of the failure mechanisms of slope failure is that the rise of pore pressure as a result of rainfall decreases the mean effective stress in the slope, which lowers the soil strength and eventually triggers landslides. In saturated granular deposits, instability under the undrained condition has been used to explain flow slides or collapses in granular soil slopes. Figure 10.4 presents typical effective stress paths in isotropically consolidated undrained triaxial tests on a granular soil with different initial void ratios. The loose soil exhibits unstable behavior if its initial stress state, p_1, is located above the critical-state line (CSL). The line that connects the peak states (p_1) of the effective stress paths is defined as the instability line (IL). The zone bounded by the IL and the CSL is termed "zone of potential instability" (Lade 1993). A medium dense soil at an initial state, p_2, exhibits stable behavior when its void ratio is smaller than the void ratio at the critical state. The shear resistance increases due to the generation of negative pore water pressure. The effective stress path, p_3, of a dense soil reaches the failure line and then goes to a lower stress point on the critical-state line. The zone between the failure line and the critical-state line is the "zone of potential instability" for the dense soil.

10.2.3.1 Drained Shearing Behavior

Drained triaxial tests were performed to investigate the critical-state characteristics of the granular soils with f_c = 50, 68, and 90%. The stress–strain behavior and volumetric strain relationships of the f_c = 50, 68, and 90% specimens with different void ratios are shown in Figs. 10.5, 10.6, 10.7, 10.8, and 10.9. The stress–strain relationships show that the increase rates of deviatoric stresses and volumetric strains decrease toward the end of the tests. The last point of test data is taken as the critical state. The soil compressibility affects its volumetric behavior in drained tests. Compared with the three types of soils under the same conditions, the f_c = 68% specimen shows the highest compressibility upon loading, producing large contractive volumetric strains. The f_c = 90% specimen has lower compressibility due to the higher stiffness of its skeleton, causing dilation under drained loading. The stress–strain relationships of the f_c = 68% specimens under the confining pressure of 15 kPa are used to illustrate the shearing behavior of soils with different void ratios. As shown in Fig. 9.7a, for the f_c = 68% specimens with e = 0.87, strain hardening is observed. The deviator stress increases monotonically with the axial strain and approaches a steady value at the end of the test. The volumetric behavior is contractive as shown in Fig. 10.8b. As the void ratio of the specimen decreases, the contractive behavior

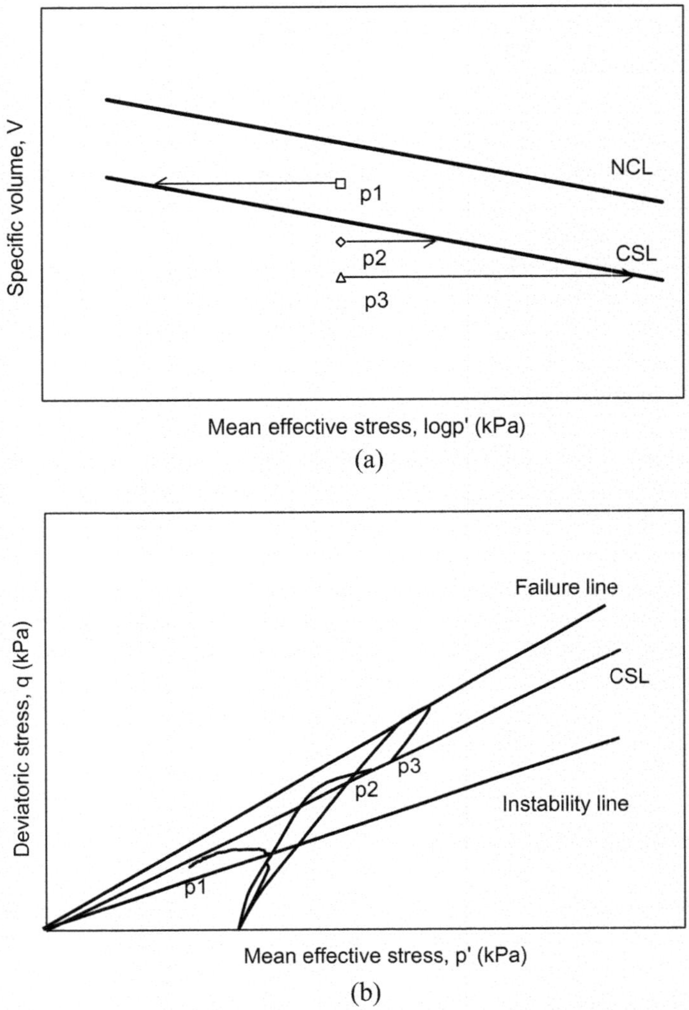

Fig. 10.4 Schematic instability behavior of saturated granular soil under undrained condition in **a** v-log p' space and **b** p'–q space

becomes weaker. When the void ratio decreases to 0.52, strain-softening behavior is observed as shown in Fig. 10.8a. The deviator stress reaches a peak value at a low axial strain. It then reduces from the maximum as the strain further increases and continues to drop but at a reduced rate toward the end of test. Dilative behavior is observed for the volumetric behavior. The specimens compress first and then swell until reaching the critical state.

The critical-state lines of the three types of soil are shown in Figs. 10.10, 10.13, and 10.15. The critical-state stresses in the p'–q space can be represented fairly well by a straight line, and a unique CSL can be defined for each type of soil. The gradients

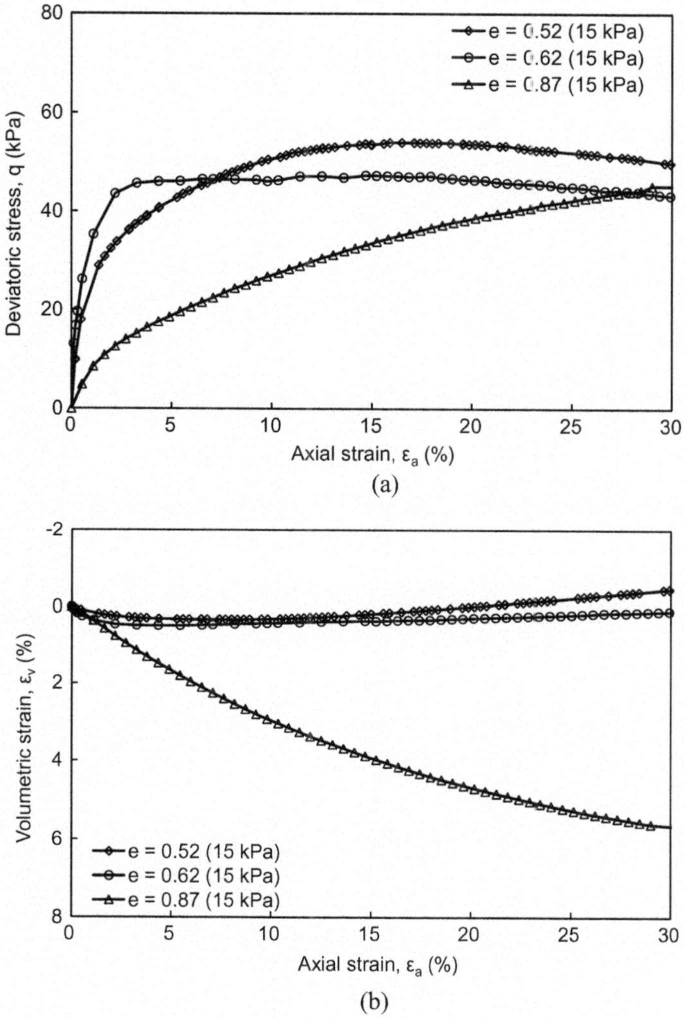

Fig. 10.5 Stress–strain relationships of the $f_c = 50\%$ specimens with different void ratios: **a** deviatoric stress and axial strain; **b** volumetric strain and axial strain

of the CSLs, M, are 1.42, 1.48, and 1.68 for the soils with $f_c = 50\%$, 68%, and 90%, respectively. The corresponding critical friction angles, ϕ'_{cs}, are 34.9°, 36.4°, and 41.1°, respectively. ϕ'_{cs} increases with the course content.

Figures 10.10b, 10.13b, and 10.15b show the CSLs in the v-log p' space for the $f_c = 50\%$, 68%, and 90% specimens, respectively. The CSL in the v-log p' space can be expressed as:

$$v = \Gamma - \lambda_{cs} \ln p' \quad (10.1)$$

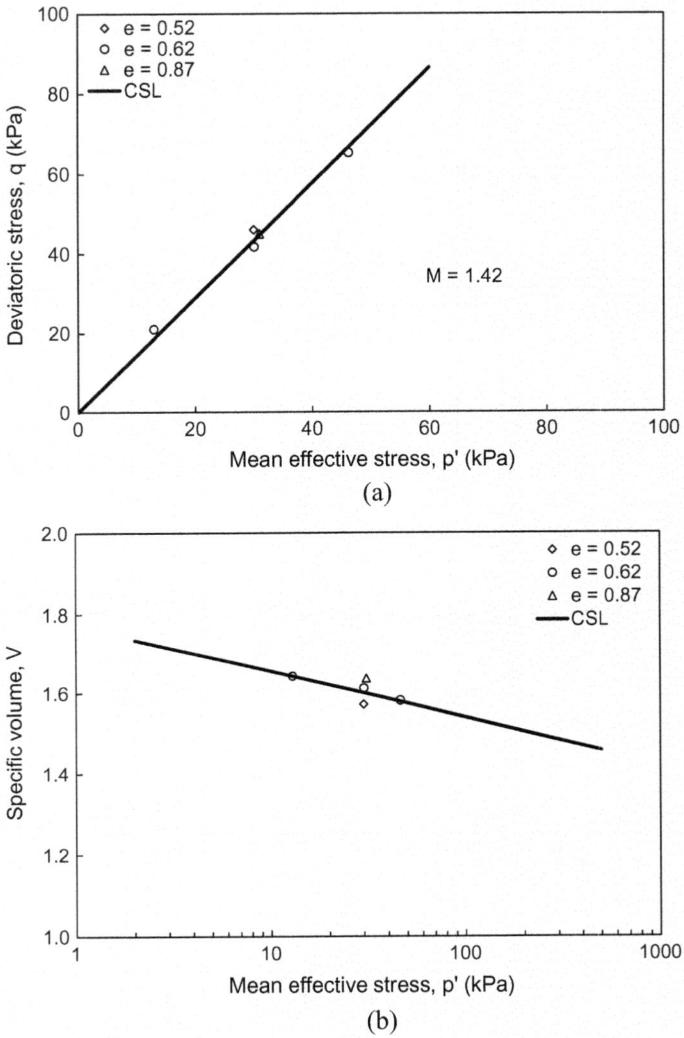

Fig. 10.6 Critical-state lines of the $f_c = 50\%$ specimens in **a** p'–q space and **b** v-log p' space

where Γ and λ_{cs} are the intercept and slope of the CSL, respectively. As shown in Fig. 10.6b, all the tests on the $f_c = 50\%$ specimens with different void ratios fall on the same critical-state locus, so do those on the $f_c = 90\%$ specimens [Fig. 10.13b]. However, the CSLs for the $f_c = 68\%$ soils of three different initial void ratios do not fall on the same line but are parallel with each other [Fig. 10.10b]. The values of Γ and λ_{cs} for the three types of soils with different void ratios are listed in Table 10.2. It can be observed that the CSLs for the transition soil ($f_c = 68\%$) have the same λ_{cs} value but different Γ values, which means the location of the CSL depends on its void ratio. This behavior is termed as transitional behavior (Nocilla et al. 2006;

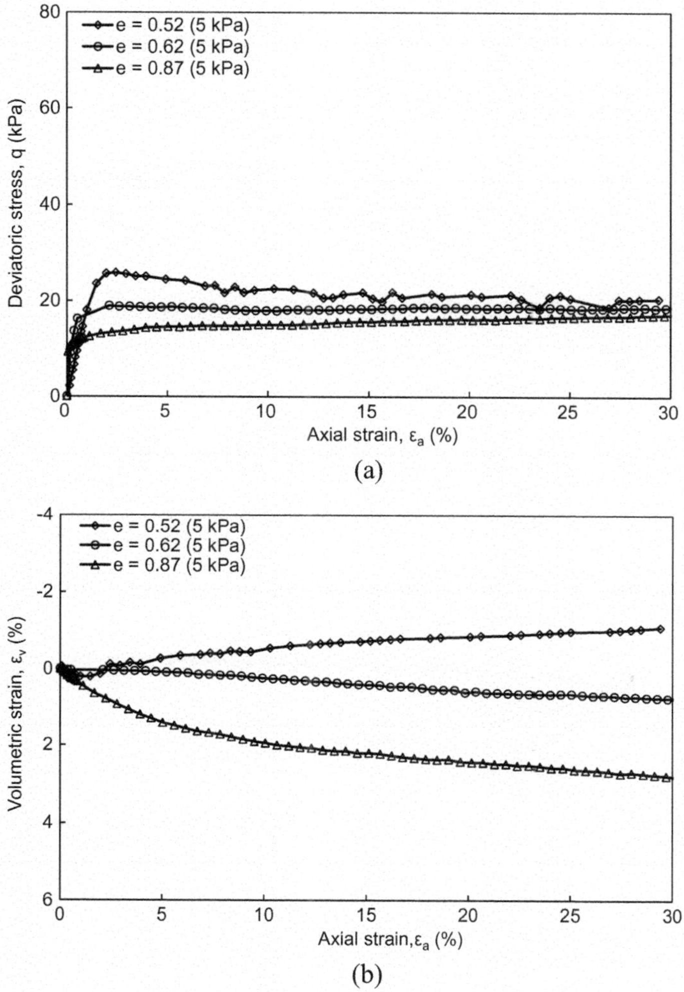

Fig. 10.7 Stress–strain relationships of the $f_c = 68\%$ specimens with different void ratios under 5 kPa: **a** deviatoric stress and axial strain; **b** volumetric strain and axial strain

Martins et al. 2001; Altuhsfi et al. 2010; Ferreira and Bica 2006). As shown in Table 9.2, the values of λ_{cs} are 0.05, 0.059, and 0.045 for the soils with $f_c = 50\%$, 68%, and 90%, respectively. This is in agreement with the instability in the microstructures, with the microstructure of the $f_c = 68\%$ soil being most compressible.

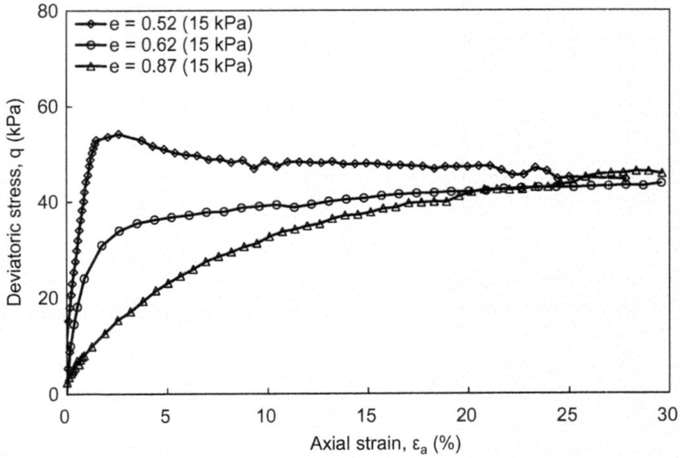

(a)

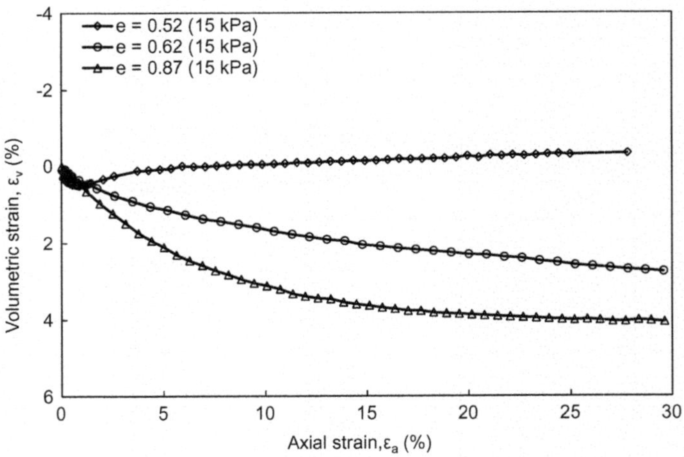

(b)

Fig. 10.8 Stress–strain relationships of the $f_c = 68\%$ specimens with different void ratios under 15 kPa: **a** deviatoric stress and axial strain; **b** volumetric strain and axial strain

10.2.3.2 Undrained Shearing Behavior

Undrained triaxial tests were conducted on the three types of soils under an effective confining pressure of 15 kPa. The stress–strain relationship, pore-water pressure development for the $f_c = 50$, 68, and 90% soils with different void ratios are shown in Figs. 10.14, 10.16, and 10.18. Comparing the three types of soils under the same void ratio and effective confining pressure, the $f_c = 68\%$ specimen shows the strongest contractive tendency under undrained shearing. Due to the high stiffness skeleton of the $f_c = 90\%$ specimen, the inter-particle pores are difficult to compress. Upon

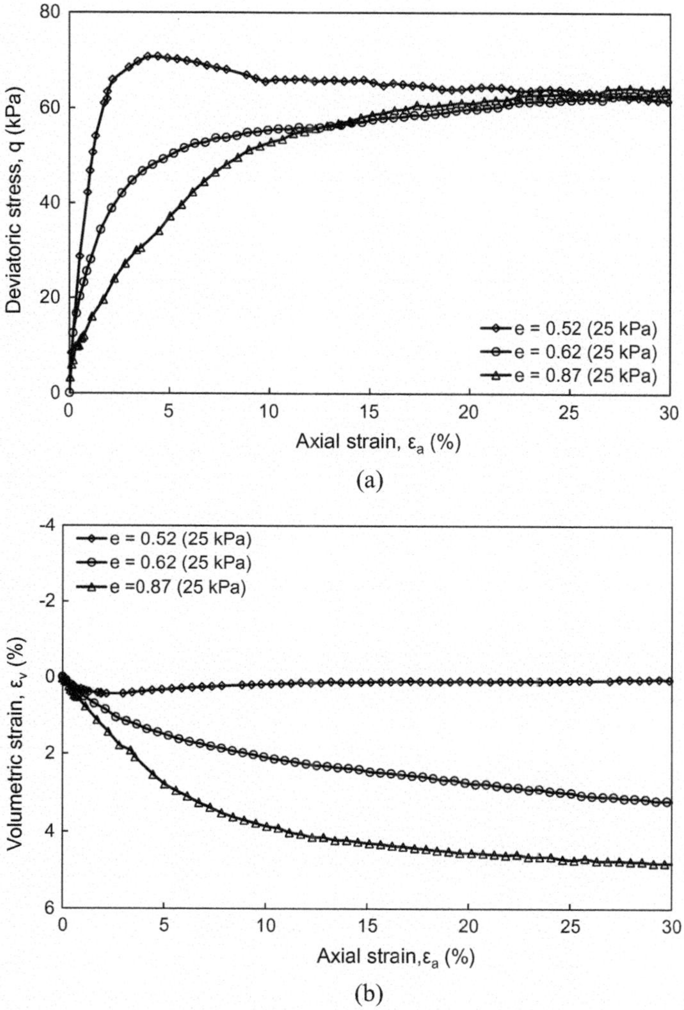

Fig. 10.9 Stress–strain relationships of the $f_c = 68\%$ specimens with different void ratios under 25 kPa: **a** deviatoric stress and axial strain; **b** volumetric strain and axial strain

undrained shearing, the coarse particles rearrange themselves and show highly dilative tendency. Void ratio significantly affects the shearing behavior. Take the $f_c =$ 68% specimen as example (Fig. 10.16), strain softening is observed for the specimens with high void ratio, and positive excess pore-water pressures are recorded during shearing, whereas strain hardening is observed for specimen with low void ratio, and negative excess pore-water pressures are developed.

Figures 10.15a, 10.17a, and 10.19a show the effective stress paths for the $f_c =$ 50%, 68%, and 90% specimens, respectively. The effective stress paths originate at the isotropic consolidation stress points on the horizontal axis. As mentioned earlier,

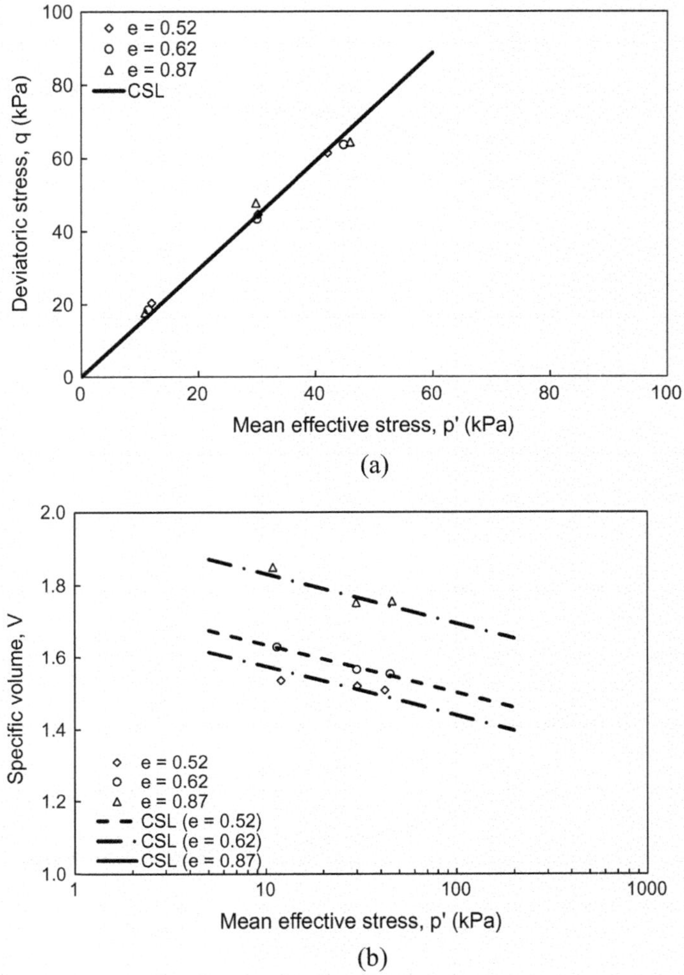

Fig. 10.10 Critical-state lines of the $f_c = 68\%$ specimens in **a** p'–q space and **b** v-log p' space

a loose specimen (i.e., $e = 0.87$) has a highly compressible skeleton. The specimen exhibits strong contractive tendency upon undrained shearing. The deviatoric stress reaches a peak. As the pore water pressure continues to increase after the soil yields, the stress path approaches the CSL with progressive reduction of deviator stress, resulting in further increases in stress ratio and plastic strains. As the void ratio decreases, the compressibility of the skeleton decreases since more stable contacts will develop. The specimen then exhibits stronger dilative tendency. As shown in Figs. 10.15a, 10.17a, and 10.19a, all the specimens seem to approach the CSLs that are obtained from the drained tests, regardless of the initial stress state or the confining pressure.

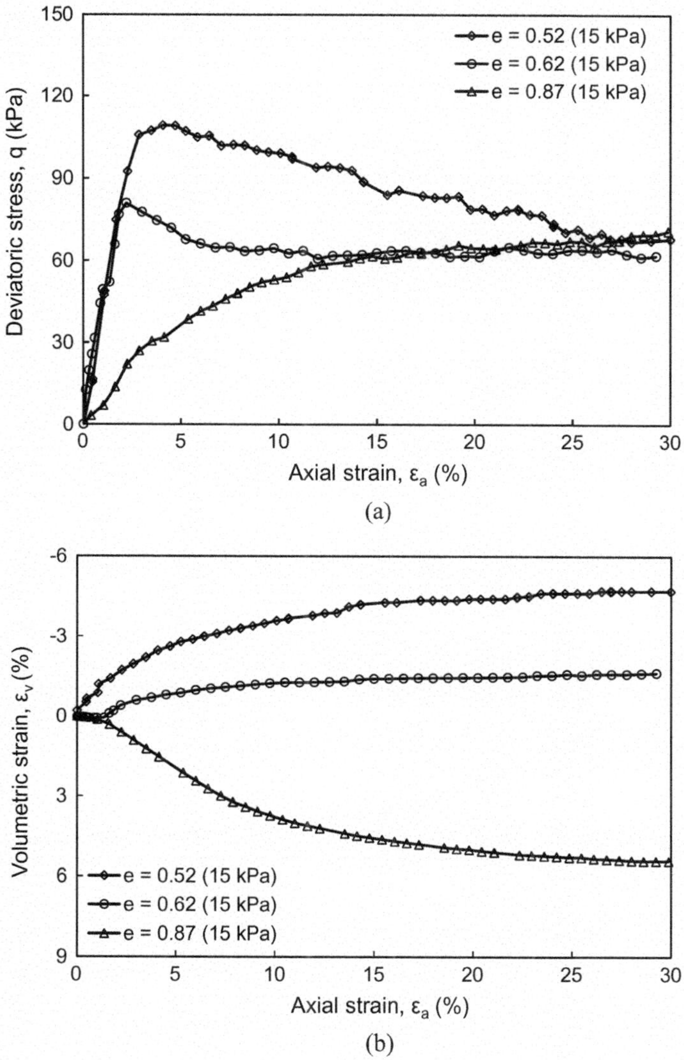

Fig. 10.11 Stress–strain relationships of the $f_c = 90\%$ specimens with different void ratios under 15 kPa: **a** deviatoric stress and axial strain; **b** volumetric strain and axial strain

Results of the undrained shearing tests on the $f_c = 50\%$, 68%, and 90% specimens are also presented in the v-log p' space in Figs. 10.15b, 10.17b, and 10.19b, respectively. The initial and final stress states in the tests are marked to show the undrained stress paths. The CSLs obtained from the drained tests are also drawn in the graphs. Under the undrained condition, the void ratio remains constant and the stress state reaches the critical-state eventually. For all the three types of soils, the CSLs obtained from the undrained tests are shown to be identical to the CSLs

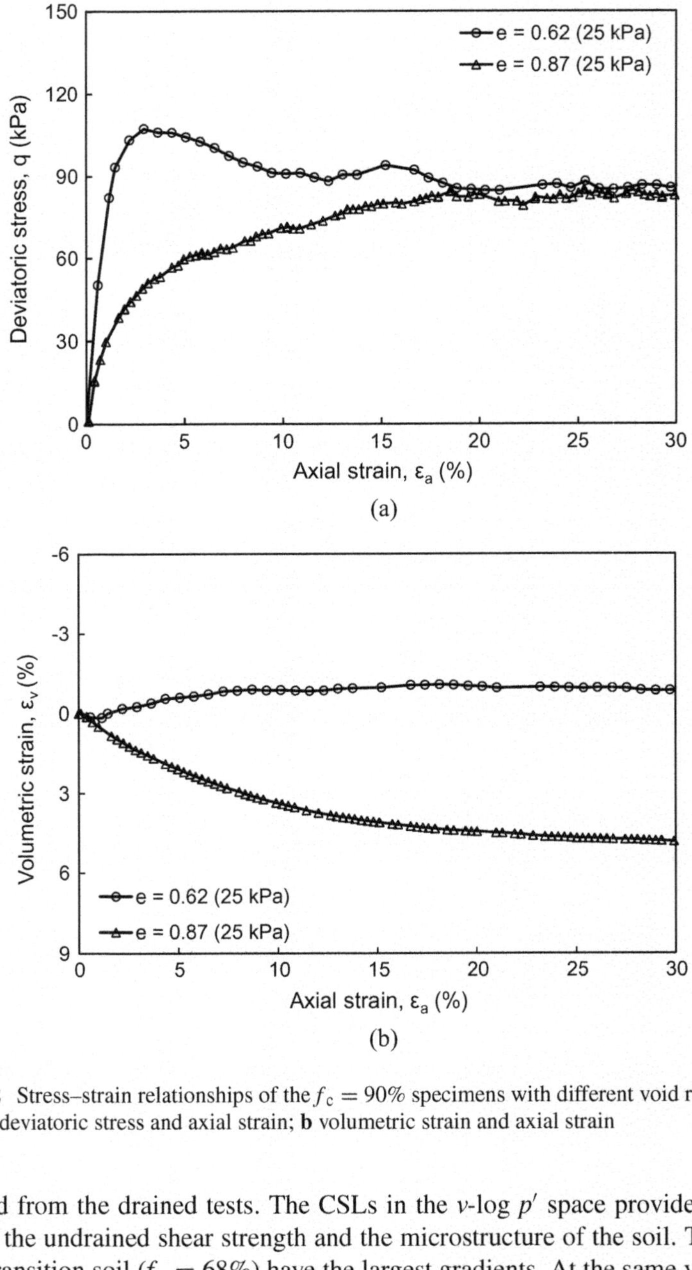

Fig. 10.12 Stress–strain relationships of the $f_c = 90\%$ specimens with different void ratios under 25 kPa: **a** deviatoric stress and axial strain; **b** volumetric strain and axial strain

evaluated from the drained tests. The CSLs in the *v*-log *p*′ space provide a bridge between the undrained shear strength and the microstructure of the soil. The CSLs for the transition soil ($f_c = 68\%$) have the largest gradients. At the same void ratio, the transition soil has the lowest undrained peak strength as it has the least stable microstructure.

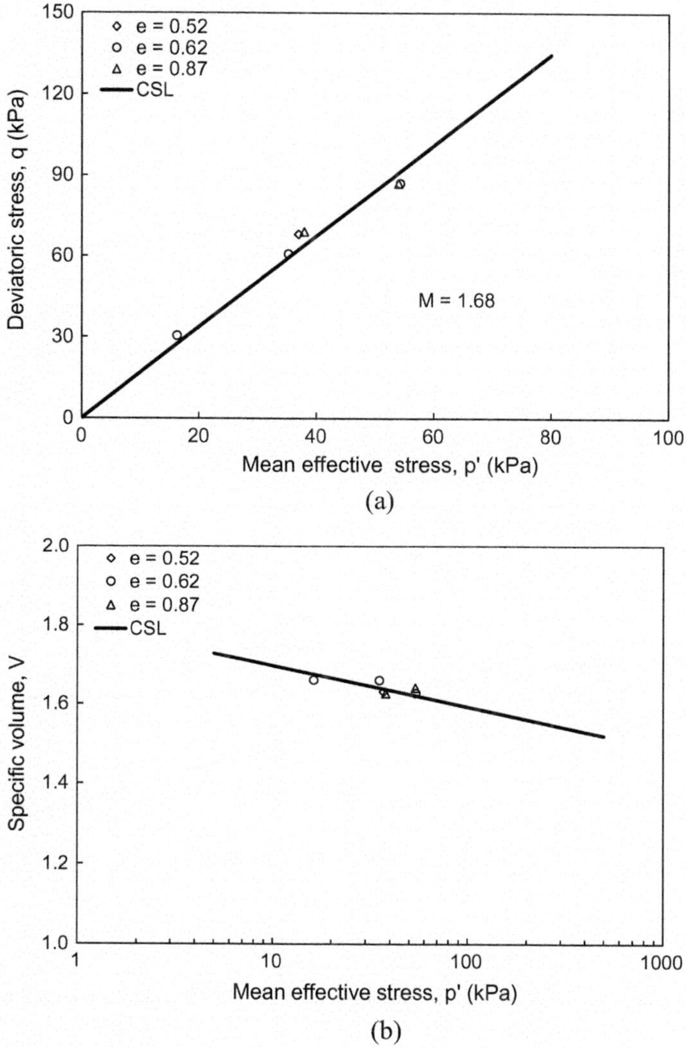

Fig. 10.13 Critical-state lines of the $f_c = 90\%$ specimens in **a** p'–q space and **b** v-log p' space

Table 10.2 Parameters of the critical-state lines in the v-log p' space

$f_c = 50\%$		$f_c = 90\%$		$f_c = 68\%$					
$e = 0.87, 0.62, 0.52$		$e = 0.87, 0.62, 0.52$		$e = 0.87$		$e = 0.62$		$e = 0.52$	
Γ	λ_{cs}	Γ	λ_{cs}	Γ	λ_{cs}	Γ	λ_{cs}	Γ	λ_{cs}
1.77	0.05	1.76	0.045	1.97	0.059	1.77	0.059	1.71	0.059

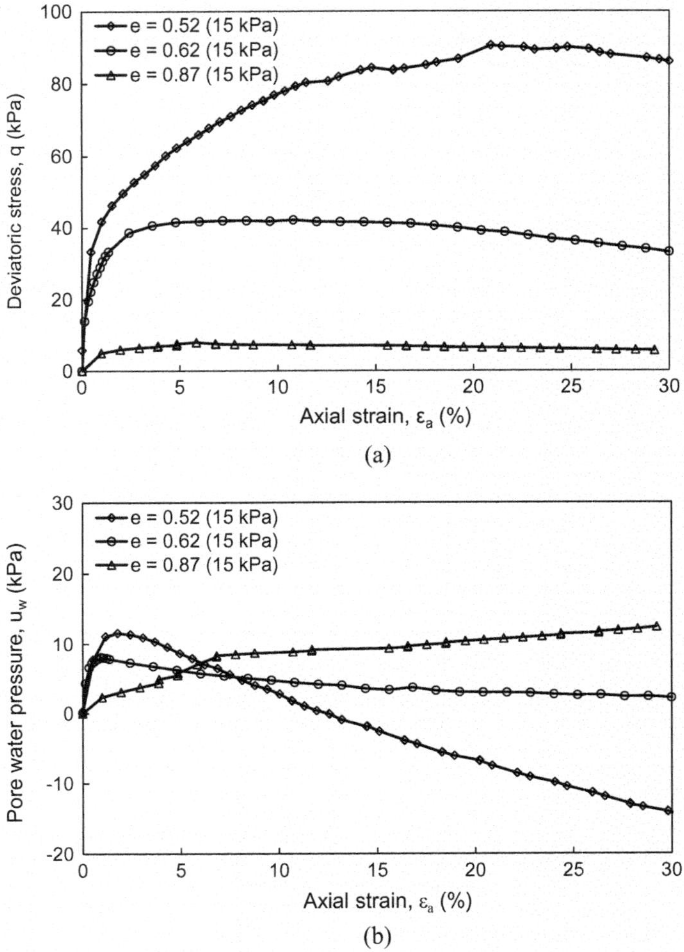

Fig. 10.14 Undrained stress–strain relationships of the $f_c = 50\%$ specimens with different void ratios: **a** deviatoric stress and axial strain; **b** pore water pressure and axial strain

10.2.3.3 Evaluation of Instability of Granular Soils Under Undrained Condition

The undrained peak strength ratio (q_{peak}/p_o') was used for instability analysis of tailing dams (Martin and McRoberts 1998), here p_o' denotes the initial mean effective stress that exists in a soil element, and q_{peak} is the undrained peak shear strength. The experimental data presented in Figs. 10.15, 10.16, 10.17, 10.18, and 10.19 show that the peak strength (onset of instability) under undrained condition depends on the void ratio of the soil. In other words, the onset of instability depends on the state of soil. The state parameter, ψ, is often used to describe the initial state of soil in terms of the difference between the initial state and the critical state in the e-log p' space (Been

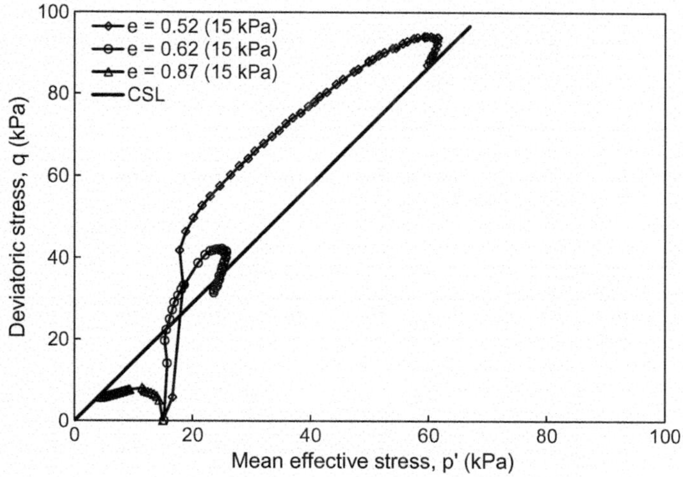

(a)

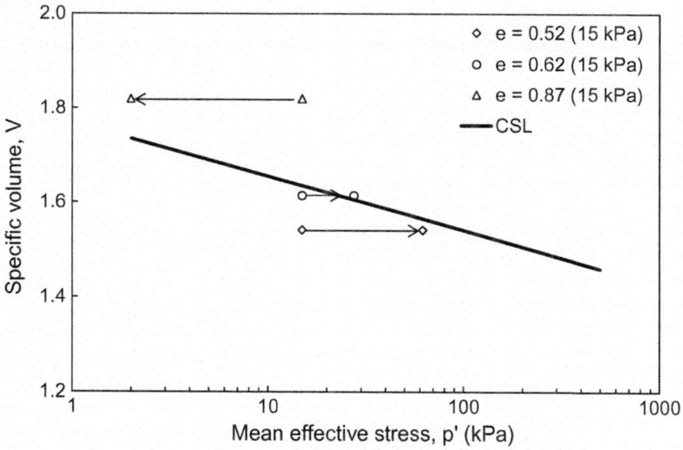

(b)

Fig. 10.15 Stress paths in the undrained tests on the $f_c = 50\%$ specimens with different void ratios in **a** p'–q space and **b** v-log p' space

et al. 1991). A relationship between q_{peak}/p_o' and ψ can be established for assessing the instability of the soil (Fig. 10.20). Under the same ψ, q_{peak}/p_o' increases with the coarse content, f_c. So, q_{peak}/p_o' is further normalized by the product of f_c and the slope of the CSL, M. Figure 10.20b shows the relationship between the normalized peak strength ratio, $q_{peak}/p_o'/(Mf_c)$, and the state parameter, ψ. After normalization, the test data for the three types of soils follow a unique relationship. The normalized relationship in Fig. 10.20b can be used to evaluate the stability of granular soils. For example, parameter M is equal to 1.48 for the soil with $f_c = 68\%$ (Chap. 7).

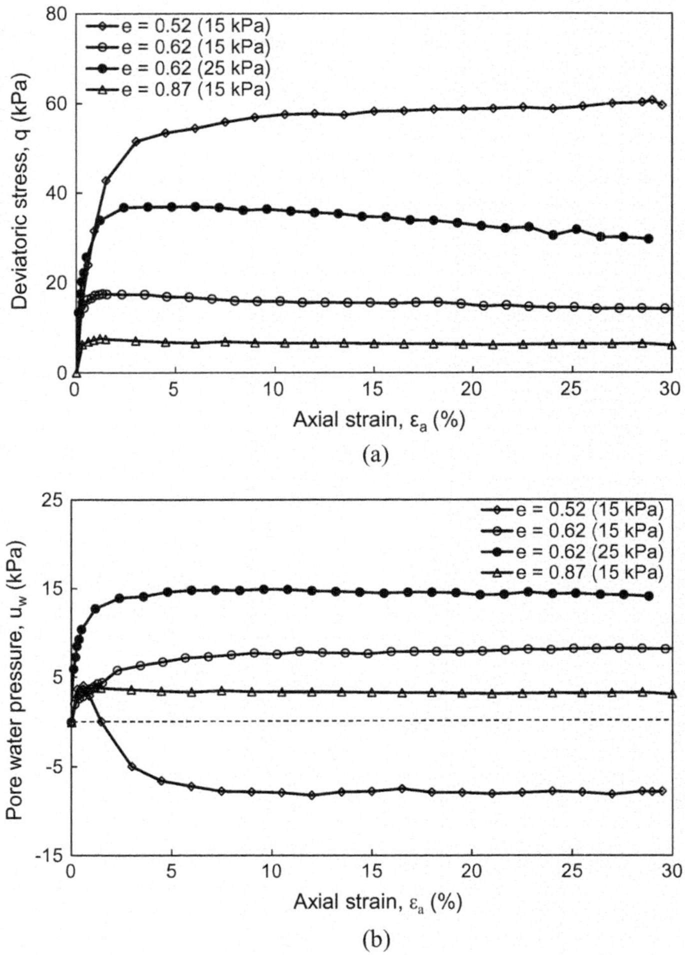

Fig. 10.16 Undrained stress–strain relationships of the $f_c = 68\%$ specimens with different void ratios: **a** deviatoric stress and axial strain; **b** pore water pressure and axial strain

The initial mean effective stress that exists in a soil element in a soil deposit can be estimated based on the depth of the soil element. The mean effective stress in a soil deposit at a depth of 1.5 m is about 15 kPa. For $\psi = -0.05$, $q_{peak}/p_o'/(Mf_c) = 4.65$, the undrained peak strength of the soil element can be estimated as 70 kPa.

The relationship between the slope of the instability line, η_{IL}, and ψ is also proposed for assessing the instability of granular soils. The values of η_{IL} and ψ from the undrained tests on the three types of soils are plotted in Fig. 10.21a. Since η_{IL}/M has more fundamental physical meanings than η_{IL} (Jefferies and Shuttle 2002), a relationship between η_{IL}/M and ψ is plotted in Fig. 10.21b to better evaluate the stability of granular soils. The line separates the soil states into stable state and unstable state in the stress space. As a result, if the stress conditions of a soil element

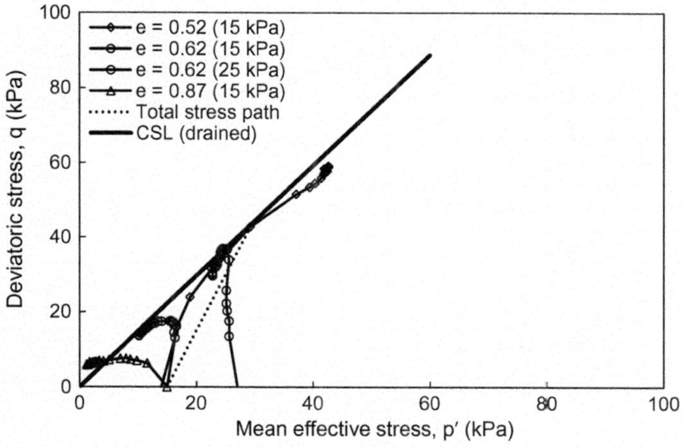

(a)

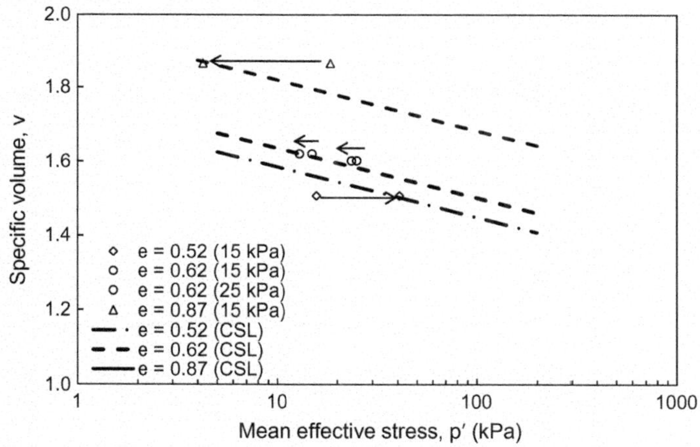

(b)

Fig. 10.17 Stress paths in the undrained tests on the $f_c = 68\%$ specimens with different void ratios in **a** p'–q space and **b** v-log p' space

reach this line, instability is to be triggered and the shear resistance will decrease rapidly to the critical-state strength. η_{IL}/M decreases with increasing state parameter. When $\psi < 0$, the soil will exhibit dilative behavior and the values of η_{IL}/M should be larger than 1.0. When the initial state of soil is at the critical state, no plastic volumetric strain will occur, $\psi = 0$, $\eta_{IL}/M = 1.0$. When $\psi > 0$, the soil will exhibit contractive behavior and η_{IL}/M should be smaller than 1.0. Knowing the state parameter of a soil element, the slope of the instability line can be determined using Fig. 10.21b.

Stark and Mesri (1992) evaluated the undrained critical strength (q_{cs}) in terms of the critical strength ratio (q_{cs}/p_o') for stability analysis that incorporates the variation of q_{cs} with effective vertical stress. The experimental data from the undrained tests

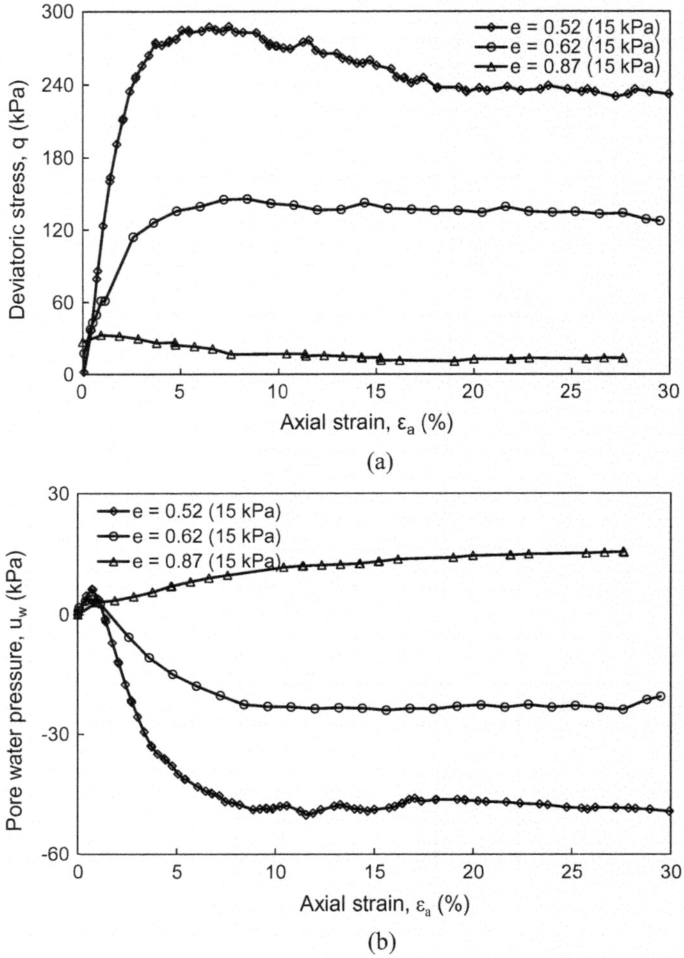

Fig. 10.18 Undrained stress–strain relationships of the $f_c = 90\%$ specimens with different void ratios: **a** deviatoric stress and axial strain; **b** pore water pressure and axial strain

on the three types of soils are presented in Fig. 10.22. The theoretical curves are calculated based on the initial void ratios of the soils and the CSLs from the drained tests. It should be noted that only one theoretical curve is obtained for the $f_c = 68\%$ specimens even though the CSLs in the v-log p' space are void ratio dependent. As shown in Fig. 10.22, the theoretical curve is in good agreement with the experimental data. The relationship between q_{cs}/p_o' and ψ is soil-type specific as the critical shear strength increases with the coarse content. After normalization by Mf_c, the relationship between $q_{cs}/p_o'/(Mf_c)$ and ψ becomes unique as shown in Fig. 10.22b.

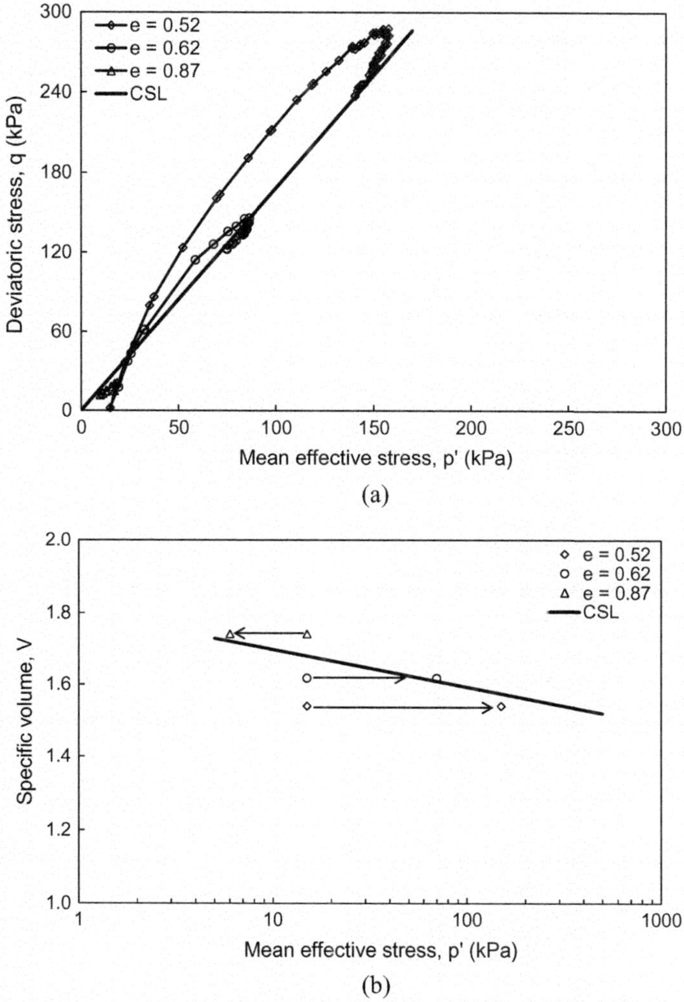

Fig. 10.19 Stress paths in the undrained tests of the $f_c = 90\%$ specimens with different void ratios in **a** p'–q space and **b** v-log p' space

10.2.3.4 Instability of Unsaturated Granular Soils During Wetting

Wetting tests were conducted on the $f_c = 68\%$ specimens to study the instability of the soil under unsaturated condition and simulate the failure mechanism induced by rainfall infiltration. As shown in Fig. 10.23, the loading-collapse curve represents the locus of yield points in the (p, s) space (Alonso et al. 1990). p_0 is the preconsolidation net mean stress under saturated condition. To the left of the yield curve, an elastic zone is postulated in which reversible deformation occurs in response to suction changes. The wetting test starts from A. When the stress path crosses the yield point

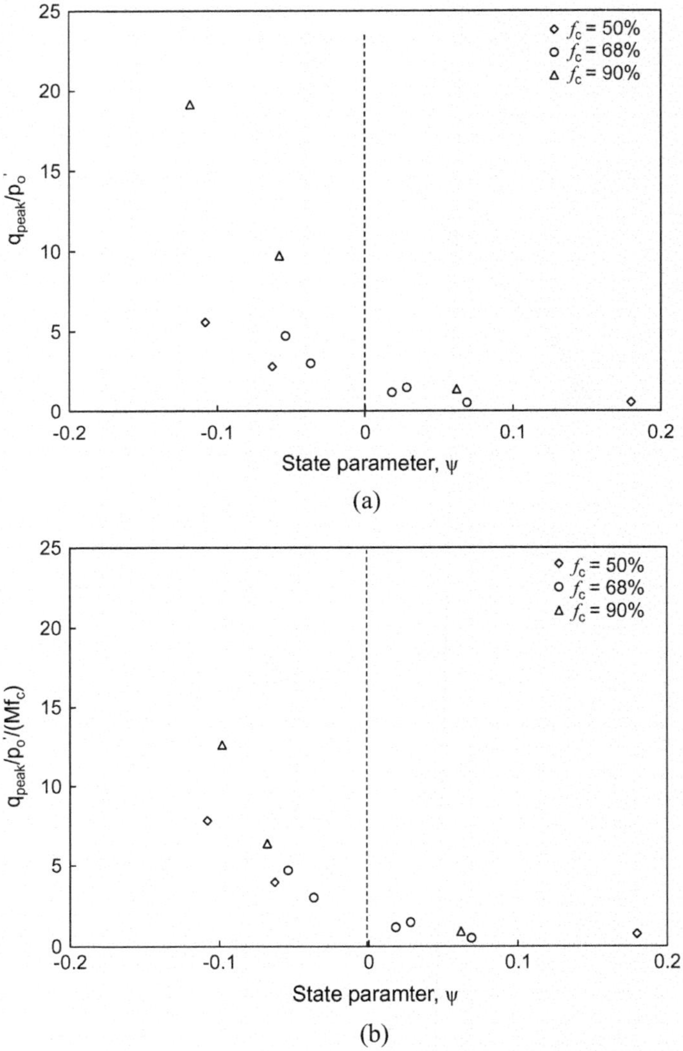

Fig. 10.20 Relationships for assessing the instability of granular soils using the undrained peak shear strength: **a** between q_{peak}/p_0' and ψ; **b** between $q/p_0'/(Mf_c)$ and ψ

(B), irreversible volumetric strain occurs and large collapse strains take place. The onset of instability in the wetting test is subsequently revealed by acceleration of both the axial and volumetric strains at yield point B.

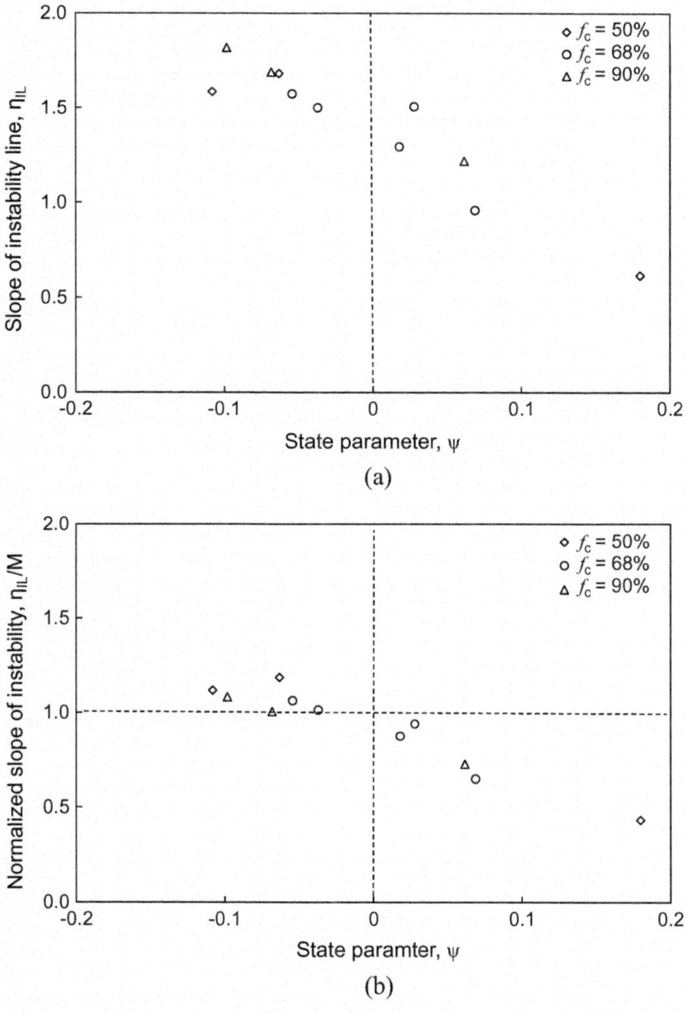

Fig. 10.21 Relationships for assessing the instability of granular soils using the slope of the instability line: **a** between η_{IL} and ψ; **b** between η_{IL}/M and ψ

10.2.3.5 Wetting Behavior of Unsaturated Granular Soils

Four unsaturated specimens were isotropically consolidated at a net confining pressure of 15 kPa and a suction of 280 kPa. These specimens were then sheared at a constant net confining pressure to four designated stress ratios, η (e.g., 0, 1.0, 1.5, and 1.85). Afterward, the specimens were wetted by decreasing the soil suction while keeping the deviator stress and net mean stress constant. Figure 10.24 shows the changes in volumetric strain, ε_v, axial strain, ε_a, and degree of saturation, ΔS_r, with suction during the wetting tests under different stress ratios. As shown in Fig. 10.24a,

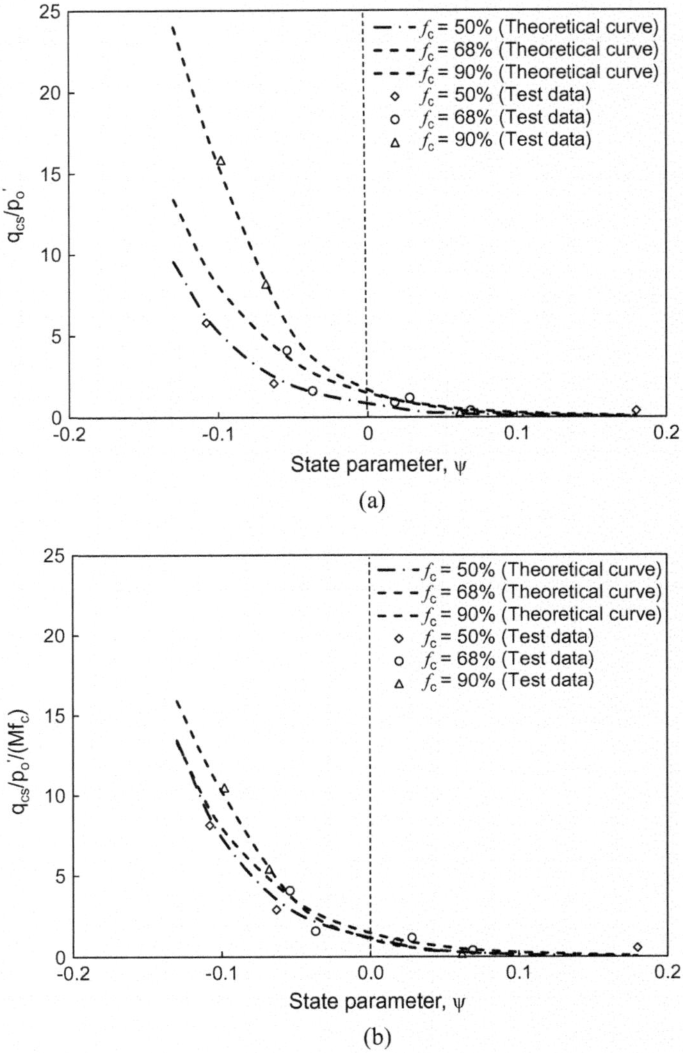

Fig. 10.22 Relationships for assessing the instability of granular soils using the undrained critical shear strength: **a** between q_{cs}/p_0' and ψ; **b** between $q_{cs}/p_0'/(Mf_c)$ and ψ

at the beginning, limited volumetric strain is mobilized with decreasing suction. The rates of increase in the volumetric strain and axial strain accelerate toward the end of test. ΔS_r is used because the initial degrees of saturation in these tests are slightly different. As expected, an increase in the degree of saturation is observed as the suction decreases [Fig. 10.24c]. The axial strain, volumetric strain, and ΔS_r respond consistently to suction changes. The tests were terminated at $\varepsilon_a = 25\%$.

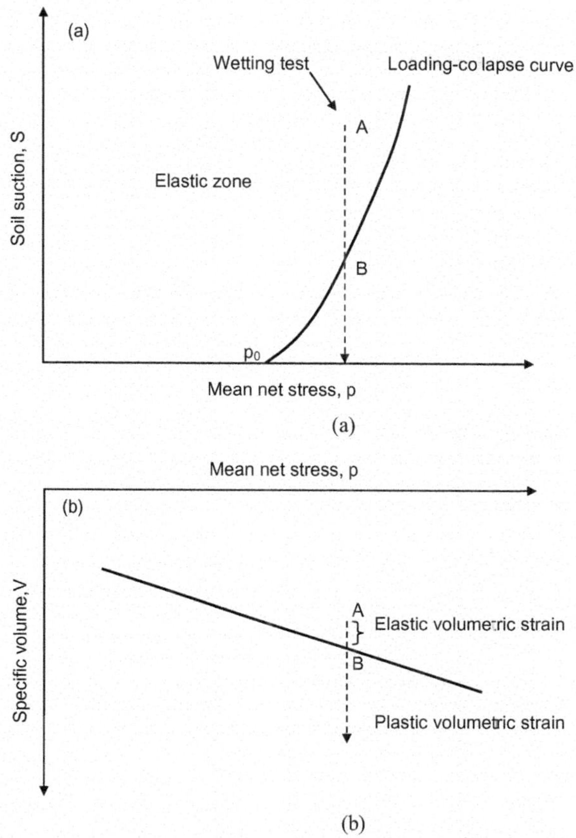

Fig. 10.23 Schematic volumetric instability behavior of unsaturated granular soil in **a** s-log p space and **b** v-log p space (Alonso et al. 1990)

The specimen wetted under the isotropic condition ($\eta = 0$) did not yield even though the suction was reduced to zero. Limited volumetric strain and axial strain were developed. The rest specimens failed with bulging at large axial strains. In Fig. 10.24, for the specimens that fail, significant volumetric strain and axial strain commence relatively suddenly at a certain suction value. This point is identified as the yield point that marks the onset of instability of the soil. The onset of instability is affected by the stress ratio. An unsaturated specimen under a higher stress ratio approaches instability at a higher suction. In other words, steep slopes are more susceptible to failure than gentle slopes when subject to rainfall infiltration. The values of soil suction and degree of saturation when instability occurred during the tests are listed in Table 10.3. For example, for the specimen under $\eta = 1.85$, instability occurred at a suction of about 100 kPa, while the degree of saturation was $S_r = 40\%$.

Figure 10.25a shows the variations of volumetric strain with suction in the wetting tests under different net confining pressures (i.e., 5, 15, and 25 kPa) but the same

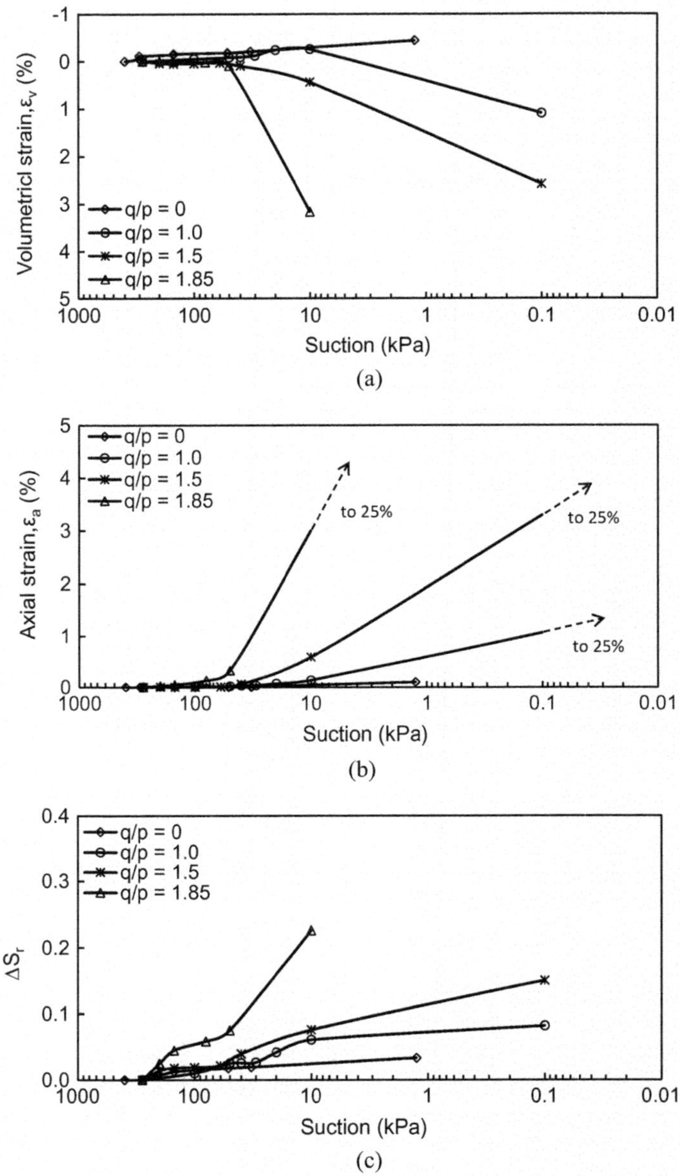

Fig. 10.24 Relationships between **a** volumetric strain and suction, **b** axial strain and suction, and **c** incremental degree of saturation and suction during wetting tests under a constant net confining pressure of 15 kPa

Table 10.3 Parameters for assessing the soil instability along the wetting path

Soil type	Net mean stress before wetting (kPa)	Stress ratio (η)	Suction (kPa)	Degree of saturation (S_r)	State parameter (ψ)	η_{IL}/M
f_c = 68%(this study)	15	0	–	–	–	No failure
f_c = 68%(this study)	26.8	1.0	20	0.64	– 0.013	0.97
f_c = 68%(this study)	30	1.5	35	0.46	– 0.032	1.01
f_c = 68%(this study)	39.5	1.87	100	0.40	0.048	0.76
f_c = 68%(this study)	9	1.5	–	–	–	No failure
f_c = 68%(this study)	45	1.5	45	0.62	0.035	0.85
CDG (Ng and Chiu 2003)	25	1.4	20	0.46	– 0.14	1.18
CDG (Ng and Chiu 2003)	50	1.4	16	0.49	– 0.057	1.14
CDG (Ng and Chiu 2003)	100	1.4	10	0.57	– 0.046	1.08
CDG (Ng and Chiu 2003)	150	1.4	11	0.66	– 0.067	1.03

stress ratio of 1.5. The specimen wetted under a net confining pressure of 5 kPa does not yield. Only limited swelling volumetric strain and axial strain are observed. For those specimens that fail in the wetting tests, relatively small volumetric strains develop before the suction reaches the yield point. After the yield point, the rates of volumetric strain and axial strain increase rapidly with decreasing suction. This is shown by sudden changes in the slopes of the curves in Fig. 10.25. The unsaturated specimens under higher applied net mean stresses approach failure at higher suctions. The volumetric contraction of the specimen increases as the net confining pressure increases.

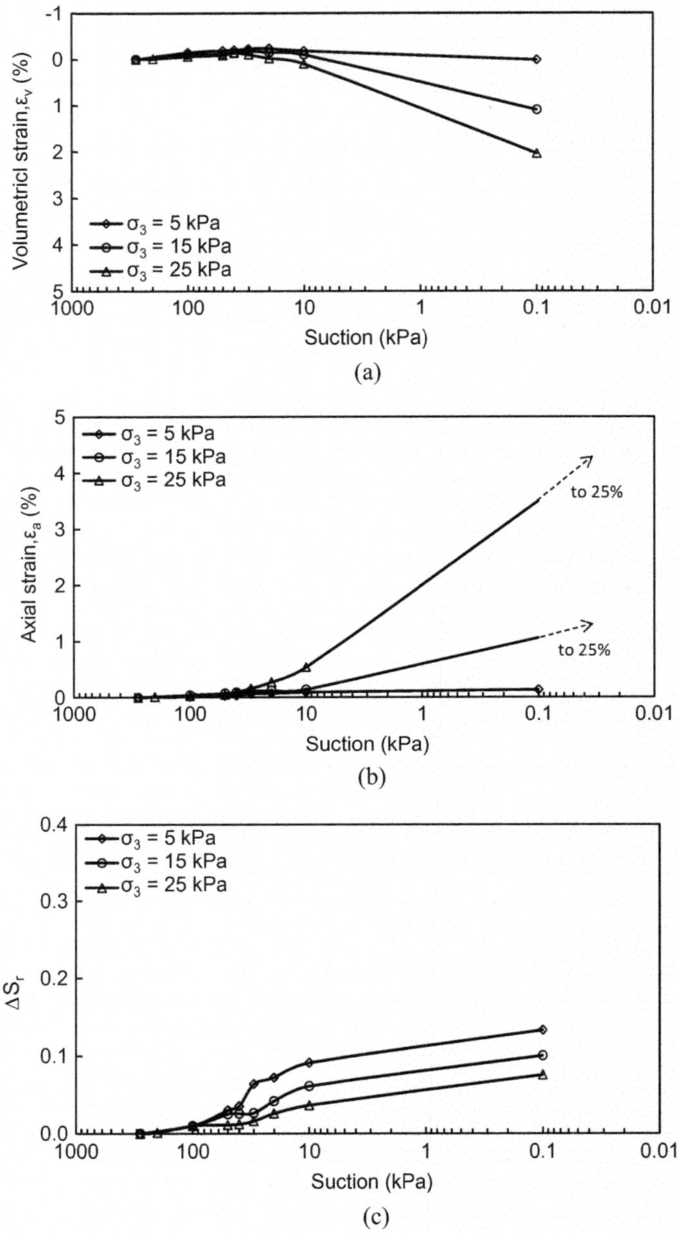

Fig. 10.25 Relationships between **a** volumetric strain and suction, **b** axial strain and suction, and **c** incremental degree of saturation and suction during wetting tests under a constant stress ratio of 1.5

10.2.3.6 Instability Analysis of Unsaturated Granular Soils

The concept of bulk water and meniscus water proposed by Karube and Kato (1994) can be used to explain the influence of suction on the wetting behavior of an unsaturated soil. Bulk water is the pore water that occupies the pore volume between soil particles, while meniscus water is the soil water existing at the contact point between soil particles. Matric suction has an influence on the meniscus water and bulk water. The effects of the meniscus water and bulk water on the soil skeleton are different. The suction within the meniscus water increases the inter-granular adhesive force that acts between the soil particles and, hence, increases the stiffness of soil skeleton. The suction within the bulk water influences both the normal and tangential forces at the particle contacts.

As the degree of saturation becomes higher and the suction becomes lower during wetting, the meniscus water at the contact points expands gradually with water absorption. The expanded meniscus water will combine and change to bulk water. The ratio of the bulk water increases, and the proportion of the meniscus water decreases.

Karube et al. (1996) described the influence of the meniscus water and the bulk water using the meniscus stress, p_{m}, and the bulk stress, p_{b}, respectively. The sum of the two stress components is equal to the suction stress, p_{s}. Meniscus stress, bulk stress, and suction stress are expressed as follows:

$$p_{\mathrm{m}} = \frac{S_{\mathrm{rm}}}{100 - S_{\mathrm{r0}}} s \tag{10.2}$$

$$p_{\mathrm{b}} = \frac{S_{\mathrm{rb}}}{100 - S_{\mathrm{r0}}} s \tag{10.3}$$

$$p_{\mathrm{s}} = p_{\mathrm{m}} + p_{\mathrm{b}} = \frac{S_{\mathrm{r}} - S_{\mathrm{r0}}}{100 - S_{\mathrm{r0}}} s \tag{10.4}$$

where s is the matric suction; S_{rm} is a degree of saturation ratio related to the meniscus water; S_{rb} is a degree of saturation ratio related to the bulk water; S_{r0} is the residual degree of saturation; S_{r0} is 0 when the soil is completely dry. So, the suction stress can be expressed as:

$$p_{\mathrm{s}} = S_{\mathrm{r}} s \tag{10.5}$$

Similar to the effective stress concept proposed by Bishop (1961) for unsaturated soils, equivalent mean effective stress p' is introduced to describe the instability behavior in a wetting test:

$$p' = p + p_{\mathrm{s}} = p + S_{\mathrm{r}} s \tag{10.6}$$

where $p = (\sigma_1 + 2\sigma_3)/3 - u_a$ is the net mean stress. The values of p, S_r, and s at the yield point in each test are listed in Table 9.3.

Figure 10.26 plots the q and p' values at the instability points in the wetting tests. According to Alonso's model (Alonso et al. 1990) in Fig. 10.23, p_0 is the yield point under saturated condition and is the minimum value of the yield stress along the loading-collapse curve. If a sample is subject to a net mean stress smaller than p_0, a change in suction will not result in plastic deformation. The line that connects point (p_0, 0) and the instability points is defined as "instability line." The instability line is not unique but varies with the stress ratio and the net confining pressure. Ng and Chiu (2003) also conducted wetting tests on completely decomposed granitic soil (CDG), which was the same type of soil as that used in the present study except the particle-size distribution. The maximum dry density and optimum water content of CDG are 1670 kg/m^3 and 20%, respectively. The test results are used to further verify the method proposed in this study. The specimens were wetted under a stress ratio of $\eta = 1.4$ and four different net mean stresses (i.e., $p = 25$, 50, 100, and 150 kPa). The corresponding values of p, S_r, and s at the yield points are also listed in Table 10.3. The values of p_0 are 15 kPa and 20 kPa for the $f_c = 68\%$ specimen and the CDG, respectively. The values of M are 1.48 and 1.549 for the $f_c = 68\%$ specimen and the CDG, respectively.

The instability conditions for the specimens subject to wetting are expressed as a relationship between the normalized slope of the instability line (η_{IL}/M) and the state parameter (ψ) in Fig. 10.27. The instability curve shown in Fig. 10.21b from the undrained tests is also plotted in Fig. 10.27. The values of η_{IL}/M obtained from the wetting tests on the $f_c = 68\%$ specimens overlap with those from the undrained tests presented previously. Hence, the relationship between η_{IL}/M and ψ can be used to

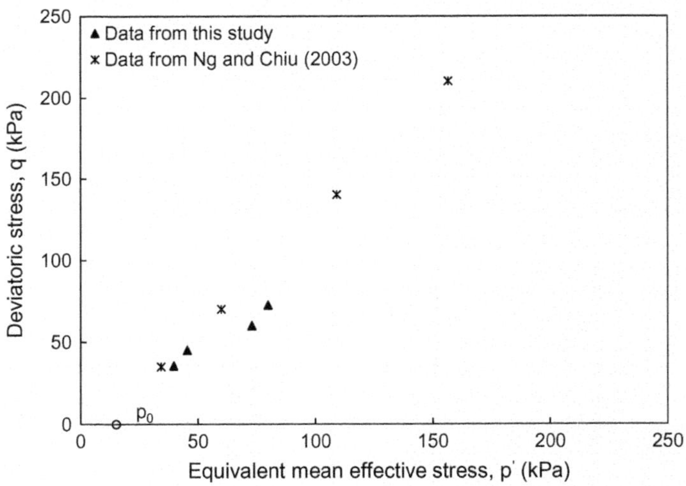

Fig. 10.26 Relationship between deviatoric stress at failure, q, and the equivalent mean effective stress, p'

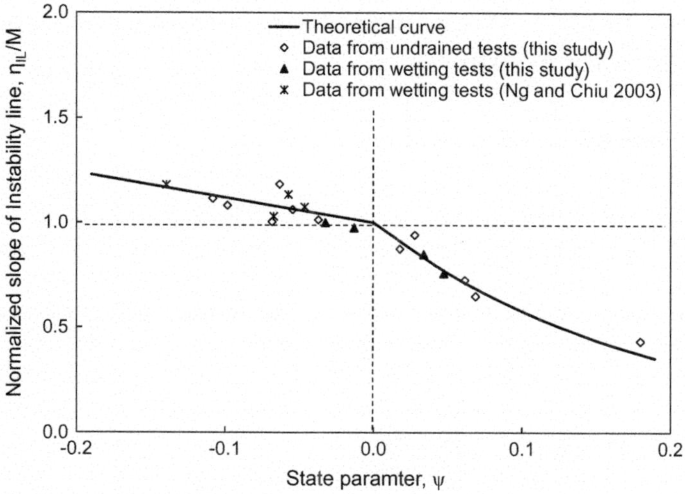

Fig. 10.27 Relationship between the η_{IL}/M and ψ for saturated and unsaturated granular soils

assess the instability of both saturated and unsaturated granular soils. Although the particle-size distribution of the CDG is different from that of the $f_c = 68\%$ specimen, the data for both soils fall on the same curve.

The relationship between η_{IL}/M and ψ has been established for saturated sand (Li and Dafalias 2000; Jefferies and Shuttle 2002; Wood et al. 2009). Wood et al. (2009) introduced a breakage criterion as a second yield mechanism for sand when considering particle crushing. The relationship between η_{IL}/M and ψ proposed by Li and Dafalias (2000) and Wood et al. (2009) is adopted to describe the instability behavior along p_1 and p_3 (Fig. 10.4), respectively:

$$\eta_{IL}/M = e^{m\psi} \quad \text{when } \psi \geq 0 \tag{10.7}$$

$$\eta_{IL}/M = 1 - k\psi \quad \text{when } \psi \leq 0 \tag{10.8}$$

where m and k are model parameters, which are $m = 5.5$ and $k = 1.2$ in this study. The theoretical curve agrees well with the experimental data (Fig. 10.27). In fact, a maximum positive ψ_{max} and a maximum negative ψ_{min} exist for each type of soil. ψ_{max} is the distance between the NCL and the CSL, $\psi_{max} = N - \Gamma$, whereas ψ_{min} is defined when the soil is at its densest state. The bounds of η_{IL}/M for a particular soil element can, hence, be obtained according to ψ_{max} and ψ_{min}. For example, for the $f_c = 68\%$ specimen with $e = 0.62$, $\psi_{max} = 0.05$, and $\psi_{min} = -0.17$, the bounds of η_{IL}/M are 0.76 and 1.204, respectively.

10.3 Case Study 1: Hydraulic Behavior of Colluvial Slopes upon a Rainstorm

10.3.1 Simulation Methodology

Evans and King (1998) suggested that natural terrain deposits with a slope angle between 25° and 50° have a high degree of susceptibility to shallow failure. The slope simulated in this study is 16 m high at an inclination of 38.6° (inclination 1H: 0.8 V). The slope geometry is shown in Fig. 10.28a. In order to simplify the problem, it is assumed that the slope is composed of a homogenous and isotropic colluvial soil. The left and right edges are located at a distance of 30 m from the slope crest and the toe, respectively, in order to avoid any influence of the boundary conditions on the seepage process within the slope area.

A commercially available software, Seep/W (Geo-slope International Ltd. 1991a), is used to simulate the infiltration process. Seep/W is a finite element software that can be used to model the water movement and pore-water pressure distribution within

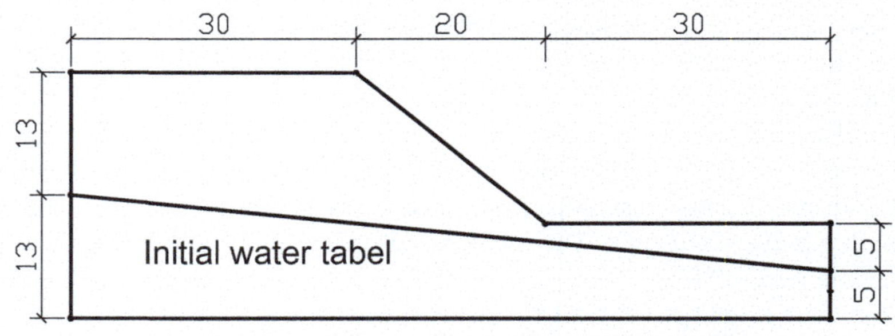

(a) Slope geometry and initial water table (unit: meter)

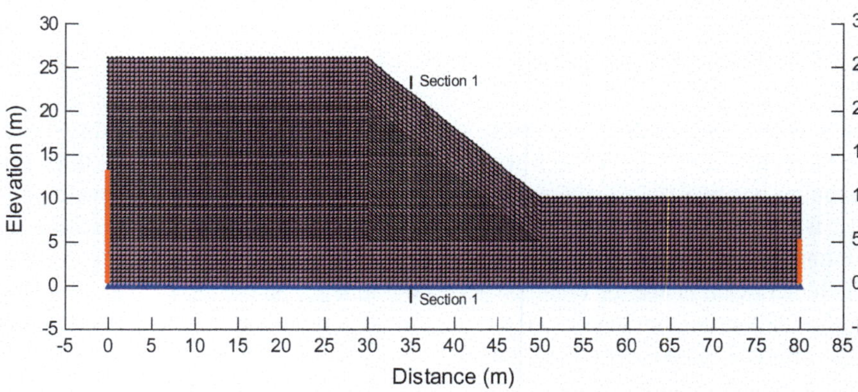

(b) Finite element mesh (boundary conditions at initial steady state)

Fig. 10.28 Slope profile for numerical simulation

porous materials such as soil and rock. Its comprehensive formulation makes it possible to analyze complex seepage problems. Seep/W has been applied to the analysis and design for geotechnical, civil, hydrogeological, and mining engineering projects.

A finite element mesh is designed to analyze the infiltration of water in the slope under rainstorms as shown in Fig. 10.28b. The infiltration processes within the slope are modeled in two dimensions. In the simulation, the required SWCC and SHCF are inputted manually. The hysteresis of the SWCC is not considered in the simulation.

For better comparison of the hydraulic behavior of the slopes composed of the five soils investigated in this thesis, all the simulations start from a hydrostatic state corresponding to the same initial water table [as shown in Fig. 10.28a]. The initial pore water pressure distributions are simulated using the steady-state method. It is assumed that the base soil has very low hydraulic conductivity, and therefore, a zero nodal flow boundary (Q) is applied along the base. Along the left and right edges, constant water head boundaries are applied in order to define the initial groundwater level and to produce the initial pore water pressure profile. The water heads are 13 m at the nodes on the left edge and 5 m at the nodes on the right edge. Hence, a hydrostatic state can be obtained for these slopes.

A 24-h rainfall event is applied at time 0. In addition to the boundary conditions defined for solving the initial hydrostatic state, the precipitation rate is modeled as a unit flux boundary (q) along the edges at the ground surface. After 24 h, the rainfall is stopped. No evaporation is taken into consideration during the no-rain period afterward.

10.3.1.1 Soil Parameters

The slopes composed of five different colluvial soils (referring to Chap. 2, Fig. 2.4) are simulated. The hydraulic properties (SWCCs and SHCFs) of these colluvial soils were measured in the laboratory (referring to Chap. 5). As shown in Fig. 10.29, the fitted SWCCs using the bimodal SWCC model (Chapter 8, Eq. 8.11) based on the measured SWCCs data are used in the simulation. Figure 10.29 shows the SHCFs employed, including three fitted curves using the Gardner (1958) model (Chap. 5, Figs. 5.27–5.33) for three fines-controlled soils (CL with sand, sandy ML, and CL with sand) based on measured data and two fitted curves using the bimodal SHCF model (Chap. 8, Eq. 8.25, Figs. 8.23 and 8.24) for the coarse-controlled soils (GW-GM with sand and SM with gravel). The hydraulic property functions are inputted manually, and the exact values from the curves in Figs. 10.29 and 10.30 are taken.

Granular material-fines mixtures are studied here, including GW-GM with sand, SM with gravel, SC with gravel, sandy ML, and CL with sand (they can be looked as mixtures containing 100, 75, 50, 25, and 0% of GW-GM with sand by weight, referring to Chap. 3.3). When the percentage by weight of the granular material in the granular material-clay mixture is greater than 72%, the shear strength of the mixture is similar to the shear strength of the granular material. When the proportion by weight of the granular material in the mixtures is less than 50%, the shear strength

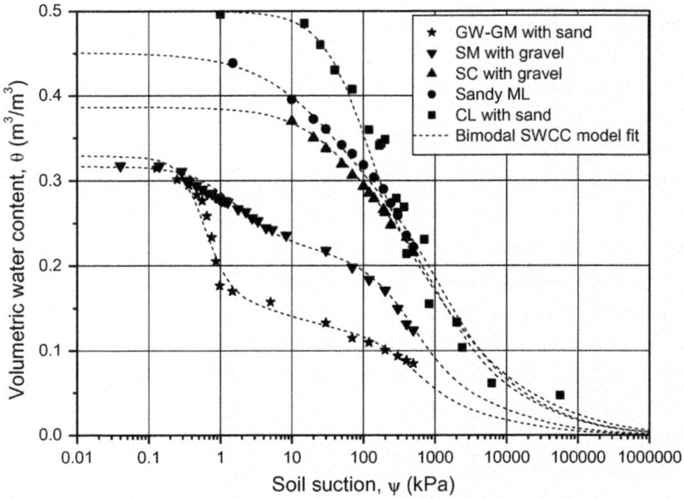

Fig. 10.29 SWCCs for five soils used in seepage analysis

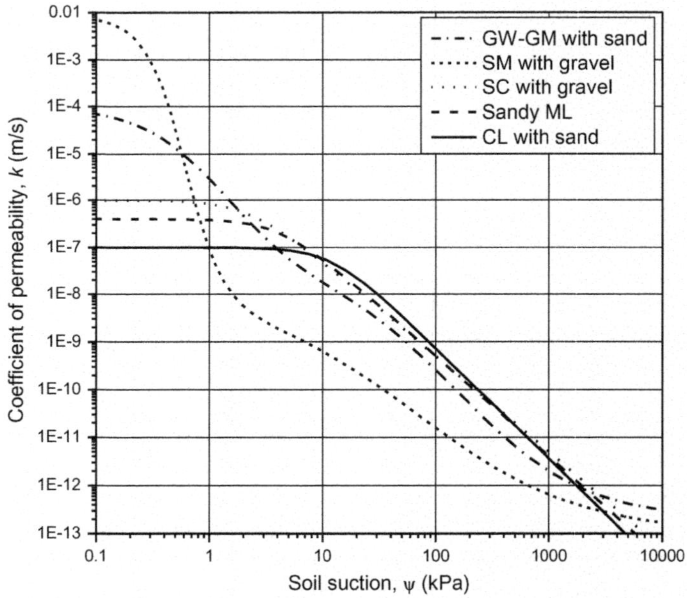

Fig. 10.30 SHCFs for five soils used in seepage analysis

of the mixtures is similar to that of the fines. For the case in which the percentage of the granular materials in the mixture is between 50 and 72%, the shear strength of the mixtures is influenced by both the granular materials and the fines.

Irfan and Tang (1993) studied the effect of coarse fractions on shear strength of colluvial soils and proposed a model to characterize the shear strength of the granite aggregate—fine-grained colluvium mixtures, as

$$\begin{aligned} \tau &= c_0' + \sigma' \tan \phi_0' && \text{for } P_g < P_c \\ \tau &= c_0' + \sigma' \tan(\Delta\phi + \phi_0') && \text{for } P_u > P_g > P_c \end{aligned} \quad (10.9)$$

where P_g is the proportion of the granular materials, P_c is the critical proportion of the granular materials at which the shear strength of granular material-fine mixtures starts to increase, P_u is the upper critical proportion of granular material, at which the shear strength is similar to the granular material, $\Delta\phi$ is the increase in friction angle resulting from an increase in granular material proportion and is equal to

$$\Delta\phi = (P_g - P_c)\alpha \quad (10.10)$$

where α is a material constant which depends on the shape, size, and density of the granular material, etc.

Same as the silty sand and granite aggregate used by Irfan and Tang (1993), the colluvial soils were drawn from natural completely decomposed granites in Hong Kong. So, the colluvial soils are considered to have shear strength behavior similar to the granite aggregate—silty sand mixtures used by Irfan and Tang (1993). As recommended by Irfan and Tang (1993), the parameters, P_c and α, are chosen as 35% and 3.5° per 10% granular material proportion, respectively. Because Irfan and Tang (1993) did not conduct experiments on soils with coarse fractions above 52% and did not find the exact value of P_u, P_u is taken as 72% as recommended by Vallejo and Mawby (2000). The fines (CL with sand) used in this study are finer than that used by Irfan and Tang (1993), i.e., a silty sand. Therefore, the CL with sand is considered to have a smaller friction angle comparing to the silty sand used by Irfan and Tang (1993). The effective friction angle of the CL with sand is chosen as $\phi_0' = 25°$, which is the same as that of the clayey silty sand used by Holtz and Gibbs (1956), but it is smaller than that of the silty sand used by Irfan and Tang (1993), i.e., 37.1°. The effective cohesion of the CL with sand is chosen as $c_0' = 5$ kPa, a recommended value for colluvial soils (Irfan and Tang 1993).

The estimated shear strength parameters of the five soils are listed in Table 10.4. The angles indicating the increase rate in shear strength related to soil suction are taken as two-thirds of the effective angles of internal friction.

Table 10.4 Soil parameters used for stability analysis

Group symbols (ASTM D2487)	Gravel proportion (%)	Dry density (g/cm^3)	Effective cohesion c' (kPa)	Effective angle of internal friction ϕ' (°)	Angle related to matric suction ϕ^b (°)
GW-GM with sand	58	1.78	5	33.1	22
SM with gravel	44	1.81	5	28.1	18.7
SC with gravel	29	1.62	5	25.0	16.7
Sandy ML	15	1.45	5	25.0	16.7
CL with sand	0	1.32	5	25.0	16.7

10.3.1.2 Rainfall Event

Maximum rolling 24-h rainfall (MR24HR) has been adopted by the GEO in establishing relations with the failure frequency of man-made slopes (Yu 2004). The current slope warning criteria are based on the use of MR24HR (Pun et al. 1999; Chan et al. 2003). A rigorous assessment of the relationship between rainfall and slope failure in the period from 1985 to 2000 was carried out by Evans (1997) and Ko (2003). Their assessment shows that the failure frequency of natural slopes increases exponentially with the normalized MR24HR, which is defined as the ratio of the MR24HR and the location-specific mean annual rainfall.

The maximum value of the MR24HRs from 1985 to 1994 in Hong Kong is 800 mm. Evans and Yu (2001) calculated the MR24HR based on the data from 100 years of observations in Hong Kong and found that the MR24HR of 10 yr. return period is 402 mm; the MR24HR of 50 yr. return period is 548 mm; the MR24HR of 100 yr. return period is 609 mm. The Geotechnical Manual for Slopes (GCO 1984) recommends that slope design and stability assessments can be carried out for the groundwater conditions that would result from a rainfall event with a ten-year return period.

In this study, the MR24HR of 10 yr. return period in Hong Kong is taken for numerical simulation. The rainfall event is uniformly distributed as 16.75 mm/h (i.e., 4.7×10^{-6} m/s) and lasts for 24 h.

10.3.2 Seepage Analysis Results

The infiltration processes are simulated using Seep/W (Geo-slope International Ltd). For better comparison of the hydraulic behavior of the five soils, all the simulations start from a steady state with the same initial water table [as shown in Fig. 10.28a]. The 24-h rainfall event is applied at time 0. After 24 h, the rainfall is stopped. In order to allow the slope to drain and to study the recovery of negative pore water pressures, the seepage simulation lasts for 111 days. Because of the same initial water

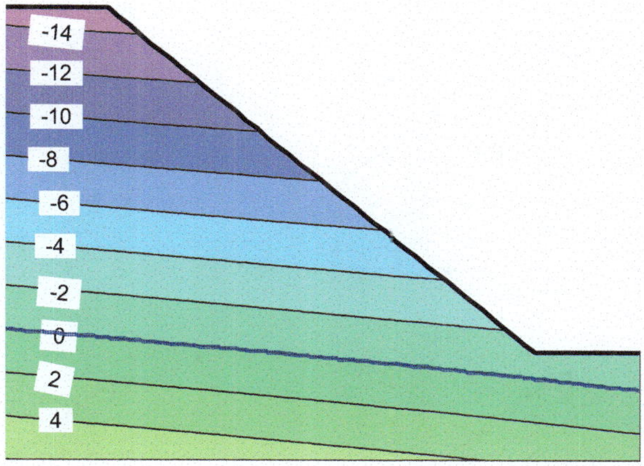

Fig. 10.31 Initial pore water pressure head (in meter) contours (at time = 0 and at a steady state)

table and initial steady state, the initial pore water pressure contours are the same for these five soils and are shown in Fig. 10.31. However, the water content contours and hydraulic conductivity contours are different for different slopes. The pore water pressure profiles change with time during the rainfall and after the rainfall. The pore water pressure contours of the five slopes during and after the rainfall are shown in Figs. 10.32, 10.33, 10.34, 10.35, 10.36, 10.37, and 10.38.

10.3.2.1 Pore Pressure Changes in Slopes Composed of Soils with High Coarse Fractions

The pore water pressure contours in the slope composed of GW-GM with sand change very quickly at the beginning of rainfall (as shown in Fig. 10.32). A thin layer below the ground surface becomes quasi-saturated (pressure head smaller than 2 m) in 0.02 days due to the rainfall infiltration [see Fig. 10.32b]. Then, the quasi-saturated zone expands rapidly and reaches a relative stable state at 0.03 days. In the remaining raining period, the quasi-saturated zone expands slowly. As shown in Fig. 10.33, the difference of the hydraulic conductivity values in the quasi-saturated zone and the saturated zone is distinct. The hydraulic conductivity in the quasi-saturated zone is similar to the rainfall intensity and ensures that the rainfall can flow through the quasi-saturated zone without increasing of the soil water content in that zone. Therefore, nearly all the water flows through the quasi-saturated zone. Because of the fairly low unsaturated hydraulic conductivity in the unsaturated zone (smaller than 10^{-9} m/s), little water flows into the unsaturated zone. After the rainfall, the quasi-saturated zone drains slowly (as shown in Fig. 10.34), while the low unsaturated hydraulic conductivity controls the drainage of the quasi-saturated zone. The pore-water pressure profiles at different times along Sect. 1 (referring to Fig. 10.28) are shown in Fig. 10.35. In the GW-GM with sand slope, only the negative pressure heads in the top layer of 0.5 m thickness dissipate during the rainfall.

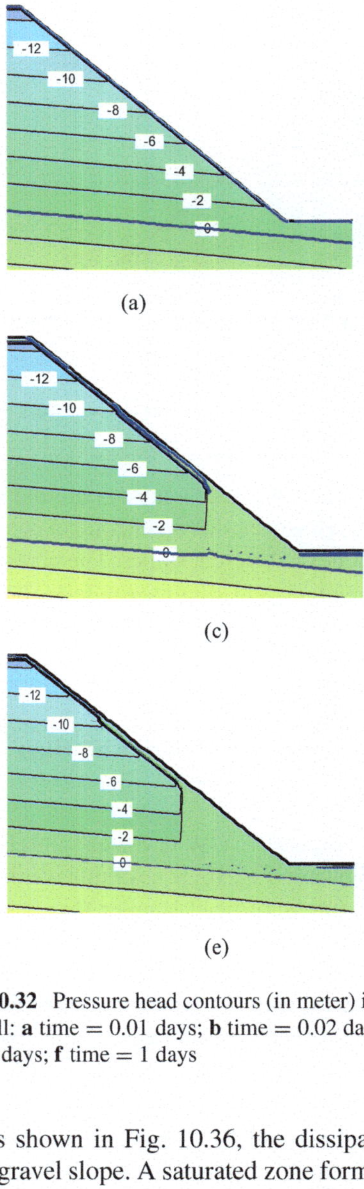

Fig. 10.32 Pressure head contours (in meter) in the slope composed of GW-GM with sand during rainfall: **a** time = 0.01 days; **b** time = 0.02 days; **c** time = 0.03 days; **d** time = 0.05 days; **e** time = 0.1 days; **f** time = 1 days

As shown in Fig. 10.36, the dissipating rate of suctions is striking in the SM with gravel slope. A saturated zone forms below the ground surface at the beginning of the rainfall and expands continuously during the rainfall. After 0.3 days, the saturated zone merges with the groundwater table. Behind the saturated zone, an expanding quasi-saturated zone follows. The soil in that quasi-saturated zone has hydraulic conductivity values matching the rainfall intensity (4.65×10^{-6} m/s). After rainfall, the saturated zone moves down (Fig. 10.37). After 6 days, the saturated zone moves to the initial water table, and the groundwater table recovers to the initial state [i.e., a straight line in Fig. 10.37b]. After rainfall, drainage from the quasi-saturated zone continues for a long period and shows a remarkable recovery of

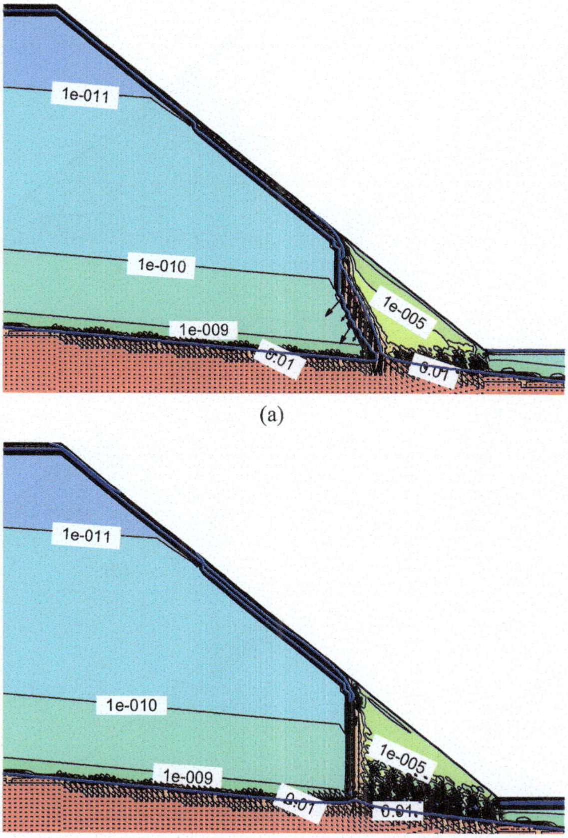

Fig. 10.33 Variation of hydraulic conductivity profiles in the GW-GM with sand slope (*unit* m/s): **a** at time = 0.02 days; **b** at time = 0.03 days

suctions to the hydrostatic steady state. After 21 days, the soil at the top surface has been recovered to pressure head values above 2 m. The pore-water profiles at different times along Sect. 1 in the SM with gravel slope (referring to Fig. 10.28) are shown in Fig. 10.38 After 0.5-day rainfalls, the negative pressure head above the initial water table dissipates and remains at values close to 0 during the rainfall. After rainfall, the pressure head above the initial water table recovers to the hydrostatic steady state gradually. For the SM with gravel slope, the striking wetting speed during rainfall and the remarkable drainage speed after rainfall are mainly contributed from the high unsaturated hydraulic conductivity of SM with gravel. As shown in Fig. 10.30, the unsaturated hydraulic conductivity of SM with gravel is in a range of 10^{-5} m/s–10^{-7} m/s as negative pressure head increases from 0.1 to 1 m and is about 20 times larger than that of the GW-GM with sand. Hence, the two slopes show quite different infiltration processes.

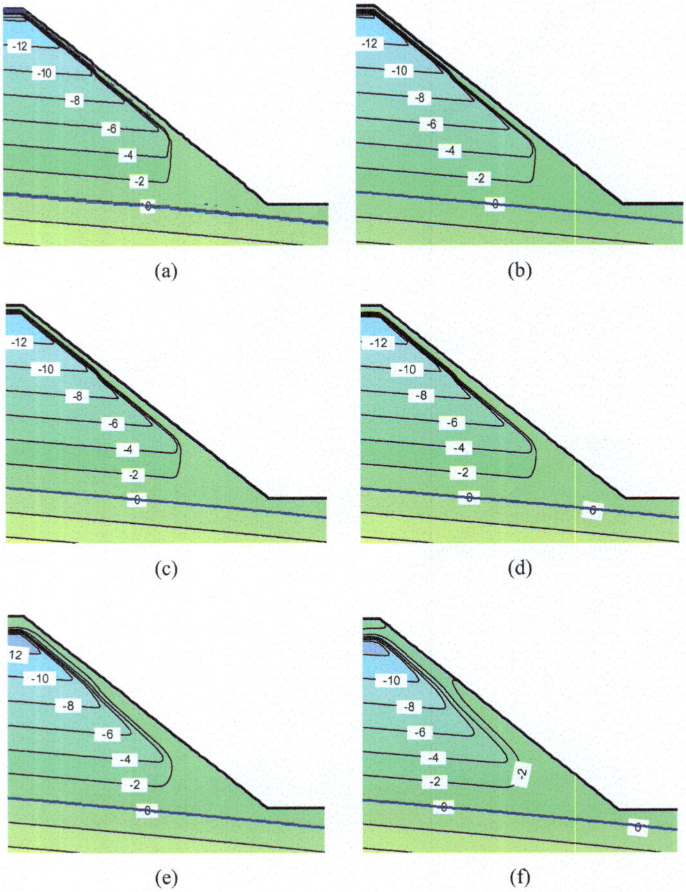

Fig. 10.34 Pressure head contours (in meter) in the slope composed of GW-GM with sand after rainfall: **a** time = 2 days; **b** time = 6 days; **c** time = 11 days; **d** time = 21 days; **e** time = 61 days; **f** time = 111 days

In summary, for colluvial soils with high coarse fractions (i.e., GW-GM with sand and SM with gravel), saturated/quasi-saturated zones form along the ground surface at the beginning of rainfall and advance into deeper zones quickly during the rainfall. The inflow during rainfall is controlled by the rainfall intensity. The advancing speed of these saturated/quasi-saturated zones is substantially controlled by the SHCF, especially the unsaturated hydraulic conductivity of the soil in that zone (i.e., soils in a suction range of 0–100 kPa). After rainfall, the saturated/quasi-saturated zones move down continuously until merging with the initial groundwater. Then, the quasi-saturated zones drain slowly and recover gradually to the hydrostatic steady state. This process is also substantially controlled by the unsaturated hydraulic conductivity. For soils with hydraulic conductivity in the order of 10^{-7} m/s, this process would continue for weeks. For soils with hydraulic conductivity in the

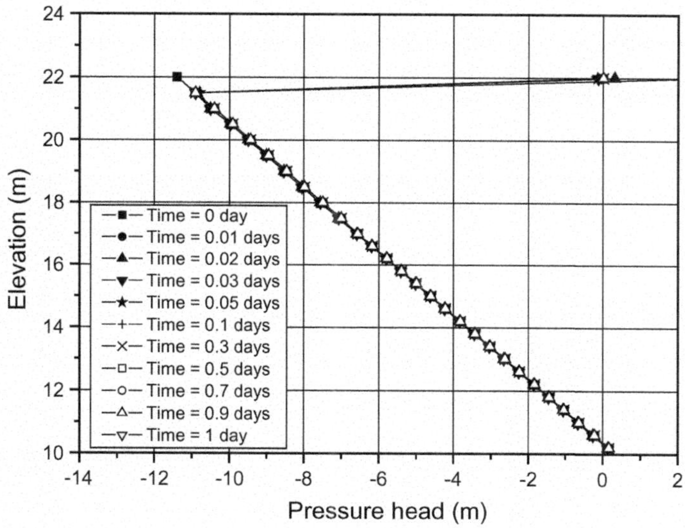

Fig. 10.35 Pore water pressure head profiles along Sect. 1 in GW-GM with sand slope

order of 10^{-8} m/s, this process would continue for months. For soils with hydraulic conductivity in the order of 10^{-9} m/s, this process would continue for years. The hydraulic behavior of slopes composed of soils with high coarse fractions upon rainfall is summarized in Table 10.5.

With different hydraulic properties, the infiltration process will be quite different. The hydraulic property functions for SM with gravel show weak bimodal features and result in high dissipation rate of negative pore-water pressures in the SM with gravel slope. On the contrary, the hydraulic property functions for GW-GM with sand show strong bimodal features and result in a much lower dissipation rate of negative pore-water pressures in the GW-GM with sand slope. If unimodal SHCFs are taken for a coarse-controlled soil and the saturated hydraulic conductivity is taken as the measured data, the dissipation rate of negative pore-water pressures will be larger than that obtained in the present simulation, in which the measured bimodal SHCFs are used. The negative pore water pressures in the whole slope zone would dissipate in one half to several hours if the unimodal functions were used.

10.3.2.2 Pore Pressure Changes in Slopes Composed of Soils with High Fines Fractions

During the rainfall, limited dissipation of pressure head is found in the SC with gravel slope, the sandy ML slope, and the CL with sand slope (as shown in Figs. 10.39, 10.40, and 10.41). Although the ground surface is saturated during the rainfall, pore water pressures change only in a limited zone near the ground surface during the rainfall. Because the rainfall intensity (4.7×10^{-6} m/s) is much larger than the

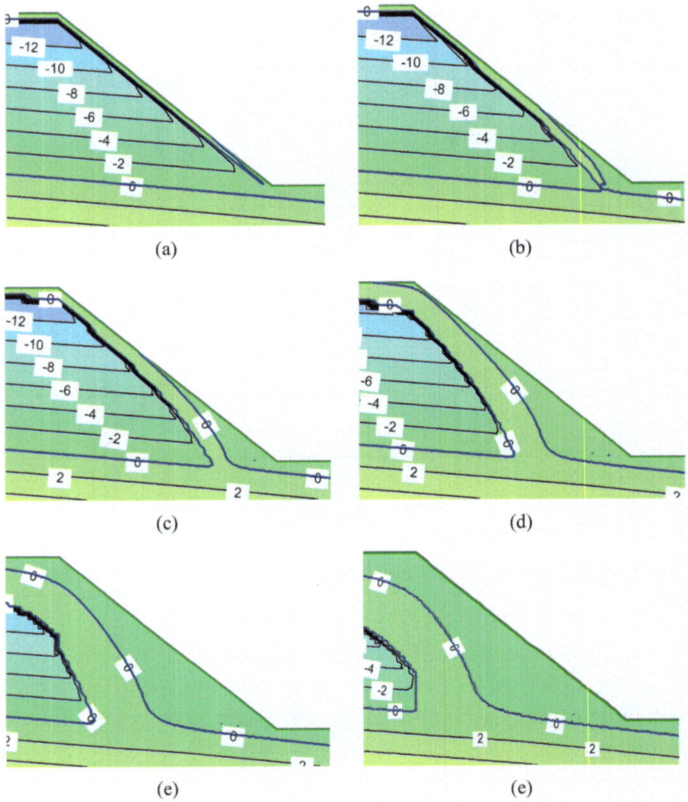

Fig. 10.36 Pressure head contours (in meter) in the slope composed of SM with gravel during rainfall: **a** time = 0.1 days; **b** time = 0.2 days; **c** time = 0.3 days; **d** time = 0.5 days; **e** time = 0.8 days; **f** time = 1 days

saturated hydraulic conductivity of SC with gravel, sandy ML, and CL with sand, most water turn to runoff from the ground surface (i.e., 78% for SC with gravel, 91% for sandy ML, and 98% for CL with sand). Therefore, the dissipation rate of suctions in these three slopes is mainly controlled by the soil saturated hydraulic conductivity and time of rainfall duration. In Fig. 10.39, the water table rises at the toe of SC with gravel during the rainfall. This demonstrates that the water infiltrated into one element during the rainfall event (i.e., 1 day), which is controlled by the saturated hydraulic conductivity of SC with gravel (i.e., 1×10^{-6} m/s), is more than the water required for saturating the element (e.g., 3% for the element at the slope toe, which has a volumetric water content of 36% at the steady state shown in Fig. 10.31). On the contrary, no rising of groundwater table is found in the sandy ML slope and in the CL with sand slope (Figs. 10.40 and 10.41). The required water for saturating one element of sandy ML and CL with sand is similar to that for SC with gravel (e.g., 6% for the element at the sandy ML slope toe and 2% for the element at the CL with sand slope toe). But, the saturated hydraulic conductivity values of sandy

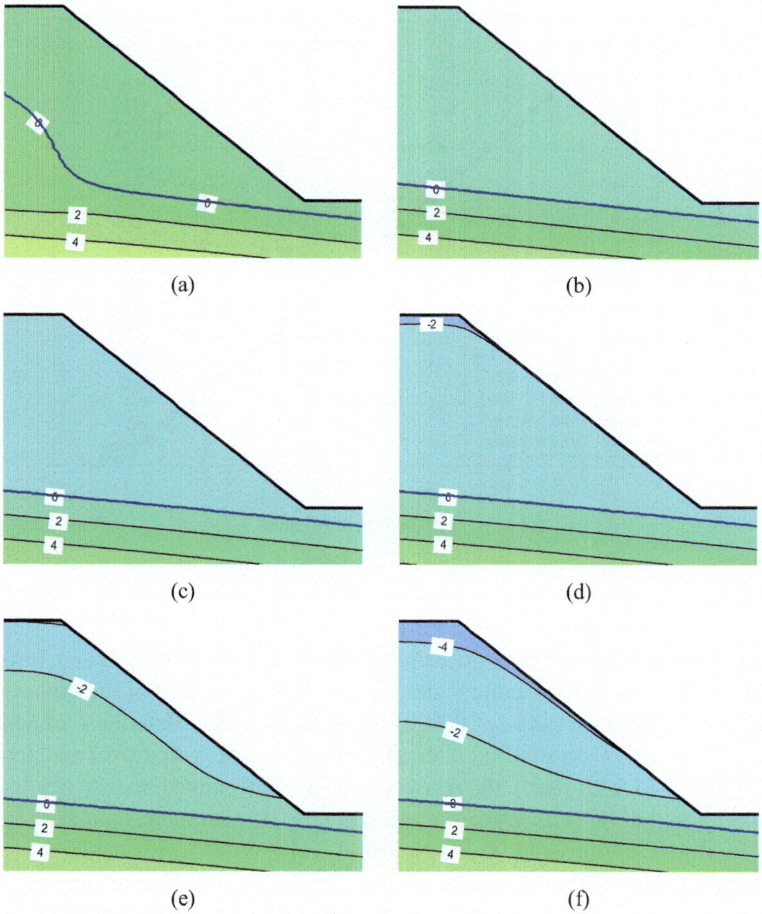

Fig. 10.37 Pressure head contours (in meter) in the slope composed of SM with gravel after rainfall: **a** time = 2 days; **b** time = 6 days; **c** time = 11 days; **d** time = 21 days; **e** time = 61 days; **f** time = 111 days

ML (3×10^{-7} m/s) and CL with sand (1×10^{-7} m/s) are much lower than that of SC with gravel. Therefore, the infiltrated water into one element is much smaller. After rainfall, the dissipation of pressure head in the SC with gravel slope, sandy ML slope, and CL with sand slope continues for months due to the low unsaturated hydraulic conductivity of these three soils.

The pore-water profiles at different times along Sect. 1 (referring to Fig. 10.28) are shown in Figs. 10.42, 10.43, and 10.44. The evolution of pressure head profiles in these three slopes, the SC with gravel slope, the sandy ML slope, and the CL with sand slope, is similar during and after the rainfall. During the rainfall, the negative pore water pressures in soils within 1 m depth decrease gradually. After rainfall, the soils within 1 m depth show gradual increase of negative pore water pressure,

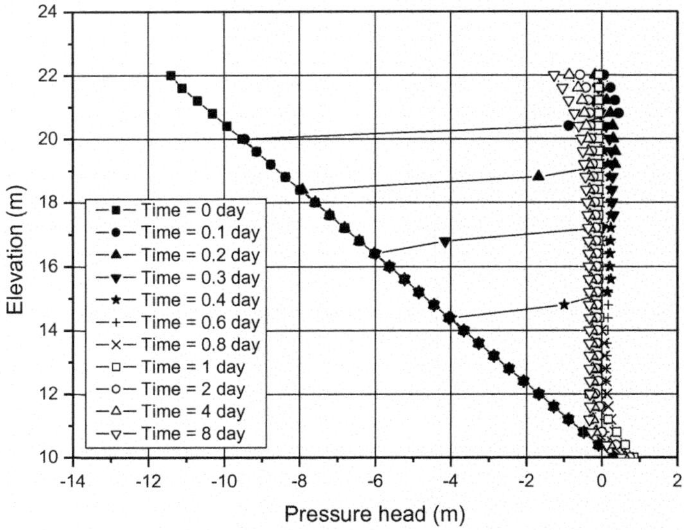

Fig. 10.38 Pore water pressure head profiles along Sect. 1 in SM with gravel slope

whereas the negative pore water pressures in soils between 1 m depth and 3 m depth continue to decrease. Because the hydraulic conductivity values of the three soils in a negative pressure head range of 0.3–10 m are much higher than these of SM with gravel and GW-GM with sand (i.e., about 5 times for SM with gravel and about 15 times for GW-GM with sand), the recover process of negative pressure heads in the slopes composed of SC with gravel, sandy ML, and CL with sand is much faster than that in the slopes composed of SM with gravel and GW-GM with sand.

In summary, for colluvial soils with high fines fraction, limited pressure head changes occur during the rainfall. The inflow is mainly controlled by the saturated hydraulic conductivity and the rainfall duration. The pore water pressure changes after rainfall are mainly controlled by the unsaturated hydraulic conductivity of the soils in the slope. The hydraulic behavior of slopes composed of soils with high fines fractions is summarized in Table 10.4.

10.3.3 Stability Analysis Results

The stability of these slopes is analyzed using Slope/W (Geo-slope International Ltd. 1991b). The settings for center of slip surface and radius of slip surface are shown in Fig. 10.45. The slip patterns change during and after the rainfall as shown in Figs. 10.46, 10.47, and 10.48, so do the factors of safety (FOS).

The FOS of critical slip pattern decreases rapidly from 1.727 to 1.583 after 0.03 days of rainfall in the GW-GM with sand slope. After 0.03 days, the FOS remains

Table 10.5 Risk of failure for colluvial slopes

Slope soil	Saturated hydraulic conductivity	Unsaturated hydraulic conductivity in slope zone [a]	Controlling parameter of inflow	Influenced depth after 1-day rainfall [e]	Rising of groundwater table	Potential slip mode
GW-GM with sand	High [b]	Low	Rainfall intensity	~ 2 m	Limited	Nil [f]
SM with gravel	High	High to low	Rainfall intensity	Whole slope	Large	Shallow seated
SC with gravel	Medium [c]	Medium to low	Saturated hydraulic conductivity	> 3 m	Some	Nil
Sandy ML	Low [d]	Medium to low	Saturated hydraulic conductivity	~ 3 m	Limited	Nil
CL with sand	Low	Medium to low	Saturated hydraulic conductivity	~ 2.5 m	Limited	Nil

[a] Corresponding to suctions smaller than 100 kPa

[b] The order of hydraulic conductivity is 10 times larger than the average 24-h rainfall intensity, which is about 10^{-6} m/s (therefore, about 10^{-5} m/s)

[c] The order of hydraulic conductivity is similar to the average 24-h rainfall intensity (therefore, about 10^{-6} m/s)

[d] The order of hydraulic conductivity is 10 times smaller than the average 24-h rainfall intensity (therefore, about 10^{-7} m/s)

[e] An average value, which is larger at the slope toe

[f] The FOS shows only about 10% decrease

about 1.583 as the water pressure profiles show limited changes in the remaining rainfall period. After rainfall, the FOS increases slowly, while the negative pore water pressures increase gradually to values at the hydrostatic steady state.

The critical slip patterns in the SM with gravel slope during rainfall are shown in Fig. 10.47. The FOS of the critical slip pattern decreases rapidly from 1.440 to 0.943 after 0.5 days of rainfall. The failure mode is shallow seated. The shallow failure zone expands quickly during the rainfall and finally covers the whole slope. If the rainfall continues for a long time, deep-seated slip may occur in the SM with gravel slope. During the rainfall, the dissipation of pore water pressures results in decrease of the soil shear strength. Also, the rising of the groundwater table in the slope causes added seepage forces. Therefore, the FOS decreases significantly.

No shallow failure or deep-seated failure is found for the slopes composed of SC with gravel, sandy ML, and CL with sand (Fig. 10.48). Because the pore water pressure changes in these three slopes are limited during and after the rainfall, the changes of FOS are also limited. It is anticipated; however, the FOS of these three slopes will decrease significantly in the case of a sustained rainfall.

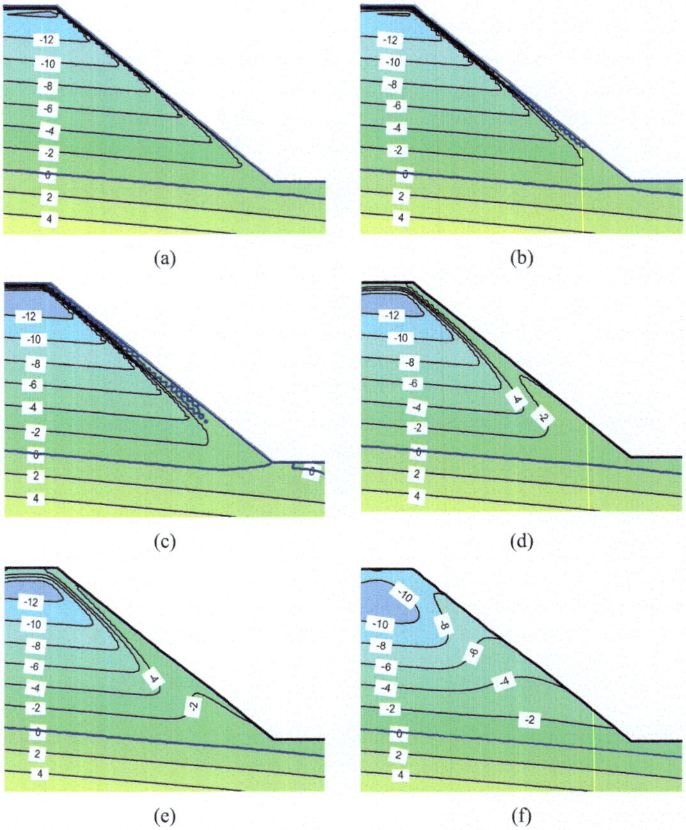

Fig. 10.39 Pressure head contours (in meter) in the slope composed of SC with gravel during and after rainfall: **a** time = 0.1 days; **b** time = 0.5 days; **c** time = 1 days; **d** time = 6 days; **e** time = 11 days; **f** time = 111 days

10.4 Case Study 2: Identification of Hazardous Loose Landslide Deposits and Scars Formed During the 2008 Wenchuan Earthquake

10.4.1 Introduction

PR303 in Sichuan Province is an important road to several national scenic parks. The Wenchuan earthquake triggered 61 slope failures, 63 rock avalanches, and 19 debris flows along a 45 km segment of the road. Over 80% of the road between Yingxiu (the epicenter) and Wolong (the Grand Panda National Park) was damaged or buried by landslide deposits. Landslide deposits formed during the earthquake are densely populated on the steep terrains. Under rainfall infiltration, such deposits may lose stability and, hence, pose hazards to the safety of the traffic and road facilities.

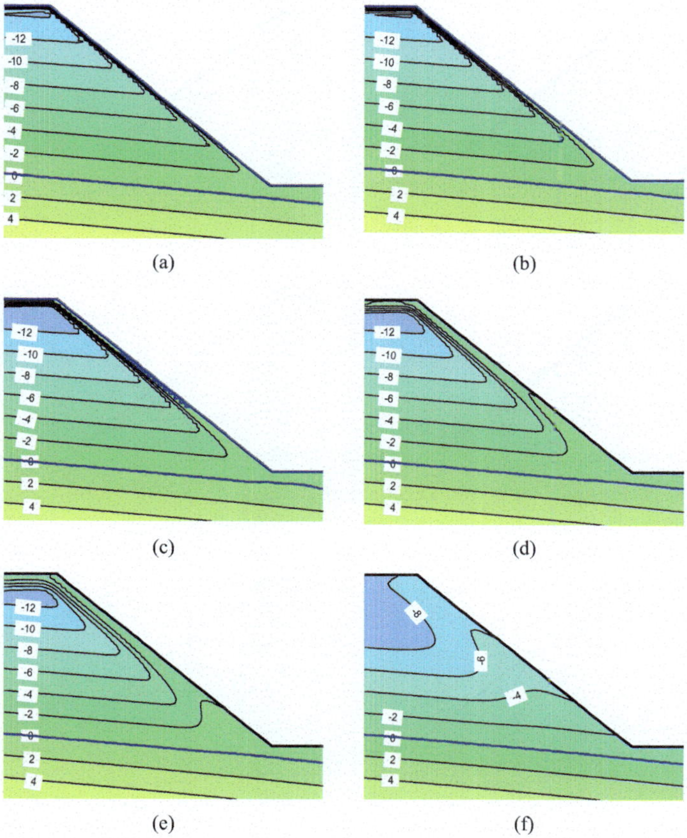

Fig. 10.40 Pressure head contours (in meter) in the slope composed of sandy ML with gravel during and after rainfall: **a** time = 0.1 days; **b** time = 0.5 days; **c** time = 1 days; **d** time = 6 days; **e** time = 11 days; **f** time = 111 days

Hazards identification is the basis for reconstruction of PR303. Some areas along PR303 are inaccessible as the terrains are very steep and high. It is unlikely to identify the sources of hazards through field reconnaissance alone. In this chapter, a method for hazards identification is demonstrated on a large earthquake-induced landslide between milestones K34 + 00 and K34 + 200 m. In this method, preliminary field surveys, past failure events, and numerical simulations are combined to identify and ascertain sources of risk near K34.

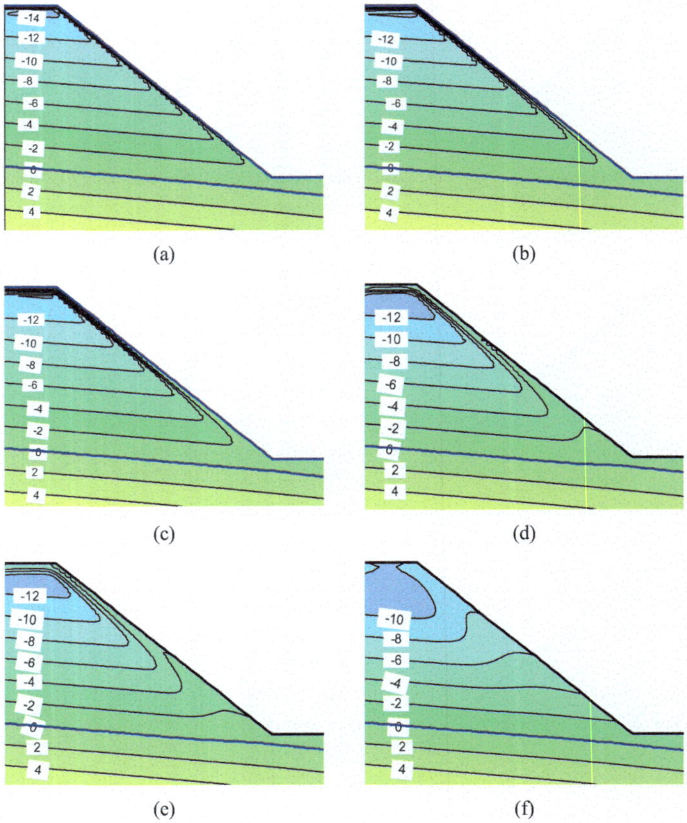

Fig. 10.41 Pressure head contours in the slope composed of CL with sand during and after rainfall: **a** time = 0.1 days; **b** time = 0.5 days; **c** time = 1 days; **d** time = 6 days; **e** time = 11 days; **f** time = 111 days

10.4.2 Geographic and Geological Settings Near K34

Many slides and boulder fall events occurred at K34 in the past two years after the Wenchuan earthquake. The geomorphic and geologic features of the K34 landslide site are illustrated in a 1:2000 digital elevation map with a resolution of 2 m (Fig. 10.49) and an aerial photograph (Fig. 10.50). The K34 terrain is typically steep. The slope angles of the landslide area range from 35 to 40°, and the depth of valley is nearly 1500 m. As the terrain is very steep and valley is very deep, it is unlikely to climb the mountains to conduct detailed surveys and identify the risk sources.

The geology of the landslide area comprises mainly of slate and sandy slate of the Triassic Xikang group, crystalline rock and phyllite of the Permo-carboniferous group and the Devonian period. A cross section along the main run-out axis is generated based on the digital elevation map and shown in Fig. 3b. Based on preliminary field surveys and aerial photographs, the sources of hazards near K34 are divided

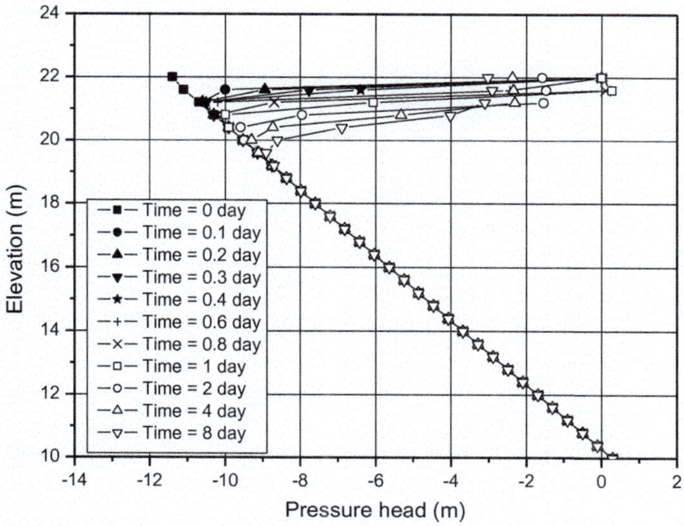

Fig. 10.42 Pore water pressure head profiles along Sect. 1 in SC with gravel slope

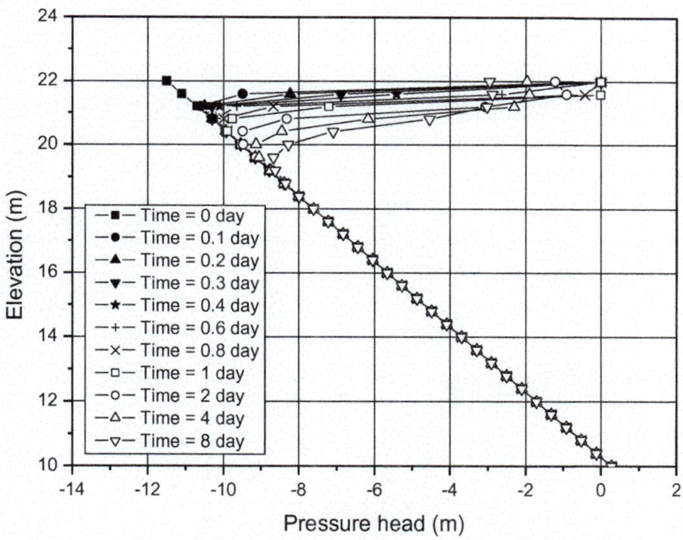

Fig. 10.43 Pore water pressure head profiles along Sect. 1 in sandy ML slope

into three parts, namely the landslide debris on the bottom, the deposits on a terrace, and the top-fissured landslide scars. The bottom landslide deposit has a volume of approximately 40,000 m^3, which buried the road. The landslide debris is stable, and a temporary road was built through the debris. Some fissured landslides scars are found 500–700 m above the road (Fig. 10.50) through survey from the opposite side

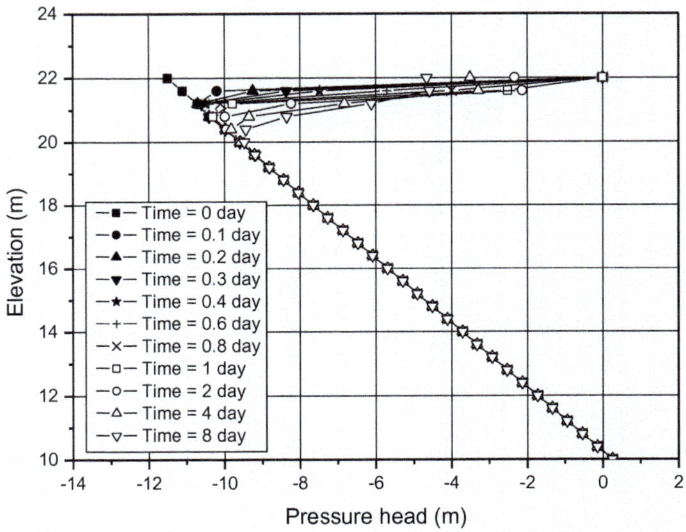

Fig. 10.44 Pore water pressure head profiles along Sect. 1 in CL with sand slope

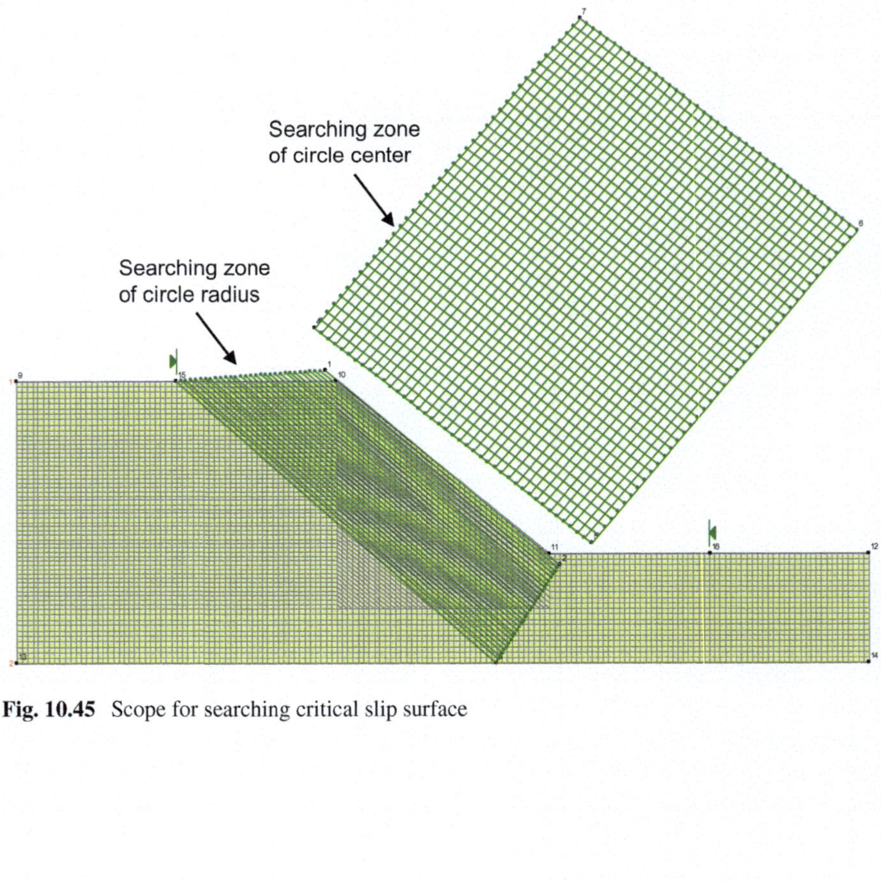

Fig. 10.45 Scope for searching critical slip surface

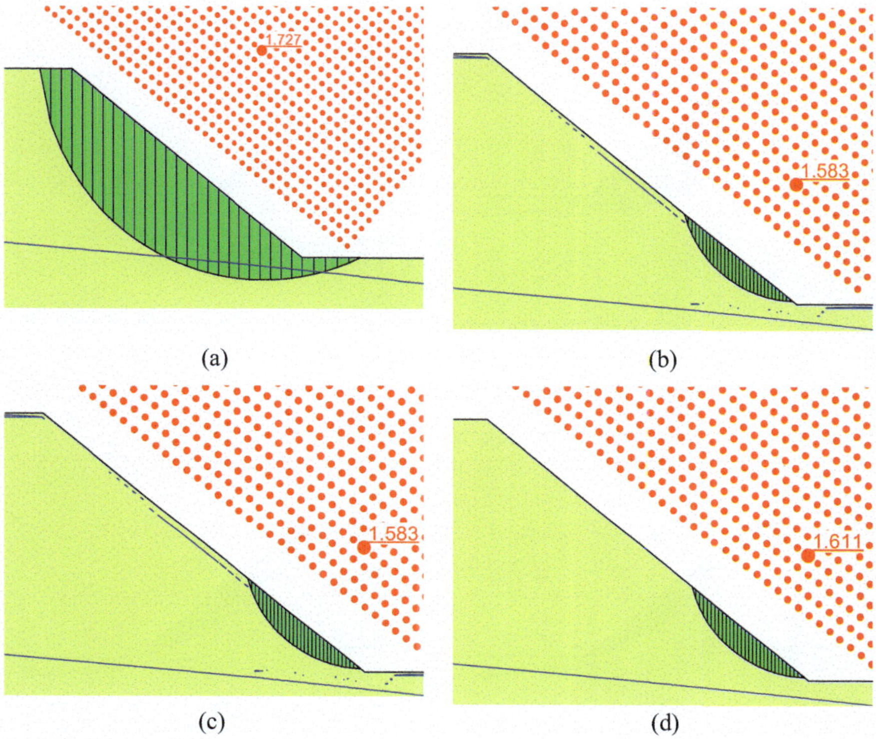

Fig. 10.46 Critical slip patterns of GW-GM with sand slope during rainfall infiltration: **a** time = 0 day (FOS = 1.727); **b** time = 0.1 days (FOS = 1.583); **c** time = 1 days (FOS = 1.583); **d** time = 11 days (FOS = 1.611)

of the valley. A terrace covered by gravelly soils with boulders is also identified 300–500 m above the road. The grain size of the deposit ranges from clays to boulders (Fig. 10.50). The thickness of the deposits is about 5–12 m. Figure 10.51a shows a photograph of the bottom part of the slope, with the run-out axis marked. The fissured scars and the terrace are not visible from the road elevation.

10.4.3 Inventory of Failures Occurred in 2009–2010

Repeated slides occurred at K34, which caused one fatality and more than 10 shutdowns of the road. We obtained some information of failure events and landslide damage from the Sichuan Provincial Department of Transportation. Additional information was obtained through our field surveys along the road and by interviewing the contractors. Table 1 summarizes our findings; some photographs of the failure events are shown in Fig. 10.52.

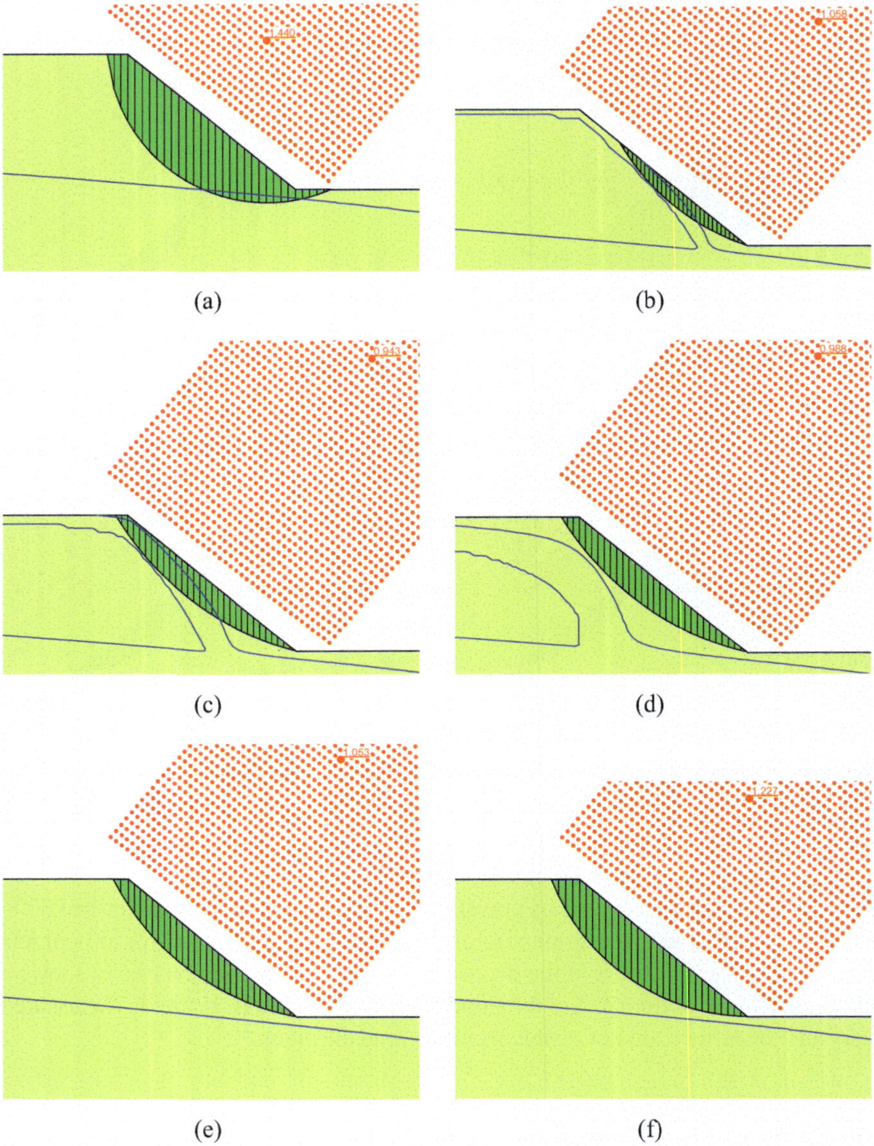

Fig. 10.47 Critical slip patterns of SM with gravel slope during rainfall infiltration: **a** time = 0 day (FOS = 1.440); **b** time = 0.3 days (FOS = 1.058); **c** time = 0.5 days (FOS = 0.943); **d** time = 1 days (FOS = 0.988); **e** time = 11 days (FOS = 1.053); **f** time = 111 days (FOS = 1.227)

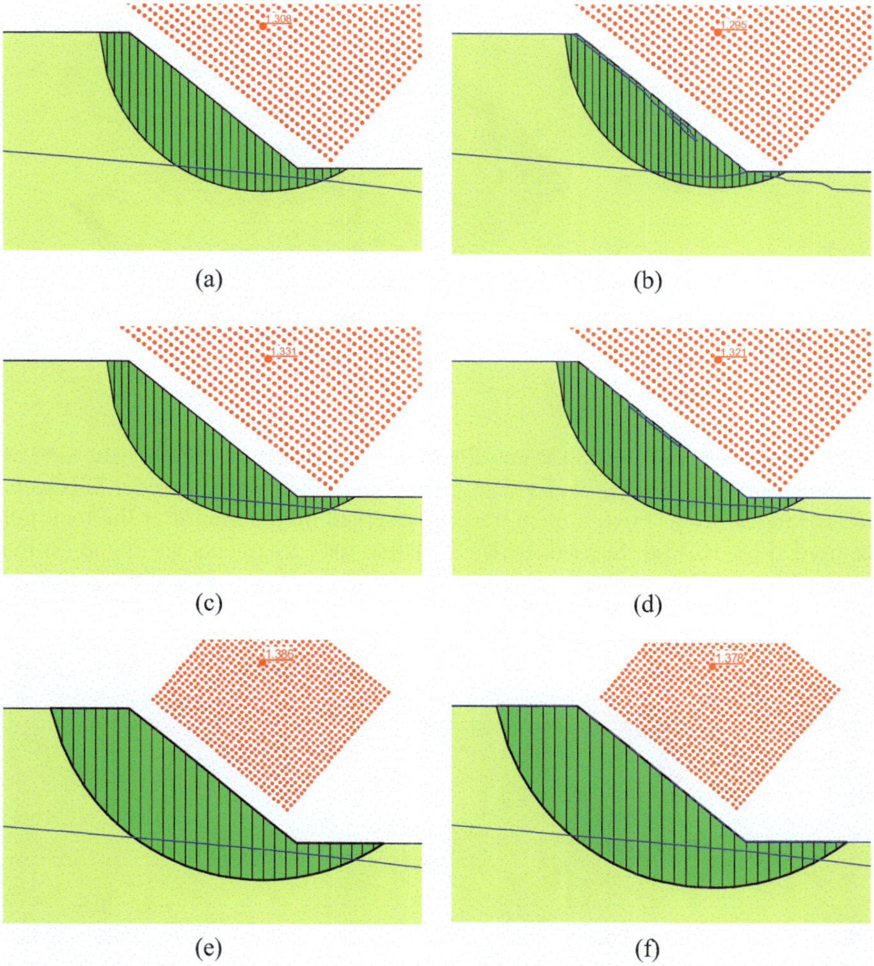

Fig. 10.48 Critical slip patterns of slopes composed of soils with high fines fractions: **a** SC with gravel, time = 0 day (FOS = 1.308); **b** SC with gravel, time = 1 days (FOS = 1.295); **c** sandy ML, time = 0 day (FOS = 1.331); **d** sandy ML, time = 1 days (FOS = 1.321); **e** CL with sand, time = 0 day (FOS = 1.386); **f** CL with sand, time = 1 days (FOS = 1.378)

10.4.4 Identification of Sources of Risk at K34

Identification of potential sources of boulder falls based on the inventory. Once the soil/rock detaches from a scar at a very high elevation, the most important factor controlling its falling trajectory is the slope geometry. The historical data in Table 10.6 provide a good indication of the sources of boulders on the terrain.

Following detachment, the landslide material proceeded at a high velocity, flying and bouncing as it traveled down [Fig. 10.53c]. A field survey was conducted after

Fig. 10.49 Digital elevation map of the terrain near K34 (1:2000)

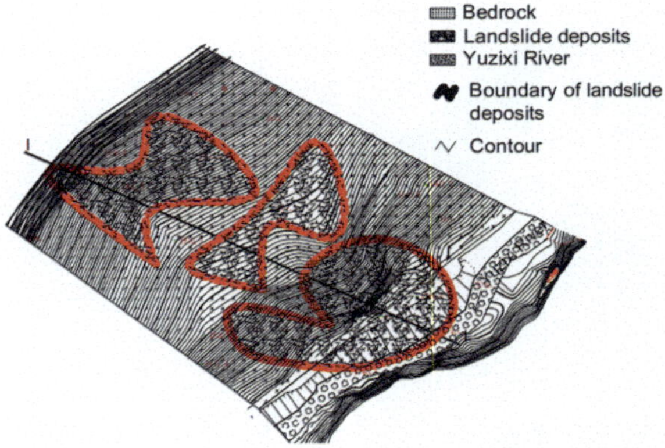

the Sept. 3, 2009, boulder fall event. Photographs taken from the opposite side of the slopes are shown in Fig. 5. Evidence has been found that the boulder falls were initiated at the scars. First, a fresh trace is observed on the scar after the rock fall occurred [Fig. 10.53a]. Second, deposits of fine rock fragments are found on the terrace [Fig. 10.53b], which are considered as segregated materials of the collapsed

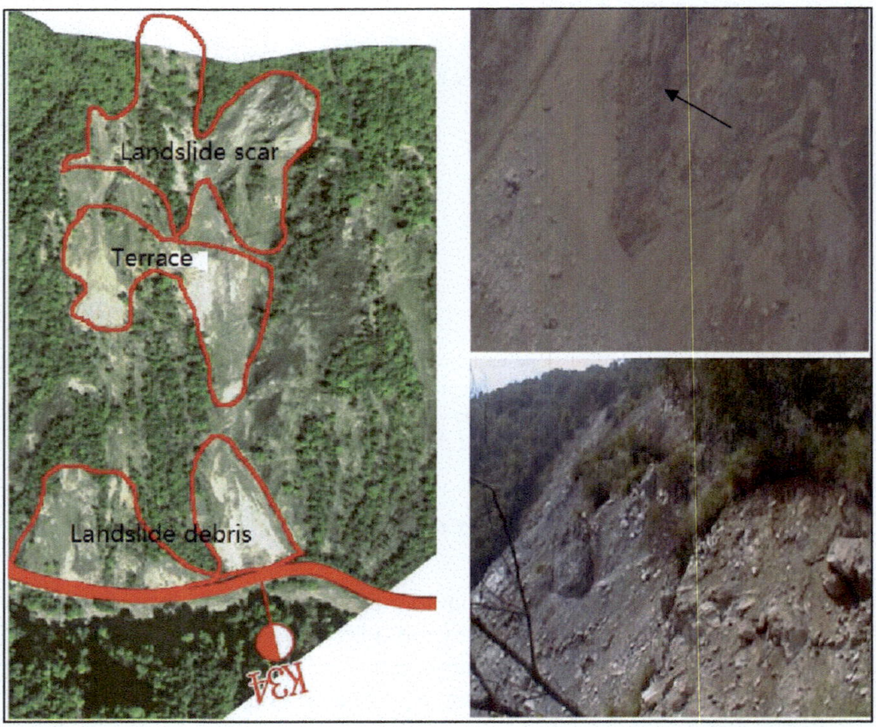

Fig. 10.50 Photographs showing **a** location of landslides, **b** fissure scars, and **c** landslide deposits

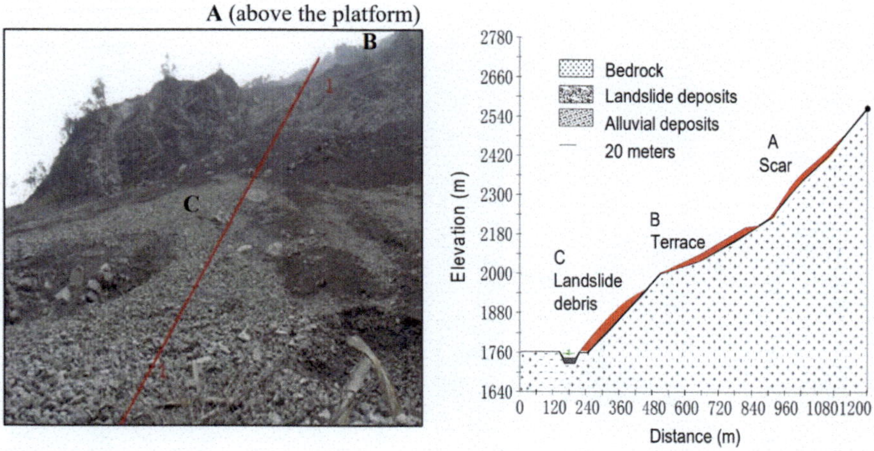

(a) (b)

Fig. 10.51 Geomorphology and geology of the landslide site: **a** photograph; **b** cross-section I-I

Fig. 10.52 Failure events at K34

Table 10.6 List of slide and boulder fall events at K34

Time of failure	Type of damage
May 01, 2009	Buried a car, temporarily interrupted the road
July 08, 2009	Buried a truck and a car, temporarily interrupted the road
July 18, 2009	A man on motorbike was killed by flying boulders
July 22, 2009	Buried a truck, damaged equipment and interrupted the road
Aug. 30, 2009–03 Sept. 2009	Interrupted the road and formed a landslide dam
May 26, 2010	Hurt four contactors and destroyed three engineering trucks
May 28, 2009–12 June 2010	Interrupted the road six times, damaged boulder protection fences
June 14, 2010	Buried the road; destroyed the boulder protection fences

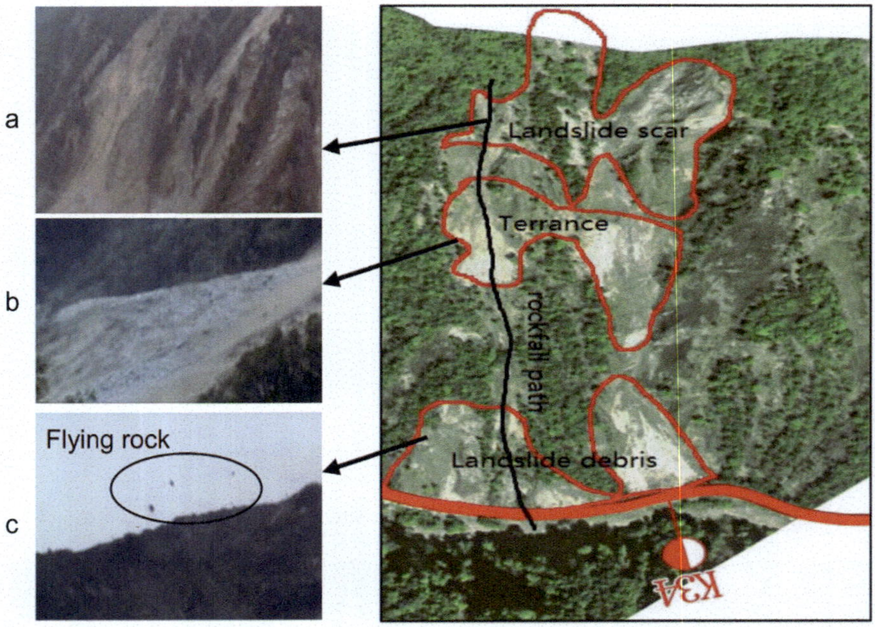

Fig. 10.53 Rock fall events on August 30 and September 3, 2009: **a** fresh trace on the scar; **b** loose deposit on the terrace; **c** flying and bouncing rocks

rocks from the scars. It is noted that, during a granular flow event, coarse particles selectively move faster and to the surface, while fine materials move slower and to the bottom.

Numerical simulation. Infiltration plays an important role in triggering of slope failures. With rainfall infiltration, the initially negative pore-water pressures in the slope will increase and the shear strength of the slope soil will decrease gradually. The slope may fail finally when the shear strength is insufficient to maintain the equilibrium of the soil.

Numerical analysis is conducted to simulate the event happened on August 30–September 3, 2009, and to confirm the likely sources of risk. All of possible hazards are modeled. A finite element program Seep/W (Geo-slope International 2001) is used to simulate the pore-water pressure responses due to rainfall infiltration, and Slope/W is used to calculate the factor of safety of the slopes.

Based on the digital elevation model, the inclination angles of the bottom landslide debris, the terrace, and the fissured scar are 40°, 24°, and 42°, respectively. A tension crack is simulated on the top of the fissured scar. The crack is modeled as a special kind of "soil," which is much more permeable than intact soil, with its hydraulic conductivity being two orders of magnitude larger than the hydraulic conductivity of the intact soil. The assumed SHCFs and soil water characteristic curve for the soil matrix and the crack, which are required in the simulation of rainfall infiltration into

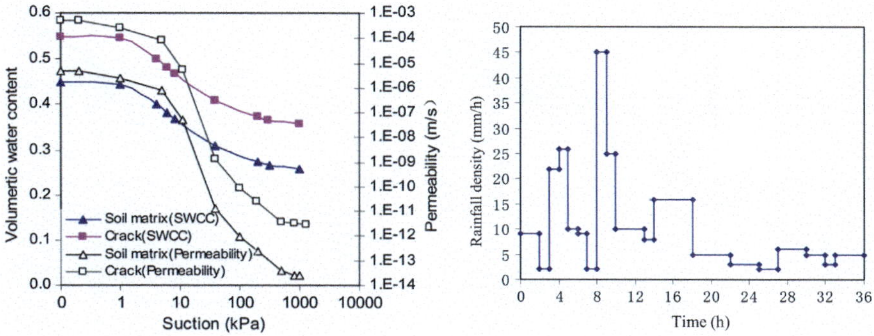

Fig. 10.54 Soil property functions and rainfall data

the terrains, are portrayed in Fig. 10.54. The shear strength of unsaturated slope soils is described by the extended Mohr Column equation (Fredlund et al. 1978):

$$\tau = c' + (\sigma - u_a)\tan\phi' + (u_a - u_w)\tan\phi^b \quad (10.11)$$

where τ is the shear stress on the failure plane at failure; c' is the effective cohesion; ϕ' is the effective angle of internal friction; ϕ^b is the angle indicating the rate of increase in shear strength with respect to matrix suction.

The design ϕ' values for soil matrix provided by Sichuan Provincial Highway Design Institute (2009) are used in the simulations, while other parameters are assumed. The parameters for the soil matrix and the crack are shown in Table 10.7. The presence of the crack reduces the soil strength. The rainfall data on August 19–20, 2010, reported by Weather China for Wenchuan County is used as a reference, as shown in Fig. 10.54. The rainfall on August 30–September 3, 2009, is not available.

Figure 10.55 shows the pore-water pressure contours in the terrace and fissured scar at the end of the rainfall event. During water infiltration, the wetting front moves from the slope surface toward deeper locations. For the terrace, the perched water table does not reach the initial groundwater table. The soil between the perched water table and the groundwater table is still unsaturated after 36 h of rainfall. For the fissured scar, the infiltrated water first passes through the crack and accumulates at the bottom of the crack. Then, the water penetrates to the soil matrix that surrounds the crack. All the soils of the slope are saturated after 36 h. Figure 10.56 shows the

Table 10.7 Soil parameters used in Slope/W

	Soil matrix		Crack	
	Mean value	Standard deviation	Mean value	Standard deviation
γ (kN/m^3)	21	3	16	3
ϕ' (degree)	30	1	0	1
c (kPa)	2	2	0	2
ϕ^b (degree)	10	1	6	1

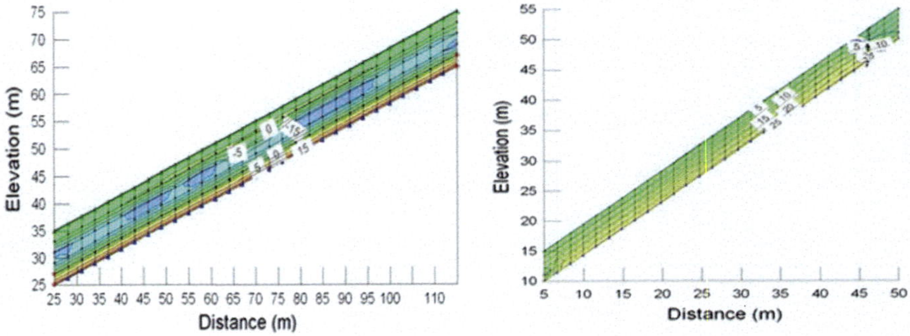

Fig. 10.55 Pore-water pressure contours (kPa) in the slopes

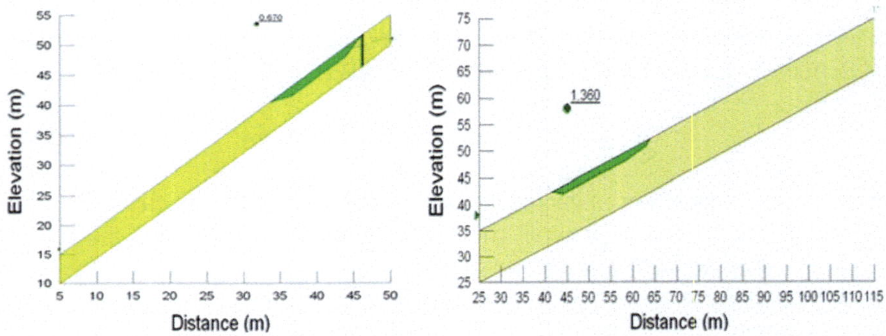

Fig. 10.56 Factors of safety of the terrace deposit and fissured scar

potential slip surfaces, and Fig. 10.57 shows the decrease of factor of safety with time. At the end of the rainfall event, the FOS is 1.14 for the bottom landslide debris, 1.46 for the terrace, and 0.65 for the fissured scar. Hence, when subject to the same heavy rainfall, the fissured scar appears to be less stable than the deposit on the terrace. The primary source of risk at K34 is the fissured scars at very high elevations.

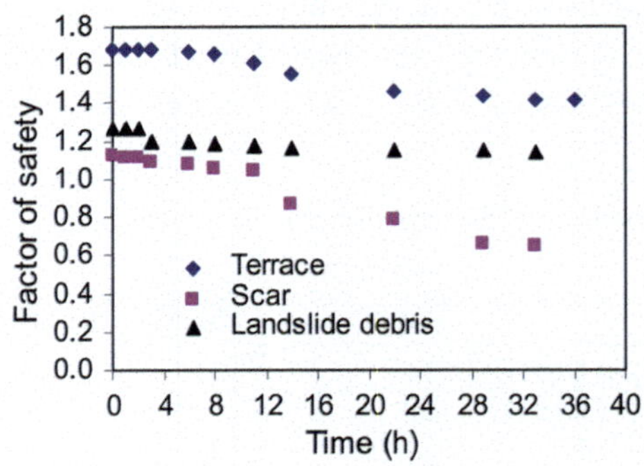

Fig. 10.57 Decrease of factor of safety with rainfall infiltration time

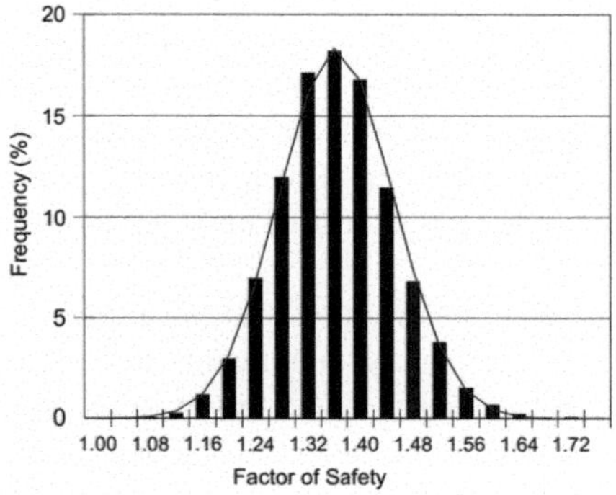

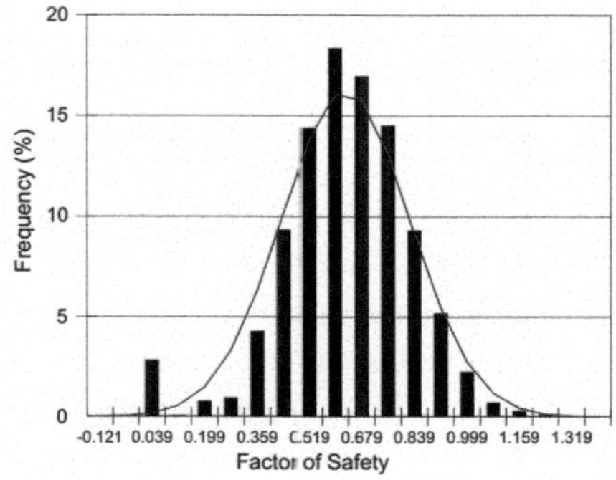

Fig. 10.58 Histograms of the factor of safety using Monte Carlo simulation

10.4.5 Reliability Analysis

Reliability of the fissured scar and the deposit on the terrace is analyzed focusing on the uncertainties in the shear strength parameters. All the parameters are assumed to be normally distributed, and the mean and standard deviation of each parameter are shown in Table 10.4. Parameters ϕ and c are negatively correlated with an assumed coefficient of correlation of $\rho = -0.5$. The Monte Carlo simulation method is used in the reliability analysis (Tobutt 1982). About 10,000 sets of soil properties are generated. The probability distributions of the factor of safety are shown in Fig. 10.58. The reliability index is obtained to indict the probability of failure. The reliability index is 4.1 for the terrace deposit and − 5.1 for the fissured scar. This supports the observation that the fissured scar is deemed to fail upon a heavy rainfall.

10.5 Summary

In this chapter, the instability in the microstructures of three types of granular soils with different coarse contents and the instability behavior of the granular soils during shearing under both saturated and unsaturated conditions are studied. Two case studies are performed to investigate the stability of slopes composed by widely-graded soils. The main findings are as follows:

(1) The potential of instability in the microstructures of the soils is in accordance with the compressibility of the soils, which initially increases with the coarse content, reaches the maximum at a transition state, and then decreases afterward. The stability of microstructure increases with decreasing void ratio. At the same void ratio, the transition soil ($f_c = 68\%$) has the highest compressibility, the least stable microstructure, and the lowest undrained peak strength among the three test soils.

(2) Unstable undrained shearing behavior is observed in loose coarse granular soils. Relationships between the state parameter ψ and the normalized undrained peak strength ratio, $q_{peak}/p_0'/(Mf_c)$, the normalized stress ratio at peak, η_{IL}/M, or normalized undrained critical strength ratio, $q_{cs}/p_0'/(Mf_c)$, are proposed to evaluate the instability of saturated granular soils under undrained shearing condition.

(3) Along a wetting stress path, limited volumetric strain and axial strain are developed before the unsaturated soil yields. Subsequently, significant volumetric strain and axial strain commence relatively suddenly. Instability in an unsaturated specimen occurs at a higher suction when it is subject to a higher stress ratio or a higher net confining pressure.

(4) Equivalent mean effective stress p' taking into account the suction stress is used to describe the instability of an unsaturated soil subject to wetting. An instability line for the unsaturated soil can also be defined. The established relationship between the normalized slope of the instability line (η_{IL}/M) and ψ can be used to assess the instability of both saturated and unsaturated coarse granular soils during shearing.

(5) The slopes composed of colluviums with high coarse fractions are very sensitive to rainfall. A saturated/quasi-saturated zone is produced at the ground surface at the beginning of rainfall, and the zone expands rapidly during the rainfall. The inflow in such slopes is mainly controlled by the rainfall intensity. Both the saturated hydraulic conductivity and the unsaturated hydraulic conductivity of soils in the slopes (normally corresponding to a suction range of 0–100 kPa) have important influences on the advancing speed of the saturated/quasi-saturated zone.

(6) The negative pore-water pressures decrease in the saturated/quasi-saturated zone, which may induce a considerable decrease of soil shear strength and, hence, a considerable decrease of FOS. Shallow-seated failure may occur in the saturated/quasi-saturated zone and expands continuously with the rainfall time.

(7) The colluvium slopes with low coarse fractions and high fines fractions are not sensitive to a short-term rainfall. Only limited dissipation of negative pore-water pressure in a thin layer near the ground surface occurs during and after a short-term rainfall. The inflow in such slopes is mainly controlled by the saturated hydraulic conductivity and the rainfall duration.

(8) A methodology for identifying sources of hazards is presented and demonstrated with a case study at K34 on PR303. This method consists of three steps: identify the all possible hazard sources based on field surveys and aerial photographs; use historic failure events to single out most likely sources of hazards; and finally, apply numerical analysis to ascertain the sources of hazards. The results of the detailed case study at K34 show that fissured scars at about 700 m above the road are the main source of repeated rock falls. The proposed method can be used to identify hazards in areas difficult to access. Reliability analysis of the fissured scar further confirms the findings from the field surveys and deterministic analyses.

References

Alonso EE, Gens A, Josa A (1990) A constitutive model for partially saturated soils. Geotechnique 40(3):405–430

Altuhsfi F, Baudet BA, Sammonds P (2010) The mechanics of subglacial sediment: an example of new "transitional" behavior. Can Geotech J 47(7):775–790

Been K, Jefferies MG, Hachey J (1991) The critical state of sands. Geotechnique 41(3):365–381

Bishop AW, Donald B (1961) The experimental study of partly saturated soil in the triaxial apparatus. 5th Int Conf Soil Mech Found Eng Vol 1:13-21

Brand EW (1981) Some thoughts on rain-induced slope failure. Proceedings of the 10th international conference on soil mechanics and foundation engineering Londen, 3:374-376

Casagrande, A. (1936). "The determination of the preconsolidation load and its practical significance." Proc., 1st Int. Conf. Soil Mech. Found. Eng., Cambridge, MA., 3, 60-64.

Chan WL, Ho KKS, Sun HW (1998) Computerised database of slopes in Hong Kong. In Proceedings of the Annual Seminar on Slope Engineering in Hong Kong, Hong Kong. A.A. Balkema Publishers, Rotterdam: 213-220

Chu J, Leong WK (2002) Effect of fines on instability behavior of loose sand. Geotechnique 52(10):751–755

Chu J, Leroueil S, Leong WK (2003) Unstable behavior of sand and its implication for slope stability. Can Geotech J 40(5):873–885

Evans NC (1997) Preliminary assessment of the influence of rainfall on natural terrain slope initiation. Discussion Note No. DN 1/97, Geotechnical Engineering Office, Hong Kong

Evans NC, King JP (1998) The natural terrain slope study. Debris avalanche susceptibility. Technical note No. TN1/98, Geotechnical Engineering Office, Hong Kong, 96

Evans NC, Yu YF (2001) Regional variation in extreme rainfall values. GEO Report No. 115. Geotechnical Engineering Office, Hong Kong

Ferreira PMV, Bica AVD (2006) Problems in identifying the effects of structure and critical state in a soil with a transitional behavior. Geotechnique 56(7):445–454

Fredlund DG, Morgenstern NR, Widger RA (1978) Shear strength of unsaturated soils. Can Geotech J 15(3):313–321

Gardner WR (1958) Some steady-state solutions of the unsaturated moisture flow equation with application to evaporation from a water table. Soil Sci 85(4):228–232

Geo-slope international Ltd. (1991a) Seep/w Define version 5.20. http://www.geo-slope.com. Calgary, Alberta, Canada

Geo-slope international Ltd. (1991b) Slope/w Define version 5.20. http://www.geo-slope.com. Calgary, Alberta, Canada

Geo-slope Ltd. (2001) Seep/W for finite element seepage analysis, user's guide

Geotechnical engineering office civil engineering department GCO (1984). Prediction of Soil Suction for Slopes in Hong Kong, GEO Report No.1/84. Geotechnical Engineering Office, Hong Kong

Holtz WG, Gibbs HJ (1956) Triaxial shear tests on pervious gravelly soils. J Soil Mech Found Div Am Soc Civ Eng 82:1–22

Irfan TY, Tang KY (1993) Effect of the coarse fractions on the shear strength of colluvium. GEO Report No. 23. Geotechnical Engineering Office, Hong Kong

Jefferies MG, Shuttle DA (2002) Dilatancy in general Cambridge-type models. Geotechnique 52(9):625–638

Karube D, Kato S (1994) An ideal unsaturated soil and the Bishop's soil. In: Proceeding of the 13th International Conference on Soil Mechanics and Foundation Engineering, New Delhi, India, Taylor and Francis, New York. 1:43–46

Karube D, Kato S, Hamada K, Honda M (1996) The relationship between the mechanical behavior and the state of pore-water in unsaturated soil. J Jpn Soc Civil Eng 535(34):83–92

Ko FWY (2003) Correlation between rainfall and natural terrain slope occurrence in Hong Kong. GEO Report No. 168. Geotechnical Engineering Office, Hong Kong

Lade PV (1992) Static instability and liquefaction of lose fine sandy slopes. J Geotech Eng 118(1):51–72

Lade PV (1993) Initiation of static instability in the submarine Nerlerk berm. Can Geotech J 30:895–904

Leong WK, Chu J, The CI (2000) Liquefaction and instability of a granular fill material. Geotech Test J 23(2):178–192

Li XS, Dafalias YF (2000) Dilatancy for cohesionless soils. Geotechnique 50(4):449–460

Martin TE, McRoberts EC (1998) Some considerations in the stability analysis of upstream tailing dams. In: Proceeding of tailings and mine waste'99. Ft. Collins, Colorado, pp 287–302

Martins FB, Bressani LA, Coop MR, Bica AVD (2001) Some aspects of the compressibility behavior of a clayey sand. Can Geotech J 38(6):1177–1186

Ng CWW, Chiu CF (2003) Laboratory study of loose saturated and unsaturated decomposed granitic soil. J Geotech Geoenviron Eng 129(6):550–559

Nocilla A, Coop MR, Colleselli F (2006) The mechanics of an Italian silt: an example of 'transitional' behavior. Geotechnique 56(4):262–271

Pun WK, Wong ACW, Pang PLR (1999) Review of Landslip Warning Criteria 1998/1999. Special Project Report SPR 4/99, Geotechnical Control Office, Hong Kong:77 http://www.cedd.gov.hk/eng/publications/geo_reports/index.htm

Sichuan Provincial Highway Design Institute (2009) Geological investigation for reconstruction of PR303 between Yingxiu and Wolong. Chengdu, China

Stark TD, Mesri GM (1992) Undrained shear strength of liquefied sand for stability analysis. J Geotech Eng ASCE 121(2):163–173

Tobutt DC (1982) Monte Carlo simulation methods for slope stability. Comput Geosci 8(2):199–208

Vallejo LE, Mawby R (2000) Porosity influence on the shear strength of granular material-clay mixtures. Eng Geol 58(2):125–136

Wood DM, Kikumoto M, Russell A (2009) Particle crushing and deformation behavior. In: Proceeding of prediction and simulation methods for Geohazard mitigation. Oka, Murakami & Kimoto. Taylor & Francis Group, London, pp 263–268

Yamamuro JA, Wood FM, Lade PV (2008) Effect of depositional method on the microstructure of silty sand. Can Geotech J 45(11):1538–1555

Yang J (2002) Non-uniqueness of flow liquefaction line for loose sand. Geotechnique 52(10):757–760

Yu YF (2004) Corrections between rainfall, slope frequency and slope information for registered man-made slope. GEO Report No.144. Geotechnical Engineering Office, Hong Kong

Zhang LM, Li X (2010) "Microporosity structure of coarse granular soils." J Geotech Geoenviron Eng 136(10):1425-1436